Functional Micro/Nanomaterials for Intelligent Sensing Devices

Xing Ma • Jinhong Guo • Wenjun Chen

Functional Micro/ Nanomaterials for Intelligent Sensing Devices

Springer

Xing Ma
School of Integrated Circuits
Harbin Institute of Technology (Shenzhen)
Shenzhen, Guangdong, China

Jinhong Guo
School of Sensing Science and Engineering
Shanghai Jiao Tong University
Shanghai, China

Wenjun Chen
School of Integrated Circuits
Harbin Institute of Technology (Shenzhen)
Shenzhen, Guangdong, China

ISBN 978-981-96-4989-1 ISBN 978-981-96-4990-7 (eBook)
https://doi.org/10.1007/978-981-96-4990-7

Jointly published with China Machine Press Co.,Ltd.

The print edition is not for sale in China (Mainland). Customers from China (Mainland) please order the print book from: China Machine Press Co.,Ltd.

The original submitted manuscript has been translated into English. The translation was done using artificial intelligence. A subsequent revision was performed by the author(s) to further refine the work and to ensure that the translation is appropriate concerning content and scientific correctness. It may, however, read stylistically different from a conventional translation.

Translation from the Chinese Simplified language edition: “功能微纳米材料与智能传感器件 ” by Xing Ma et al., © 2023 2023. Published by China Machine Press Co.,Ltd. All Rights Reserved.

This Springer imprint is published by the registered company Springer Nature Singapore Pte Ltd.
The registered company address is: 152 Beach Road, #21-01/04 Gateway East, Singapore 189721, Singapore

Contents

1 Introduction and Properties of Functional Micro and Nano Materials 1

1.1 Definition and Introduction of Functional Materials 1

1.2 Definition and Basic Characteristics of Micro and Nano Materials 2

1.2.1 Basic Characteristics of Nanomaterials 3

1.2.2 Classification of Nanomaterials 5

1.3 Physical Properties of Micro-Nano Materials 8

1.3.1 Optical Properties 9

1.3.2 Electrical Properties 10

1.3.3 Thermal Performance 12

1.3.4 Magnetic Properties 14

1.4 Sensor Applications of Functional Micro and Nanomaterials 15

1.4.1 Based on Optical Signals 16

1.4.2 Based on Electrical Signals 20

1.4.3 Based on Magnetic Signals 22

1.4.4 Based on Chemical Signals 24

1.4.5 Based on Other Signals 25

1.5 Trend Analysis 27

Bibliography 28

2 Surface Enhanced Raman Technology and SERS Sensor Detection Applications 31

2.1 Surface Enhanced Raman Technology 31

2.1.1 Introduction 31

2.1.2 Current Research Status of Surface-Enhanced Raman Technology 32

2.1.3 Surface Enhanced Raman Technology Mechanism 33

2.2 Raman Enhancement Substrate Materials 37

2.2.1 Metal Structure Nanoparticles SERS Substrate 37

2.2.2 Thin Film Structure SERS Substrate Materials 38

2.2.3 Core-Shell Structure SERS Substrate Materials 39
2.3 Current Status of Surface Enhanced Raman Sensing Detection ... 40
2.3.1 Detection of Biochemical Molecules 40
2.3.2 Cell Detection 46
2.3.3 Environmental Pollutant Detection 49
2.4 Raman Detection Technology Based on Microfluidic Platform ... 51
2.4.1 Introduction to Microfluidic Technology 51
2.4.2 Advantages of Microfluidic SERS 52
2.4.3 SERS Detection Microfluidic Channel Design 54
2.4.4 Diversity of SERS Substrates in Microfluidic Devices and Their Detection Applications 55
2.4.5 Challenges and Development Trends of Microfluidic SERS Technology 58
Bibliography 60

3 Lateral Flow Immunoassay Probe Materials and Sensing Detection Applications 65
3.1 Introduction to Immunoassay Technology 65
3.1.1 Overview of Immunoassay 65
3.1.2 Structure of Lateral Flow Immunoassay Strip 66
3.1.3 Principle of Immunoassay 66
3.2 Immunoassay Nanoprobe Materials and Sensors 68
3.2.1 Colloidal Gold Nanoprobes and Immunoassay Sensors ... 68
3.2.2 Fluorescent Nanoprobes and Immunochromatographic Sensors Fluorescent 69
3.2.3 Carbon Nanoprobe and Immunochromatographic Sensor 72
3.2.4 Magnetic Nanoprobe and Immunochromatographic Sensor 73
3.3 POCT Application Based on Immunochromatography 74
3.3.1 Multi-Index Antigen/Antibody (Human Immune Index) Detection 74
3.3.2 Bacterial Detection 77
3.3.3 Virus Detection 78
3.3.4 Molecular Detection (Drugs, Food Toxins, etc.) 79
3.3.5 Heavy Metal Detection 80
3.4 Development Trends Introduction 80
Bibliography 81

4 Optical Nanoprobes and Their Biomedical Detection Applications 83
4.1 Luminescence Mechanism of Optical Nanoprobes 83
4.1.1 Fluorescence and Phosphorescence 83
4.1.2 Fluorescence Resonance Energy Transfer 84
4.1.3 Upconversion Luminescence 85
4.1.4 Chemiluminescence 87

4.2 Classification of Optical Nanoprobe Materials 90
4.2.1 Time-Resolved Fluorescence Nanoprobes..... 90
4.2.2 Long Afterglow Phosphorescent Nanoprobes 96
4.2.3 Chemiluminescent Nanoprobes..... 98
4.2.4 Upconversion Luminescent Nanoprobe..... 101
4.2.5 Luminescent Energy Transfer Nanoprobe 102
4.2.6 Luminescence Quenching Nanoprobes 103
4.3 Biomedical Detection Application of Optical Nanoprobes 103
4.3.1 Optical Nanoprobes for In Vitro Detection 103
4.3.2 Optical Nanoprobes for In Vivo Imaging..... 105
4.4 Chapter Summary 106
Bibliography 107

5 Application of Magnetic Nanoparticles in Sensor Detection 111
5.1 Introduction to Magnetic Nanoparticles 111
5.1.1 Overview of Magnetic Nanoparticles 111
5.1.2 Sensing Characteristics of Magnetic Nanoparticles 112
5.1.3 Structural Features Affecting Magnetic Performance 112
5.1.4 Synthesis Methods of Magnetic Nanoparticles 114
5.2 Application of Magnetic Nanoparticles in Sensor Detection 120
5.2.1 MNPs-Based Electrochemical Sensors 122
5.2.2 Optical Sensors Based on MNPs..... 125
5.2.3 Piezoelectric Sensors Based on MNPs 126
5.2.4 Based on MNPs Magnetic Resistance Sensor 127
5.3 Chapter Summary 136
Bibliography 136

6 Sensors Based on One-Dimensional Nanomaterials 139
6.1 Synthesis of One-Dimensional Nanomaterials 139
6.1.1 Vapor Growth Method..... 140
6.1.2 Liquid Phase Growth Method 142
6.1.3 Porous Template Method..... 145
6.1.4 Electrochemical Deposition..... 146
6.2 Properties of One-Dimensional Nanomaterials..... 147
6.3 One-Dimensional Nanomaterial Sensors..... 148
6.3.1 Chemical Gas Molecule Sensing..... 148
6.3.2 Biomolecule Sensors 154
6.4 Chapter Summary 156
Bibliography 158

7 Two-dimensional Nanomaterials and Smart Sensor Devices 163
7.1 Types of Two-dimensional Sensing Materials..... 164
7.1.1 Graphene Family 164
7.1.2 Transition Metal Dichalcogenides (TMDs)..... 165
7.1.3 Transition Metal Carbonitrides (MXene) 166
7.1.4 Flaky Metal Oxides (MOXs) 167

7.1.5 Metal-Organic Framework Materials (MOF) 169
7.1.6 Others 169
7.2 Preparation of Two-Dimensional Sensing Materials 170
7.2.1 Exfoliation Method 170
7.2.2 Chemical Vapor Deposition (CVD) 173
7.2.3 Wet Chemical Synthesis 173
7.2.4 Atomic Layer Deposition (ALD) 175
7.2.5 Selective Extraction Method 176
7.2.6 Other Methods 176
7.3 Two-Dimensional Material-Based Smart Sensor Devices 177
7.3.1 Mechanical Sensing 177
7.3.2 Biosensing 180
7.3.3 Gas Sensing 185
7.3.4 Photo-detector 186
7.4 Chapter Summary 189
Bibliography 190

8 Liquid Metal Micro-Nano Processing and Flexible Electronic Sensors 193
8.1 Basic Properties of Liquid Metal 193
8.1.1 Gallium-Based Liquid Metal 194
8.1.2 Toxicity 195
8.1.3 Oxidation and Wettability 195
8.2 Preparation Methods of Liquid Metal Micro/Nanoparticles 196
8.2.1 Template Method 197
8.2.2 Jetting Method 197
8.2.3 Microfluidic Method 199
8.2.4 Slicing Method (Shearing Method) 200
8.2.5 Ultrasonic Method 201
8.2.6 Physical Vapor Deposition (PVD) 202
8.3 Methods of Preparing Liquid Metal Flexible Devices 203
8.3.1 Injection Method 203
8.3.2 Photolithography 205
8.3.3 Additive Method 207
8.3.4 Subtractive Method 210
8.4 Application of Liquid Metal Flexible Sensors 212
8.4.1 Liquid Metal Pressure Sensor 212
8.4.2 Liquid Metal Strain Sensor 216
8.4.3 Other Applications of Liquid Metal Flexible Sensors 221
8.5 Chapter Summary 228
8.5.1 Control of Liquid Metal Oxide Film 228
8.5.2 Stability of the Connection Between Liquid Metal and Devices 228
8.5.3 Resolution of Liquid Metal Circuits 228
8.5.4 Biocompatibility of Liquid Metal 229
Bibliography 229

9 Intelligent Sensor Devices Based on Active Micro-Nano Probes of Micro-Nano Motors 237
9.1 Micro-Nano Motors 237
9.1.1 Background Introduction of Micro-Nano Motors 237
9.1.2 Driving Mechanism of Micro-Nano Motors 238
9.1.3 Applications of Micro-Nano Motors 243
9.2 Principles and Applications of Micro-Nano Motors 246
9.2.1 Intelligent Sensing Based on Micro-Nano Motor Motion Behavior Analysis 246
9.2.2 Intelligent Sensing of Micro-Nano Motors Based on Optical Detection Methods 251
9.2.3 Intelligent Sensing of Micro-Nano Motors Based on Electrochemical Analysis Methods 258
9.2.4 Other Micro-Nano Motor Intelligent Sensing 263
9.3 Advantages of Micro-Nano Motor Intelligent Sensing 264
9.3.1 Enhanced Material Exchange in Low Reynolds Number Environment by Micro-Nano Motor Motion 264
9.3.2 Precision Targeting and Controllable Detection of Substances to Be Tested at the Micro-Nano Scale 264
9.3.3 Achieving Automated Detection by Controlling Detection Probes 265
9.3.4 Combination of Micro-Nano Motors and Smartphones . . . 265
9.4 Chapter Summary 265
Bibliography 266

Chapter 1
Introduction and Properties of Functional Micro and Nano Materials

1.1 Definition and Introduction of Functional Materials

Functional materials typically refer to materials with their own unique properties and functions, such as materials with optical, acoustic, electrical, magnetic, or mechanical functions. Functional materials come from a wide range of sources, including ceramics, metals, polymers, and organic molecules, and there are many types, which can be classified according to chemical composition, physical properties, and application fields, as shown in Fig. 1.1.

Functional materials can be designed and synthesized for certain specific functions, with appropriate surface morphology and specific performance. They are the core of the current new material research field. The study of functional materials is the foundation of today's scientific and technological development and is also the focus and hotspot of new material research and development in countries around the world. In recent years, functional materials have developed rapidly, and scientists are committed to researching and developing various functional materials, from single-function materials to complex multi-function materials, function conversion materials, and various new forms and new concept functional materials. With the in-depth study of functional materials, people can now explain the functional characteristics of individual inorganic and organic component materials, as well as the reasons for the emergence of various new properties (such as magnetism, electricity, and optics). Functional materials are widely used in various fields due to their excellent magnetic, catalytic, electrical, optical, and good mechanical properties. For example, Bismuth-Tellurium-Selenium (Bi-Te-Se) composites are a type of topological insulator with high thermoelectric performance, which can be used in energy conversion devices; Nickel-Titanium (NiTi) shape memory alloys have important application value in the medical fields such as orthodontics, heart stents, and laparoscopy; Some functional materials can directly complete the required special application functions without any external stimulation, such as carbon nanotubes

X. Ma et al., *Functional Micro/Nanomaterials for Intelligent Sensing Devices*,
https://doi.org/10.1007/978-981-96-4990-7_1

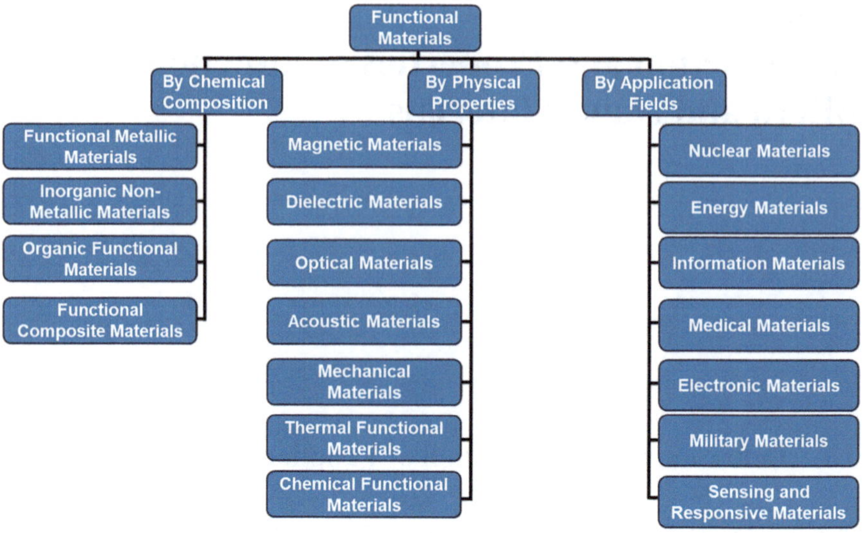

Fig. 1.1 Classification of functional materials

used for drug delivery; At the same time, there are also intelligent functional materials that respond to external stimuli such as electrical, thermal, magnetic, optical, and mechanical stimuli. In recent years, functional materials have been prepared in different forms such as powder, thin film, nanoparticles, bulk, and strip to adapt to different application scenarios, and have expanded the comprehensive functional categories of materials (physical functions, chemical functions, and biological functions), making their functionality more diverse.

1.2 Definition and Basic Characteristics of Micro and Nano Materials

With the development of micro and nano technology, the ability of humans to understand and transform the natural world has extended to the microscopic field by manipulating atoms, molecules, or atomic and molecular clusters to prepare the required materials. Micro and nano materials refer to ultrafine powders or particulate materials with a particle size less than 10 μm. Among them, materials with a particle size of 0.001–0.1 μm, that is, 1–100 nm, are called nano materials, those between 0.1–1 μm are called submicron materials, and powder materials with a particle size larger than 1 μm are called micron materials, as shown in Fig. 1.2.

Nanomaterials refer to materials that possess at least one dimension within the nanoscale range (1–100 nm) in three-dimensional space, or are composed of such units as their fundamental building blocks. This size is approximately one ten-thousandth the diameter of a human hair. At the micro-nano scale, the arrangement

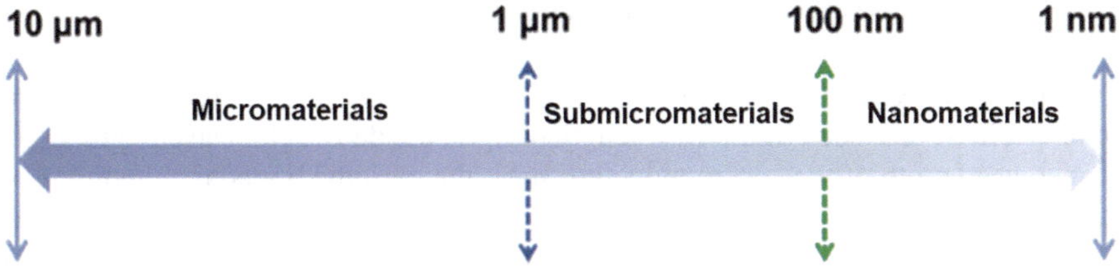

Fig. 1.2 Size range of micro and nano materials

Table 1.1 Correspondence between nanoparticle size and the number of surface atoms

Particle size (nm)	Number of atoms	Percentage of surface atoms (%)
20	2.5×10^5	10
10	3.0×10^4	20
5	4.0×10^3	40
2	2.5×10^2	80
1	3.0×10^1	99

of surface molecules, electron distribution, and crystal structure of materials have significant differences compared to macroscopic bulk materials. This can lead to changes in their structure, giving them some special properties, such as surface effects, small size effects, quantum size effects, macroscopic quantum tunneling effects, etc. This also gives nanomaterials some excellent physical, chemical, and surface properties, such as light, sound, electricity, magnetism, and catalytic properties. Further, by controlling the design synthesis of nanomaterials (including particle size, morphology, surface properties, and functional modification), it is possible to effectively "customize" the mechanical, electrical, optical, catalytic, and other properties of the materials, making them adaptable to different application fields.

1.2.1 Basic Characteristics of Nanomaterials

1.2.1.1 Surface Effect

For macroscopic bulk materials, the size of the material is much larger than the diameter of the atom, and the effect of surface atoms can be ignored. However, when the particle diameter is reduced to the nanoscale, the number of surface atoms cannot be ignored, and the material will be affected by the surface atoms, causing changes in surface properties. Due to the sharp increase in the number of surface atoms, the high specific surface area and high surface energy make these surface atoms extremely unstable and exhibit high chemical activity, bringing about some special effects, which are called surface effects. Table 1.1 shows the relationship between the size of nanoparticles and the number of their surface atoms.

1.2.1.2 Volume Effect (Also Known as Small Size Effect)

When the particle size of the material is close to or even smaller than the physical characteristic size such as the wavelength of light, the de Broglie wavelength, or the coherence length of the superconducting state, the periodic boundary conditions of the particle crystal are broken, and the density of surface atoms of the amorphous particles will decrease, causing the light, sound, electricity, magnetism, heat, mechanics, and other characteristics of the material to present new small size effects. That is, when the particle size is reduced to the nanometer level, the volume effect of nanoparticles causes a fundamental change in the properties of the material. For example, the light absorption of nanoparticles increases, and under certain light excitation, a plasmon resonance frequency shift phenomenon will occur; the conductivity of copper particles will disappear when the size is reduced to the nanometer level; the melting point of metals decreases with the decrease in size; the mechanical properties of nanomaterials improve; magnetic materials transition from an ordered state to a disordered state; the superconducting state transitions to a normal state, etc. For magnetic iron materials, bulk iron materials are silver-white conductors with metallic luster and strong ferromagnetism. As the material size is reduced to the nanometer level, the color of iron phase nanomaterials turns black and loses its metallic luster, the coercivity increases, it has paramagnetism, and the ferromagnetism disappears (as shown in Fig. 1.3).

1.2.1.3 Quantum Size Effect

When the size of materials decreases to the nanoscale, the electronic energy levels near the Fermi level of metallic materials will change, transitioning from quasi-continuous to discrete energy levels. For semiconductor materials, the energy gap will widen, a phenomenon known as the quantum size effect of nanomaterials. When the change in energy level spacing is greater than the changes in thermal

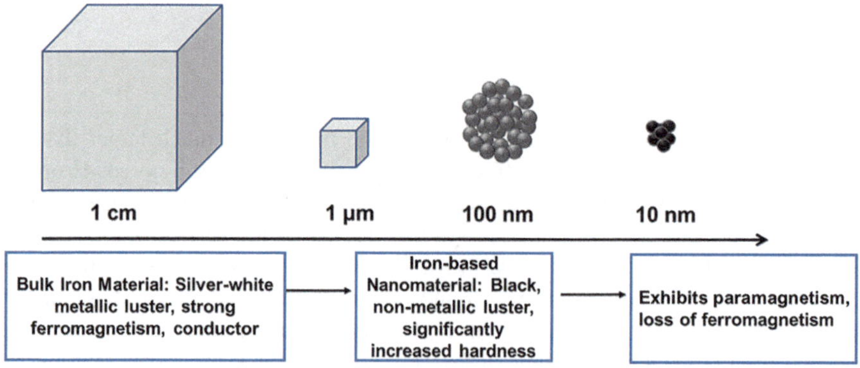

Fig. 1.3 Example of the small size effect of Fe material

energy, magnetic energy, electrostatic energy, and photon energy, the quantum size effect will cause the optical, electrical, magnetic, and mechanical properties of nanoparticles to significantly differ from the macroscopic properties of the material. In addition, due to the wave nature of electrons in the discrete quantized energy levels of nanoparticles, these particles will exhibit special properties, such as optical nonlinearity, photocatalytic activity, and unique catalytic properties.

1.2.1.4 Macroscopic Quantum Tunneling Effect

The tunneling effect refers to the ability of microscopic particles to penetrate potential barriers. It was later discovered that some macroscopic physical quantities, such as the magnetization strength and magnetic flux of particles, also have this tunneling effect, hence the term "macroscopic quantum tunneling effect". This effect limits the spatial size and time limit of information storage for tapes and disks, and is the basis for microelectronics research, establishing the limit for further miniaturization of microelectronic devices.

1.2.2 Classification of Nanomaterials

Different classification principles will yield different results. For example, according to the type of bonding: it can be divided into metallic nanomaterials, ionic semiconductor nanomaterials, semiconductor nanomaterials, and ceramic nanomaterials, etc.; according to their chemical composition: it can be divided into inorganic nanomaterials, organic nanomaterials, nanocomposite materials, and biological nanomaterials, etc.; according to their uses: it can be divided into optical nanomaterials, magnetic nanomaterials, optoelectronic nanomaterials, etc.

1.2.2.1 Classification Based on Basic Structure

According to the basic structure of nanomaterials, they can be further divided into the following four categories (as shown in Fig. 1.4):

1. **Zero-dimensional (0D) nanomaterials** refer to those whose dimensions in all three spatial dimensions are at the nanometer level. Typical zero-dimensional nanomaterials mainly include: atomic clusters, nanoparticles, quantum dots, or so-called artificial atoms, etc.
2. **One-dimensional (1D) nanomaterials** refer to those with two dimensions at the nanoscale, or those with a radial size less than 100 nm and a length dimension far greater than the radial size, similar to fibrous structures. Typical one-dimensional nanomaterials mainly include: nanotubes, nanorods, nanowires, nanoribbons, and coaxial nanofibers. For example, carbon nanotubes (CNT) are

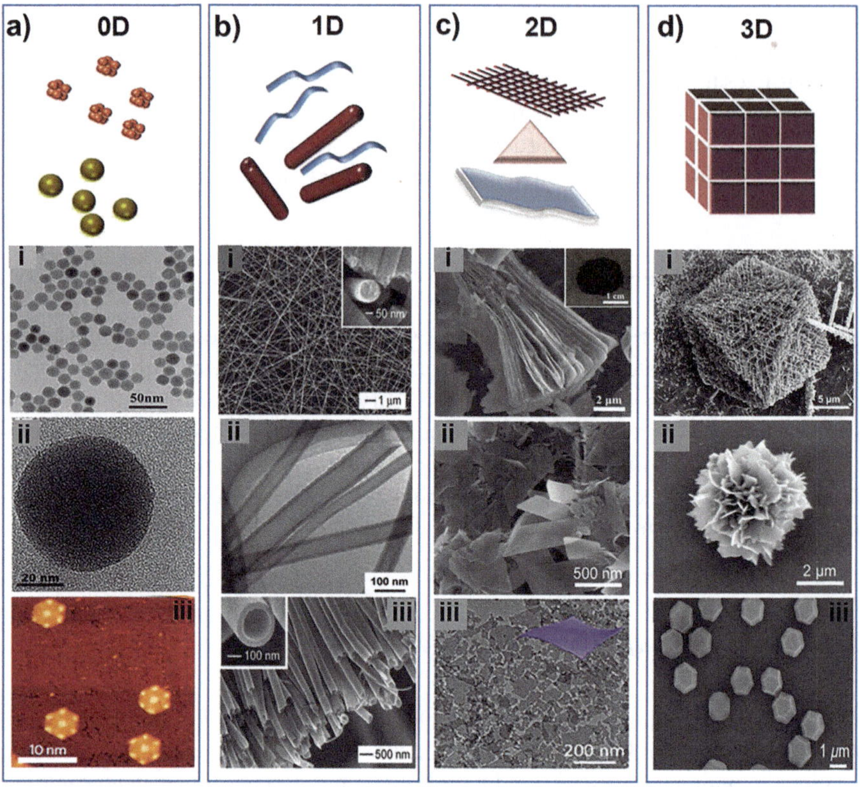

Fig. 1.4 Classification of nanomaterials based on basic structure: (**a**) zero-dimensional nanomaterials; (**b**) one-dimensional nanomaterials; (**c**) two-dimensional nanomaterials; (**d**) three-dimensional nanomaterials

typical representatives of tubular structures, as well as non-carbon materials such as SiO_2, TiO_2, BCN, WS_2, Nanotubes, etc. The definitions and distinctions of nanorods and nanowires are somewhat vague. Generally, short and straight one-dimensional cylindrical shapes are referred to as nanorods, such as metal or metal oxide nanorods; while longer, straight or curved one-dimensional linear materials are referred to as nanowires, such as metal (AgNWs, AuNWs) and non-metal (Si), oxide (ZnO) or sulfide (CdS) nanowires. The cross-section of a nanobelt is quadrilateral, with a wide range of widths. The most studied nanobelts are ZnO, Ga_2O_3, SnO_2 and other oxide nanobelts. Coaxial nanofibers have a core/shell structure, including inorganic/inorganic and coaxial nanofibers and inorganic/organic coaxial nanofibers.

3. **Two-dimensional (2D) nanomaterials** refer to materials that have one dimension at the nanoscale in three-dimensional space, usually characterized by their large lateral size (>100 nm) and thin atomic layer thickness (<5 nm). Typical 2D nanomaterials include ultrathin films, multilayer films, superlattice materials,

etc. For example, graphene, an ultrathin 2D nanomaterial, has received great attention in various application fields such as energy, biology, and catalysis due to its excellent physicochemical and electronic properties. The structural characteristics of 2D nanomaterials can confine electrons within an ultra-thin region, allowing them to move within the plane. Due to strong planar covalent bonding behavior, they have strong mechanical strength, flexibility, and optical transparency.

4. **Three-dimensional (3D) nanomaterials** refer to bulk materials composed of the above-mentioned nanostructure units in three-dimensional space, such as nanoflowers, nanocubes, etc.

1.2.2.2 Classification by Chemical Composition

Based on the chemical composition of the materials, functional nanomaterials commonly include the following types:

1. **Carbon materials** mainly include: carbon nanotubes (CNT) and carbon fibers, fullerenes, graphene/oxidized graphene, carbon quantum dots. For example, carbon nanotubes have good conductivity, structural stability, high strength, and electron affinity, and can be used in electronic devices, such as transistors, sensors, capacitors, electrodes, organic light-emitting diodes, solar cells, and wastewater treatment research fields.
2. **Metal nanomaterials** mainly include various metal nanoparticles, for example: gold nanoparticles (AuNPs) widely used in biochemical research, silver nanoparticles (AgNPs) with antibacterial activity, platinum nanoparticles (PtNPs) and palladium nanoparticles (PdNPs) with catalytic properties, as well as copper nanoparticles (CuNPs), nickel nanoparticles (NiNPs), etc. It also includes metal oxides, such as Fe_3O_4, ZnO, CuO, TiO_2. Etc. Among them, magnetic nanomaterials are prepared from magnetic metals, such as cobalt, nickel, iron and their respective oxides, such as $Fe_3 O_4$ (magnetite) and $Fe_2 O_3$ (hematite), cobalt ferrite and chromium dioxide. When the size of magnetic particles is reduced to a certain size, they become paramagnetic nanoparticles, which have stronger magnetism than other conventional preparations and can be used for diagnostic and therapeutic purposes in the medical field. For example, $Fe_3 O_4$ is a well-known biocompatible magnetic nanomaterial, and its potential uses in cancer treatment, stem cell manipulation and sorting, drug delivery, gene therapy, and magnetic resonance imaging (MRI) have been widely studied and verified.
3. **Ceramic nanomaterials** are a type of inorganic non-metallic nanomaterials, such as alumina ($Al_2 O_3$), zirconia (ZrO_2), and silica (SiO_2) nanomaterials, which have high melting points, high hardness, high wear resistance, and oxidation resistance. They can be used as auxiliary treatments for radiation therapy and bone repair, and in energy supply and storage fields. Ceramic nanoparticles often exist in amorphous, polycrystalline, dense, porous or hollow forms, and these

nanoparticles are often used in catalysis, photocatalysis, dye photodegradation, and imaging.

4. **Semiconductor nanomaterials** are nanomaterials with properties between metals and non-metals, such as GaN, GaP, InP, ZnO, ZnS, CdS, CdSe, and CdTe, etc., which can be widely used in photocatalysis, electrical equipment and other research fields. Semiconductor nanoparticles have a wide band gap, and their performance changes significantly after band gap adjustment. Therefore, they are very important materials in photocatalysis, optics, and electronic devices. For example, various semiconductor nanoparticles are found to be very effective in water decomposition applications due to their suitable band gaps and band gap positions.
5. **Polymer nanomaterials** are essentially organic nanomaterials, mainly including supramolecular compounds (cyclodextrins, crown ethers, dendrimers, calixarenes) and natural or synthetic high molecular compounds. Depending on the preparation method, they can be nanocapsules or nanospheres. Spherical NPs have a structure similar to the matrix, while capsule NPs have a core-shell structure. They have advantages such as drug molecule controlled release and protection, specificity in targeting and binding to diseases in treatment and bioimaging, biodegradability, and biocompatibility.
6. **Nanocomposite materials** refer to the combination of two or more types of nanofunctional materials. For example, carbon-metal nano-hybrid: carbon materials conjugated with metal nanomaterials. Organic-inorganic nano-hybrid: usually organic molecule-coated nano-hybrids, when inorganic materials act as a matrix and couple with long-chain polymers, drug molecules, specific proteins, DNA or long organic molecule chains, they form the so-called organic molecule-coated nano-hybrids. Surface functionalization of inorganic nanoparticles is an important means in the field of nanotechnology, which can customize the performance and application of hybrid materials. For example, in the construction of functionalized nanomaterials, the surface of inorganic nanoparticles usually needs to be functionalized to regulate the interaction between nanoparticles and organisms, effectively reduce toxicity, improve their diagnostic and therapeutic performance, and some organic or inorganic materials can be used to create a shell on the surface of nanoparticles, thus forming a composite material with broad application prospects.

1.3 Physical Properties of Micro-Nano Materials

Micro-nano materials, as a new type of material, usually have internal structural units at the micron or nanometer scale. Especially when their structural units are at the nanometer level, due to their large specific surface area and surface energy, nanomaterials exhibit many unusual optical, electrical, thermal, and magnetic characteristics, making them have a wide range of application prospects.

1.3.1 Optical Properties

Under the excitation of high-energy particles or light, the electrons in the internal structural units of micro-nano materials will transition between different states, causing light absorption and radiation. This electron transition can occur between different energy levels, within the same energy level, or between discrete energy levels. When the particle size of its structural unit is close to the Bohr radius, the de Broglie wavelength of the electron, and the superconducting coherent wavelength, the size effect of its small particles will be very obvious. At the same time, due to the large specific surface area, the behavior of electrons and atoms in the internal and surface states of the particles shows a great difference, which has a great impact on the optical properties of micro-nano materials, and even gives them new characteristics that macroscopic bulk materials do not have. This is mainly reflected in the following aspects.

1.3.1.1 Broadband and Strong Absorption

The bulk metals we see in daily life can present different metallic lusters due to their ability to absorb and reflect visible light. However, when the size is reduced to the nanometer level, various metal nanoparticles generally appear black. This is mainly due to the size effect and surface effect, which make nanoparticles have extremely strong absorption ability for visible light, and the reflection ability is very low, such as the reflection of Pt nanoparticles is 1%, and the reflection of Au nanoparticles is less than 10%. In addition, some nano-oxides, nitrides, and nano-semiconductors have obvious infrared vibration absorption in the infrared region, showing typical broadband strong absorption characteristics, so they can be applied to infrared sensors. And the composite materials based on some metal nano-particles have large third-order optical nonlinearity and short response time. Using this characteristic, it can be used as a high-efficiency photo-thermal, photo-electric conversion material, which can efficiently convert solar energy into thermal energy.

1.3.1.2 Blue Shift or Red Shift

When the particle size is reduced to the nanometer level, the optical absorption band of the material generally exhibits a blue shift compared to the bulk state, that is, the absorption band shifts towards the direction of short wavelengths. For example, the peak values of the infrared absorption frequencies of nanosilicon nitride particles and bulk silicon nitride are 949 cm^{-1} and 935 cm^{-1} respectively, showing a clear blue shift. This is mainly caused by the broadening of the energy band caused by the quantum size effect and the change in surface energy of the nanoparticles. In addition, sometimes a red shift of the optical absorption band relative to coarse-grained materials is observed. This is due to the narrowing of the energy band gap caused by the dielectric confinement effect and the dipole effect induced by chemically modified nanoparticles.

1.3.1.3 Nonlinear Optical Effects

Due to their own characteristics, the absorption changes induced by light excitation in micro-nano materials can usually be divided into two parts: the rapid nonlinear part produced by the free electron-hole pairs caused by light excitation; and the slow nonlinear process of carriers affected by trap effects. Among them, CdS nanoparticles are the most studied. Due to changes in the band structure, the migration, transition, and recombination processes of carriers in nanocrystals all show different rules from conventional materials, thus having different nonlinear optical effects.

1.3.2 Electrical Properties

The electrical properties of micro-nano materials are the basis for their application in nanoelectronic devices and have always been a hot topic. The resistivity of micro-nano materials is an important indicator of their electron conduction ability in their electrical properties, and the conductivity of materials is inversely proportional to their resistivity. It has been found that the resistivity of micro-nano materials is much higher than that of similar coarse-grained materials. For example, the resistivity of nanocrystalline Cu with a grain size of 7 nm at 275 K is 7–20 times that of conventional coarse-grained Cu. This is partly due to the higher volume fraction of grain boundaries in nanocrystalline Cu compared to coarse-grained Cu, and partly due to grain boundary electron scattering and the short mean free path (λ) of electrons, i.e., the λ of nanocrystalline Cu is approximately 4.7 nm, while that of coarse-grained Cu is approximately 44 nm. In addition, the reduction in grain size also increases the degree of localization, reduces the concentration of carriers, and thus increases the specific resistivity. Therefore, metals with small resistances under normal conditions will have increased resistances at the nanoscale, the temperature coefficient of resistance will decrease, and even negative numbers may appear; the resistance of oxide insulators, on the other hand, will decrease at the nanoscale, becoming semiconductors or conductors. Nanoelectronic devices made using the tunneling quantum effect and Coulomb blockade effect of nanoparticles have the characteristics of ultra-high speed, ultra-capacity, ultra-miniature low energy consumption, and may completely replace current conventional semiconductor devices in the near future.

Dielectric properties are one of the many special properties of functional micro-nano materials. The dielectric properties of materials can be used for energy conversion and are widely used in electronics, energy storage, and other fields. Micro-nano dielectric materials are also known as dielectrics, which are characterized by their ability to polarize. Under the action of an external electric field, the positive and negative charge centers within the molecule undergo relative displacement, forming an electric dipole moment, a phenomenon known as polarization. Dielectric materials transmit, store, or record the effects and influences of electric fields by induction rather than conduction. Dielectric materials will exhibit polarization, loss, conduction, and breakdown under the action of an external electric field. The main performance indicators are dielectric constant, dielectric loss, polarization intensity, breakdown field, and energy storage density.

Dielectric constant: Under the action of an external electric field, dielectric materials can generate induced charges to weaken the intensity of the electric field. The size of the dielectric constant is equal to the ratio of the external electric field in a vacuum to the final electric field in the dielectric material. The dielectric constant mentioned in the study usually refers to the relative dielectric constant εr, calculated as follows:

$$D = \varepsilon_0 \varepsilon_r E \tag{1.1}$$

where, D is the electric displacement; εo is the vacuum dielectric constant, with a value of 8.85×10^{-12} F/m; E is the applied electric field.

Dielectric loss: In the complex form of the dielectric constant ε(ω)=ε′(ω)-jε″(ω), the imaginary part ε″(ω) is caused by the polarization process not keeping up with the changes in the external field. The loss tangent *tan* is an important indicator of dielectric loss, equal to the ratio of the imaginary part of the dielectric constant to the real part of the dielectric constant:

$$\tan\delta = \frac{\varepsilon''}{\varepsilon'} \tag{1.2}$$

Polarization intensity: The vector sum of the molecular electric dipole moment per unit volume is the polarization intensity, which reflects the degree of polarization of the dielectric material. The calculation formula is:

$$P = \Sigma p / \Delta \mathrm{V} \tag{1.3}$$

where, p is the electric dipole moment; ΔV is the volume element. Breakdown field: Under the action of a very large electric field, dielectric materials will gradually lose their original dielectric properties and become conductors. The electric field strength at the time of breakdown is called the breakdown field, often analyzed using Weibull statistics:

$$F(x) = 1 - e^{-\left(\frac{x}{\alpha}\right)^{\beta}} \tag{1.4}$$

where, F(x) is the cumulative failure probability of the tested sample; x is the experimental breakdown field; α is the characteristic breakdown field value when the cumulative failure probability is 63.2%; β is the breakdown field distribution parameter.

Energy storage density: The energy storage density (Ue) is a measure of the material's ability to store electrical energy, expressed as follows:

$$U_e = \int EdD = \frac{1}{2}\varepsilon_r \varepsilon_0 E^2 \tag{1.5}$$

where, E is the applied electric field strength; D is the electric displacement; ε_r is the dielectric constant; ε_0 is the vacuum permittivity. For the vacuum dielectric constant. Therefore, the energy storage density value can be improved by improving the dielectric constant and breakdown electric field of the dielectric material.

1.3.3 Thermal Performance

The thermal performance of micro-nano materials is also extremely important in applications. The high proportion of interfaces and the special structure of atomic energy levels in micro-nano materials give them properties different from conventional bulk materials and single molecules. The small size effect of micro-nano materials gives their thermodynamic properties a special nature. Various thermodynamic properties of micro-nano materials such as heat capacity, melting point, thermal expansion coefficient, and thermal stability all show size and shape effects, so the thermodynamic characteristics of micro-nano materials are significantly different from those of conventional bulk materials in all aspects. When micro-nano materials are used in various applications, the measurement of these parameters is also important, such as in nano-electronic devices and gas sensors.

The melting point is the temperature at which the atoms, ions, or molecules of a crystalline material change from a periodic ordered state to a disordered state. A large amount of research shows that the melting point of nano-particles is much lower than that of conventional powders. Tkagail first discovered the phenomenon of a decrease in the melting point of metal thin films. Later, using various preparation techniques, different types of metal particles, embedded particles, or thin film crystals were obtained, all of which were found to have a melting point lower than the equilibrium melting point of bulk materials. Wronski calculated the relationship between the particle size and melting point of nano Au particles. When the particle size is larger than 10 nm, the melting point does not change significantly with the change in particle size; but when the particle size is less than 10 nm, the melting point drops sharply. The melting points of other metals such as In, Sn, Pb, Bi, Cd, Al, Ag decrease with the decrease in size, especially below 30 nm. Melting starts from the surface of the material, characterized by an increase in the mobility of surface atoms or molecules. At temperatures far below the melting point of bulk materials, the diffusion coefficient of these atoms approaches the quasi-liquid value. This is because nano-particles have a high surface area and high surface energy, so the activation energy required for surface atoms to melt is lower than that of the corresponding bulk materials.

The thermal expansion of solids is related to the non-linear vibration of the lattice. If the crystal lattice vibrates linearly, no expansion will occur. The expansion coefficient under a certain pressure P can be easily obtained from the free energy of the system, and the relationship between system expansion and temperature is:

$$\frac{V - V_0}{V_0} = \frac{\gamma \bar{E}}{KV} - \frac{P}{K} \tag{1.6}$$

In the formula, $K = V_0 \left(\frac{d^2U}{dV^2} \right)_{V0}$ is the volume elastic modulus at a volume of V_0; V_0 is the original volume; V is the volume after thermal expansion; P is the pressure; E is the system energy; γ is the Grüneisen constant; it can be expressed as:

$$\gamma = \frac{dln v_i}{dln V} \tag{1.7}$$

In the formula, $v_i$represents the frequency of lattice nonlinear vibration. When the crystal vibrates linearly, γ=0; when the temperature changes, thermal expansion occurs due to the nonlinear vibration of the lattice. When the temperature of a nanocrystal changes, the nonlinear vibration can be divided into two parts: nonlinear thermal vibration and thermal vibration of the grain boundary components. Often, the latter's nonlinear thermal vibration is more significant, and it can be said that the grain boundaries, which account for a large percentage of the volume, play a leading role in the thermal expansion of nanocrystals. Taking nanocrystalline Cu (average particle size 8 nm) as an example, the thermal expansion coefficient in the temperature range of 110–293 K is 31×10^{-6} K^{-1}, while the thermal expansion coefficient of single crystal Cu is 16×10^{-6} K^{-1}. It can be seen that the thermal expansion of nanocrystalline materials mainly comes from the contribution of grain boundary components. The measurement results of the grain boundary expansion of Cu and Au (nanoscale) by the dilatometer and X-ray diffractometer show that the contribution of the grain boundary to thermal expansion is 4 times that of the bulk phase, which indirectly explains why the thermal expansion coefficient of nanocrystals with a large percentage of grain boundary volume is higher than that of similar polycrystalline conventional materials.

The Debye characteristic temperature is an important physical quantity introduced by the lattice vibration theory, which is a representation of the binding force between atoms. It is closely related to the elasticity, hardness, melting point, specific heat, thermal expansion coefficient, electrical conductivity, and thermal conductivity of solid materials. Cheng Benpei used the X-ray diffraction method to measure the TiO_2 nanopowder of different particle sizes, and found that the square of the Debye temperature is directly proportional to the reciprocal of the particle size; Zhang Wei used the double-temperature X-ray diffraction method to measure the Debye temperature of nanosilver particles of different particle sizes, and found that the Debye temperature of most samples is higher than the characteristic value of bulk silver crystals, and there is a critical size. When the particle size is smaller than the critical value, the Debye temperature decreases with the decrease in particle size. When the particle size of the sample is larger than the critical value, its Debye temperature is lower than the characteristic value corresponding to the critical size.

1.3.4 Magnetic Properties

Magnetism is one of the basic properties of matter, and the magnetic properties of materials are determined by their constituent substances, which can be mainly divided into ferromagnetism, ferrimagnetism, antiferromagnetism, paramagnetism, and diamagnetism. Magnetic nanomaterials, as a branch of nanomaterials, also possess properties such as small size effect, quantum size effect, and surface effect, which in turn affect the magnetic properties of the material, making magnetic nanomaterials exhibit magnetic properties that magnetic bulk materials do not possess.

1.3.4.1 Coercive Force

Coercive force refers to the phenomenon where the magnetization strength of a magnetic material does not return to zero after it is saturated and the external magnetic field is reduced to zero. Only by applying a magnetic field of a certain size in the opposite direction to the original magnetization field can the magnetization strength be returned to zero. This magnetic field is called the coercive field, also known as coercivity. The size of the coercivity is most obviously affected by changes in grain size. For roughly spherical grains, the coercivity increases as the grain size decreases, reaching a maximum value and then decreasing as the grain size continues to decrease. The grain size at which the coercivity reaches its maximum is equivalent to the size of a single magnetic domain. For blocky magnetic materials, the material appears to have the maximum coercivity when the grain size is at the single domain size. For superparamagnetic nanomaterials with grain sizes less than or equal to the size of a single magnetic domain, the magnetic orientation force generated by exchange interactions is not enough to resist the interference of irregular thermal motion, and the overall magnetization direction of the material becomes disordered. Therefore, it is no longer necessary to apply a reverse magnetic field to return the material's magnetization strength to zero, and superparamagnetic nanomaterials appear to have zero coercivity.

1.3.4.2 Saturation Magnetization

Saturation magnetization refers to the maximum magnetization strength that a magnetic material can reach when magnetized in an external magnetic field. The saturation magnetization of superparamagnetic nanomaterials seems to be related to the size of the material, but the main reason is due to the influence of surface effects different from bulk materials. In an ideal state, the direction of the magnetic moment of the atoms inside the magnetic bulk material is parallel to the direction of the external magnetic field. However, for magnetic nanomaterials, the direction of the magnetic moment of their surface atoms is affected by surface effects and deviates from the direction of the external magnetic field, forming a state of surface spin disorder similar to spin glass. Therefore, the surface atoms of magnetic nanomaterials are difficult to achieve a high magnetization strength. As the size of the magnetic nanomaterials decreases, the proportion of surface atoms to total atoms increases sharply, and more atoms show a state of surface spin disorder due to the influence of surface effects, and the saturation magnetization of the material also decreases with the decrease in material size. When the size of the magnetic nanomaterials is extremely small, the proportion of surface atoms is extremely high, and the vector sum of the material's magnetic moment is very small due to the influence of surface effects. The magnetization strength of the material is linearly related to the magnetic field strength, and the material will show paramagnetism.

1.3.4.3 Superparamagnetism

Superparamagnetism refers to the phenomenon where magnetic materials do not show hysteresis when magnetized above a certain critical temperature and below a critical size, and the corresponding coercivity and residual magnetization strength are both 0. Moreover, the magnetization curves at different temperatures will coincide after conversion. The origin of the superparamagnetic state is mainly due to the fact that at small sizes, when the anisotropy energy is reduced to be comparable to the thermal motion energy, the magnetization direction is no longer fixed in an easy magnetization direction but changes irregularly, resulting in superparamagnetism. In addition, the critical sizes at which different types of micro and nano magnetic materials exhibit superparamagnetism are also different.

1.4 Sensor Applications of Functional Micro and Nanomaterials

Functional micro and nanomaterials play a significant role in various fields, including targeted drugs, bioimaging, and sensor detection. Particularly in the field of sensor applications, they are not only powerful tools for enhancing biosensing technology but also drive the development of existing detection methods. Micro and nanomaterials have a high specific surface area, unique electrochemical characteristics, catalytic activity, and physicochemical properties. These characteristics of micro and nanomaterials are crucial for the generation and amplification of sensor signals and the improvement of analytical capabilities. Scientists combine different micro and nanomaterials, such as metal nanoparticles (NPs), quantum dots (QDs), silicon dioxide nanoparticles (SiO_2 NPs), graphene (Gr), and carbon nanotubes (CNTs) with biosensors, creating various sensor devices based on different sensing principles. Currently, they can be classified into the following categories based on the sensor signal: (1) Optical signals (colorimetric, fluorescent, luminescent, and Raman, etc.); (2) Electrochemical signals; (3) Magnetic signals; (4) Elemental composition signals; (5) Other signals (thermal and acoustic).

1.4.1 Based on Optical Signals

1.4.1.1 Fluorescent Micro and Nanomaterials

Organic dyes are the most commonly used fluorescent probes, but their low absorption coefficient and weak emission signal reduce the sensitivity of detection. Moreover, some organic dyes, such as fluorescein isothiocyanate, have strong phototoxicity. Therefore, the toxicity of organic dyes and their instability in biological systems limit the application of fluorescent dyes in the field of biomedical sensor detection. With the development of nanotechnology, there is a need to develop fluorescent probes with brighter luminescence, better biocompatibility, higher photostability, and hydrophilicity for biosensing applications. In fact, fluorescent nano-probes can overcome some limitations of traditional organic dyes, improving quantum yield and detection sensitivity. Additionally, the surface functionalization of these nanomaterials provides their multifunctionality. Fluorescent nanostructures can be divided into two major categories based on size and optical properties: metal nanoclusters (NCs) and quantum dots. Metal NCs are defined as small metal clusters (such as Au, Ag, Cu, and Pt) with diameters less than 2 nm. Quantum dots include heavy metal quantum dots (Cd or PbS) and non-heavy metal quantum dots (carbon quantum dots, graphene quantum dots, and polymer quantum dots).

The fluorescence quenching mechanism of micro and nanomaterials is effectively used for the determination of analytes. There are generally two different quenching methods: static quenching and dynamic quenching. Fluorescence static quenching refers to the formation of non-fluorescent complexes between the fluorescent group and the quencher. Dynamic quenching methods, such as Förster Resonance Energy Transfer (FRET), are the energy transfer process between a pair of photosensitive molecules, where the donor fluorescent group initially in the electronic excited state transfers energy to the acceptor fluorescent group, as shown in Fig. 1.5. Dynamic quenchers provide a non-radiative pathway for the loss of excited state energy. Gold nanoparticles (Au NPs), graphene quantum dots (GQDs), and carbon nanotubes (CNT) are often used as dynamic quenchers in FRET. More details about fluorescence and other optical nano-detection can be found in Chap. 4 of this book.

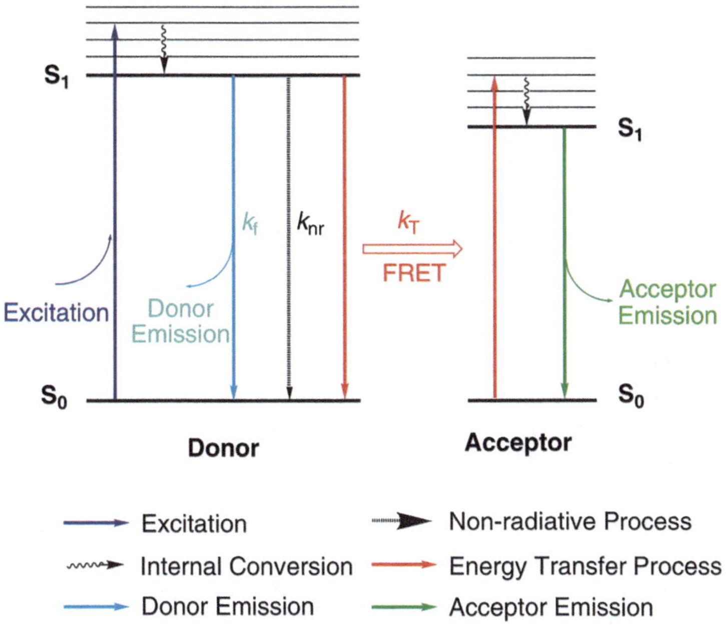

Fig. 1.5 Basic principle of FRET resonance energy transfer

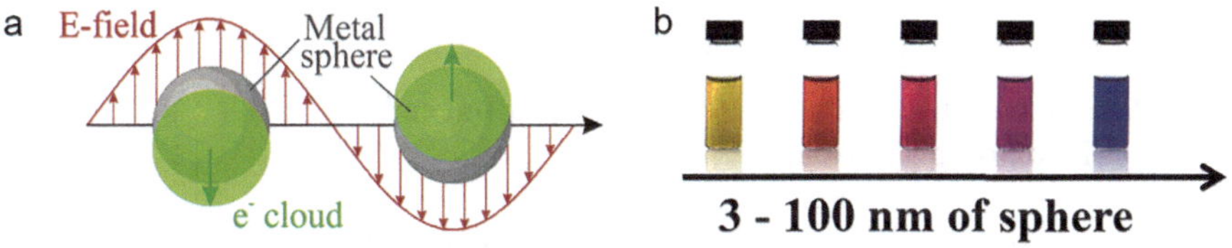

Fig. 1.6 (**a**) Schematic of Local Surface Plasmon Resonance; (**b**) Color changes corresponding to Au NPs of different sizes

1.4.1.2 Micro and Nanomaterials for Colorimetric Determination

In recent years, high-sensitivity, low-cost nanoparticle colorimetric probes have attracted widespread attention in biosensing detection. Especially due to their unique optical and electronic properties, interest in anisotropic gold and silver nanoparticles has greatly increased. When incident light shines on the surface of metal nanoparticles, the incident light interacts with the metal nanoparticles. When the frequency of the light's electromagnetic field equals the frequency of the electron's plasma oscillation, resonance occurs. This is known as Surface Plasmon Resonance (SPR), as shown in Fig. 1.6a. Because the generated surface plasmon waves are confined near the surface of the precious metal nanostructures, it is also known as Local Surface Plasmon Resonance (LSPR). The spectral characteristics of

LSPR are determined by the physical parameters of the system, such as the size, shape, material composition of the nanoparticles, and the refractive index of the surrounding environment. For Au NPs, due to their strong color effect, they meet the resonance conditions under visible light wavelengths. Au NPs with a size of 10 nm have a maximum absorption wavelength of about 520 nm in water due to the effect of LSPR. Au NPs of different sizes and shapes present different colors and have different characteristic absorption bands (Fig. 1.6b).

Colorimetric detection based on nanomaterials usually relies on changes in optical properties caused by their assembly (or aggregation), morphological transformation, and surface chemical reactions. According to the colloidal interface behavior of nanoparticles, colorimetric sensing strategies can be summarized into two types: (1) One is based on the method of analyte-induced nanoparticle aggregation. For example, as shown in Fig. 1.7a, the aggregation of appropriately sized metal-free nanoparticles (Au/Ag) causes surface plasmon coupling between particles, leading to visible color changes. The color of Au NPs is red in a dispersed state, but turns to blue-purple in an aggregated state. In fact, Au NPs can be induced to aggregate by a variety of target analytes, and the color change from red to blue-purple comes from strong surface plasmon resonance. The color change can be detected by the naked eye or measured with a UV-visible spectrophotometer. (2) Another strategy mainly utilizes the detection of morphological changes and surface chemical reactions of individual particles. For example, the interaction between mercury and gold can cause morphological changes in gold rods, resulting in color changes, achieving the purpose of colorimetric determination of mercury ions in water (Fig. 1.7b).

1.4.1.3 Micro-Nano Materials for Surface Enhanced Raman Spectroscopy

Molecules adsorbed on the rough surface of precious metals, transition metals, or semiconductor substrates can significantly enhance their Raman scattering signal, a phenomenon known as Surface Enhanced Resonance Raman Scattering (SERS) (as

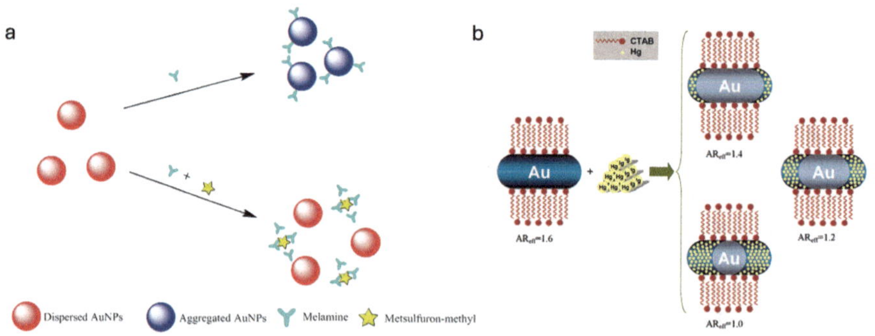

Fig. 1.7 (**a**) The detection of analyte-induced aggregation of Au NPs to achieve detection purposes; (**b**) The detection of analyte-induced morphological changes in gold nanorods for color change analysis

shown in Fig. 1.8). There are currently two mechanisms to explain the amplification of the Raman signal: Electromagnetic Enhancement (EMH) and Chemical Enhancement (CE). The electromagnetic enhancement mechanism mainly believes that the surface plasmon vibration between the rough surface of the metal or nanoparticles greatly enhances the local electromagnetic field. These enhanced electromagnetic fields have enhancement factors from 10^6. The inequality is up to 10^{11}. The chemical enhancement mechanism mainly considers the enhancement caused by the change in molecular polarizability, which is caused by the resonance of metal-molecule charge transfer. The enhancement factor of this mechanism is about 10^2–10^3. Metal nanoparticles, especially Ag and Au, are the most ideal materials for designing SERS substrates, although transition metal nanoparticles also have this ability, but the enhancement effect is limited. In recent years, graphene and its derivatives have been reported as a new type of SERS substrate, which can effectively perform Raman scattering on molecules adsorbed on the graphene surface. The SERS enhancement effect of zero-dimensional (0D) nanomaterials highly depends on the type of material (such as Au or Ag NPs), size (about 10–100 nm), shape (such as spherical, triangular, star-shaped or flower-shaped NPs) and structure (such as core-shell or hollow shell). These special structure nanomaterials provide a variety of good SERS signal enhancement substrates. One-dimensional nanomaterials, such as rod-shaped, shuttle-shaped, linear or tubular nanocrystals, have a large surface area and can generate many SERS detection hotspots. In addition, the SERS detection platform based on two-dimensional nanomaterials, including graphene or

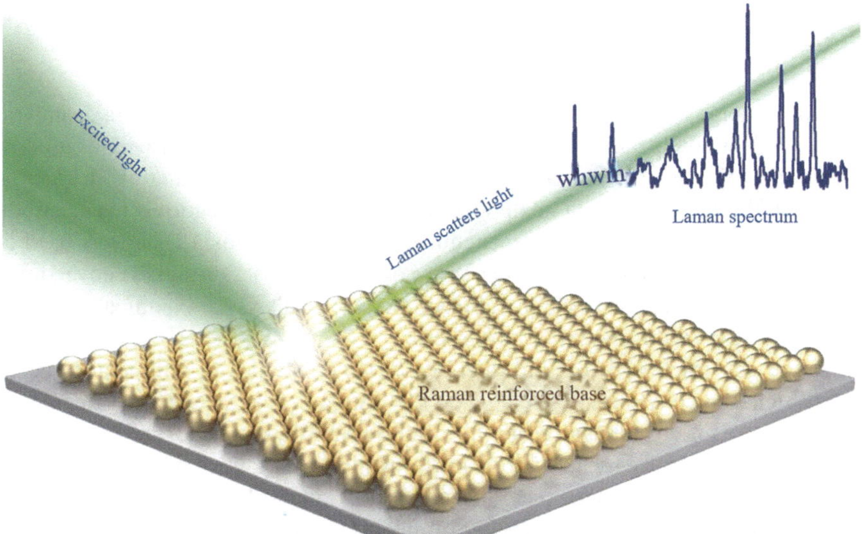

Fig. 1.8 SERS sensing technology refers to the analysis and detection of the inelastic light signals scattered by molecules adsorbed on the surface of nanostructured metal materials (such as silver or gold nanoparticles)

graphene oxide, MoS_2, black phosphorus and metal nanoparticles combination, has a larger contact area, can adsorb a large number of molecules, and can form more SERS hotspots. In addition, by changing the number of layers or shape of two-dimensional nanomaterials, a three-dimensional SERS structure can be formed, which also has an ideal Raman signal enhancement effect. More about SERS detection applications can be found in Chap. 2 of this book.

1.4.1.4 Micronanomaterials that Produce Other Optical Sensing Signals

Compared with fluorescence, the energy of chemiluminescence is provided by the chemical reaction process. Chemiluminescence can be divided into direct luminescence and indirect luminescence. Unlike direct luminescence, indirect luminescence has an energy transfer process. According to the different sources of energy, chemiluminescence can be divided into ordinary chemiluminescence, biochemical luminescence, and electrochemical luminescence. Generally speaking, chemiluminescence is based on organic structures (luminol, esters, etc.) through multiple oxidation reactions to generate excited state products, and the shorter wavelength light emitted when returning to the ground state. Some nanomaterials with enzyme-like catalytic activity (Au NPs, Ag NPs, metal-organic framework materials) are often used as catalysts for chemiluminescence reactions to improve detection signals and reduce detection limits.

1.4.2 Based on Electrical Signals

Due to the large surface area, high binding site activity, chemical stability, catalytic and excellent conductivity of micronanomaterials, they have been widely used in electrochemical detection. According to the role of micronanomaterials in electrochemical detection, they can be divided into the following categories: (1) electrode materials used to construct electrochemical sensing platforms; (2) carriers used to assist molecular probes to complete a series of complex reaction signal conversions; (3) tracers for direct electrochemical reactions; (4) catalysts for chemical reactions to promote the establishment of biosensors.

In the construction of the bio-interface of nanomaterials, it is crucial to accurately and efficiently immobilize aptamers or antibodies on the surface of nanomaterials. These micro and nano materials, such as AuNPs, graphene, and CNTs, have a large specific surface area and good biocompatibility. They are often used as electrode materials in electrochemical detection, providing a large number of binding sites by immobilizing a large number of aptamers and antibodies. The number of immobilized aptamers and antibodies will directly affect the sensitivity of the sensor. Transition metal oxides are rarely used as electrode materials due to their poor conductivity. As sensor electrode materials, nanomaterials face many challenges, one of which is how to prepare bio-functionalized micro and nano materials with aptamers, antibodies, and enzymes in a mild manner.

Compared with enzymes, the surfaces of nano-catalysts have abundant active sites, which helps them generate a large number of amplified signals. Therefore, nano-catalysts can be used in biosensors to solve some legacy issues related to the inherent environmental instability of biomaterials. Nanomaterials, as a signal amplification strategy assisted by nano-enzymes, have been widely used in biomolecular detection in recent years. For example, as shown in Fig. 1.9, Fe_3O_4 nanoparticles have intrinsic enzyme-like activity similar to natural peroxidase. The Fe_3O_4/chitosan-modified glassy carbon electrode is used for H_2O_2 and glucose detection. This electrode has the advantages of low application potential, small background current, and fast response to H_2O_2. In addition, as an inorganic nanomaterial, it has better stability to solution pH and temperature than natural enzymes. In addition, new nanomaterials such as $MoSe_2$ and black phosphorus have unique electrocatalytic activity and are also used to generate sensitive electrochemical signals, such as for magnetic immunodetection of proteins.

The measurement method based on the current generated by the electrochemical oxidation or reduction of electroactive substances has attracted widespread attention due to its high sensitivity. Nanoparticles composed of thousands of atoms can in principle be electrochemically oxidized or reduced. For example, AuNPs, AgNPs, and metal sulfide quantum dots can be used as labels for ultra-sensitive electrochemical bioassays. Gold nanoparticles are electrochemically oxidized in hydrochloric acid to generate electroactive $AuCl_4^-$, which emits a detectable signal after being reduced. Sulfide quantum dots dissolve in HNO_3 medium to obtain Cd^{2+}, Pb^{2+}, and Zn^{2+}, it can be detected with high sensitivity using the dissolution voltammetry method.

Nanomaterial electrodes can also be integrated into field-effect transistor (FET) type biosensors. FET type biosensors usually rely on the change in resistivity after the sensing material surface binds to the target object for detection and analysis. Silicon nanowires (Si NWs) are the most common materials for developing FET type biosensors. Since all the current flows through the cross-section of the nanowire, FET type biosensors based on Si NWs are very sensitive to minor changes in the

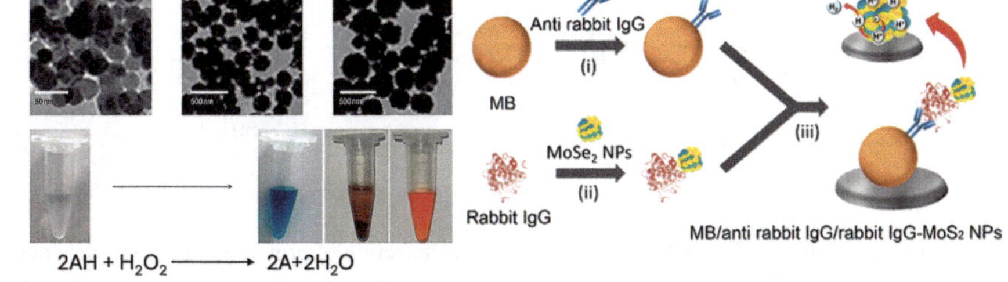

Fig. 1.9 (**a**) Peroxidase activity displayed by Fe_3O_4 nanoparticles; (**b**) Construction of magnetic immunodetection method with $MoSe_2$ labeled rabbit IgG showing hydrogen evolution reaction (HER) activity

surrounding environment. Moreover, they can easily be integrated into multi-channel biosensing, providing high-throughput platform technology. As shown in Fig. 1.10, the conductivity of biotin-modified silicon nanowires significantly increases after the addition of streptavidin, and this Si NWs-based FET sensor system can be used for the detection of a series of biomarkers such as DNA, proteins, and even viruses. Due to the high conductivity of carbon nanomaterials (carbon nanotubes, graphene, etc.), even a very small amount of target objects binding to the surface of nanomaterials will change the electrical properties of the nanomaterials, so they are also widely used in the construction of FET type biosensors.

1.4.3 Based on Magnetic Signals

Firstly, magnetic materials ranging in size from a few nanometers to tens of nanometers can be synthesized in a controlled manner. Secondly, the movement of magnetic nanoparticles can be effectively manipulated by external magnetic forces, providing new technical means for many biomedical applications including biosensing, diagnosis, and drug delivery. Because the proton magnetic moment signal around magnetic nanoparticles can be captured by resonant absorption, magnetic nanoparticles have the ability to enhance magnetic resonance contrast and can be used for magnetic signal detection and analysis. Magnetic nanomaterials, due to their unique properties and broad application prospects, have aroused widespread research interest as nanoprobes in the detection of bacteria, viruses, proteins, and other active biomolecules.

In the field of sensing detection, magnetic nanoparticles are widely used in the development of biological detection reagents because they can separate different reactants under magnetic field conditions. Using magnetic nanoparticles in biological detection can simplify the separation or washing steps involved in the detection process. For example, adding magnetic nanoparticles to the conventional quantum

Fig. 1.10 SiNW-based FET sensor used for real-time detection of protein binding. (**a**) Schematic of biotin-modified SiNW (left) and streptavidin subsequently binding to the SiNW surface (right); (**b**) The change in conductivity of biotin-modified SiNW over time, where region 1 corresponds to the buffer solution, region 2 corresponds to the addition of 250 nM streptavidin, and region 3 corresponds to pure buffer solution

dot coded microsphere detection system, combined with a magnetic field and microfluidic device to achieve its automated detection, can simplify the entire detection process.

At the same time, the change in optical and/or magnetic properties of magnetic nanoparticles during assembly allows them to be used as nanoprobes for rapid and sensitive detection of related biomarkers. During the assembly process of magnetic nanoparticles, the spin relaxation time (T_2 Relaxation time, δT_2, allows nanoparticles to act as Magnetic Relaxation Switches (MRS) (Fig. 1.11). This unique magnetic phenomenon provides a great opportunity to design magnetic nanoprobe capable of detecting various biomarkers, such as nucleic acids, proteins, antibodies, viruses, stem cells, and circulating tumor cells.

Giant Magnetoresistance (GMR) probes, such as Spin Valve (SV) probes, can detect trace amounts of magnetic nanoparticles. Therefore, magnetic nanoparticles can be used as detection labels, and the GMR effect can be used to detect the target.

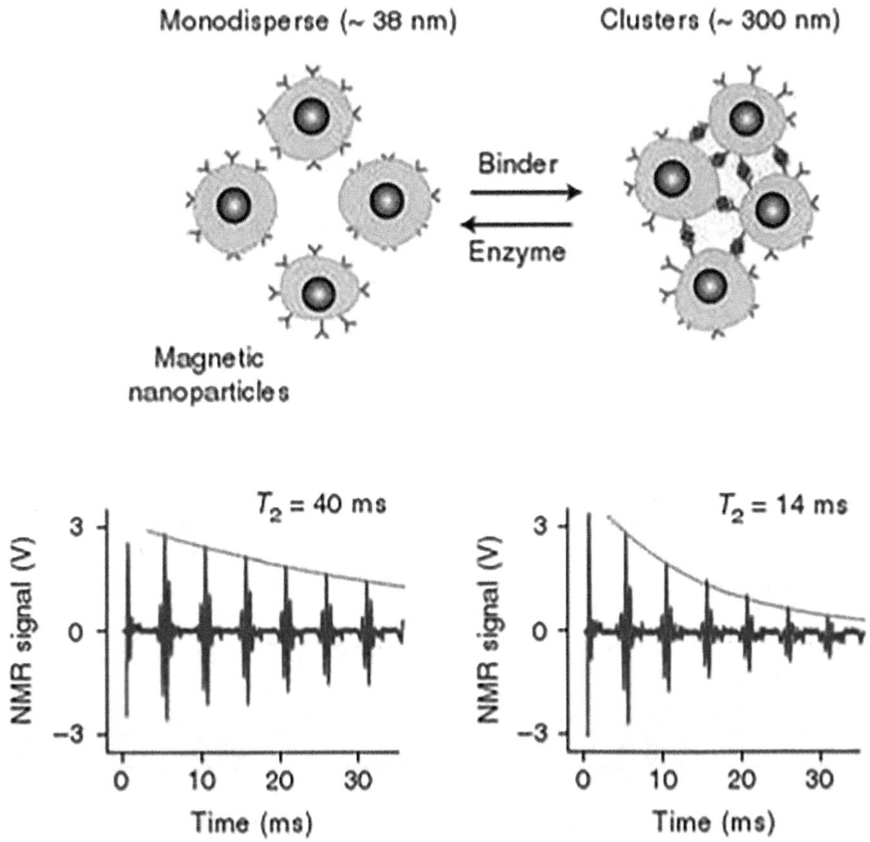

Fig. 1.11 When monodispersed magnetic nanoparticles self-assemble through the target detection object, it leads to a reduction in spin relaxation time (T_2). Adding avidin to the solution of biotin-modified magnetic nanoparticles reduces T_2 from 40 ms to 14 ms

For example, as shown in Fig. 1.12, the analyte is captured on the probe surface by antibodies, the analyte is connected to the magnetic nano-label through biotin-streptavidin interaction, and finally, the GMR probe is used to detect the magnetic nano-label, thereby quantifying the analyte. This method of quantifying analytes based on magnetic nano-labels can be used for protein detection in clinically relevant real serum samples, and can also be used for the detection of various proteins when issues such as stability and reproducibility are further resolved. In addition, the magnetic nanoprobe technology is insensitive to the matrix of various media, making it possible to directly apply this technology to various complex clinical sample environments for in vitro diagnosis, making it a good candidate technology for multifunctional and portable molecular diagnosis in clinical environments. More about the application of magnetic signal sensing detection can be found in Chap. 5 of this book.

1.4.4 Based on Chemical Signals

Mass Spectrometry (MS), which classifies ions based on their mass-to-charge ratio (m/z), is a highly sensitive and accurate analytical technique. However, traditional mass spectrometry techniques are usually time-consuming and single-channel. With the emergence of the most representative elemental mass spectrometry technique—Inductively Coupled Plasma Mass Spectrometry

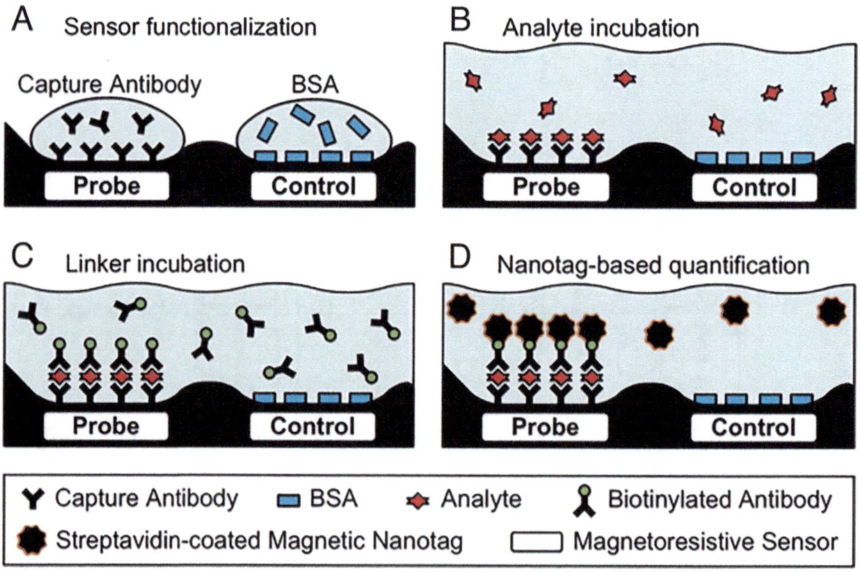

Fig. 1.12 Protein detection method based on magnetic nano-labels

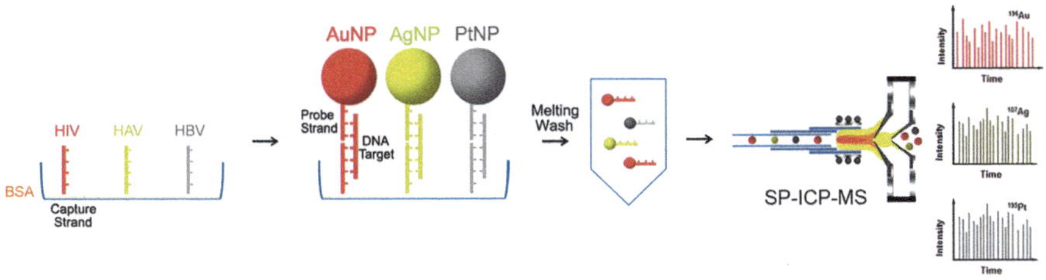

Fig. 1.13 SP-ICP-MS multi-channel DNA detection based on metal nanoprobe

(ICP-MS), and the development of exogenous elemental labeling techniques, mass spectrometry has been widely used for multi-channel bio-detection. Exogenous elemental labels developed for ICP-MS include lanthanide metal chelates, nanoparticles containing metal elements (such as Au NPs, Ag NPs, PbS NPs, and QDs), and polymer-related elemental labels. For example, as shown in Fig. 1.13, using three types of metal nanoparticles (Au NPs, Ag NPs, and Pt NPs) as labels, and using Single Particle mode Inductively Coupled Plasma Mass Spectrometry (SP-ICP-MS), low concentration HIV, HAV, and HBV three DNA targets can be detected simultaneously. In addition to nucleic acid targets, the use of multi-element metal nanoparticle labels in ICP-MS also makes multi-channel detection of low-abundance protein targets possible.

Subsequently, ICP-MS was combined with flow cytometry to develop mass spectrometry flow cytometry detection, which is particularly suitable for multi-channel bio-detection. Since large-scale cell technology is largely unaffected by spectral overlap, it allows for the simultaneous detection of many more parameters than traditional fluorescence flow cytometry.

1.4.5 Based on Other Signals

Nanoparticles that can generate heat are commonly found in photothermal therapy applications for tumor tissues. Existing research has begun to utilize the thermal properties of metallic nanoparticles for disease detection. Typically, they can be realized by marking with heat-generating micro-nano materials, such as polydopamine nanoparticles, gold nanoparticles, two-dimensional black phosphorus, and Ti_3C_2 MXene. The increase in temperature is directly proportional to the number of tracer particles, and this photothermal (PT) response allows for the quantitative detection of analyte concentration in the sample. This method, combined with thermal imaging technology, is often used in lateral flow assays. In this system, antibodies marked with photothermal materials generate heat signals under near-infrared irradiation, which can be amplified and detected by a heat detector (Fig. 1.14). Currently, compared to traditional visual lateral flow assays, the use of heat signal detection can increase the sensitivity of lateral flow assays by 32 times. Lateral flow

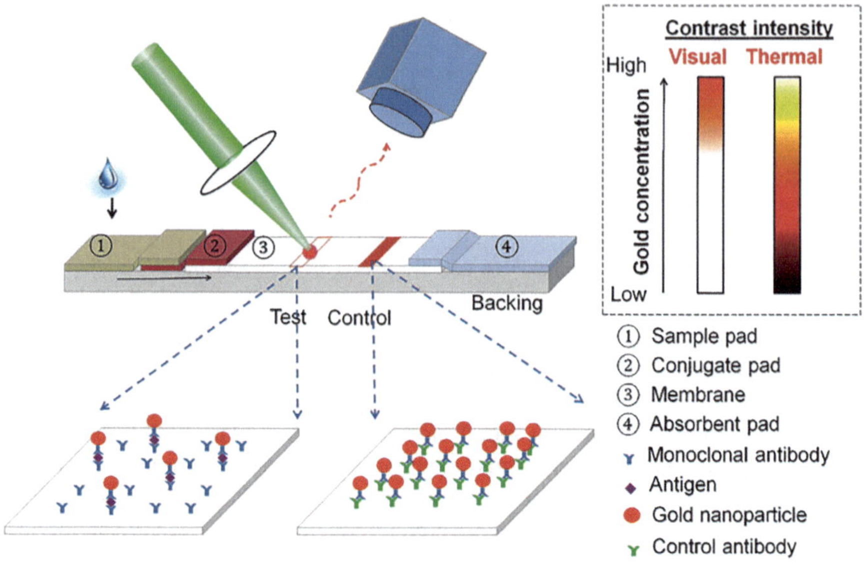

Fig. 1.14 Gold nanoparticle-antibody antigen complexes bind with monoclonal antibodies adhered to the test paper, retaining gold nanoparticles in the test area and causing a visible color change in the test strip. At low antigen concentrations, when there are not enough bound gold nanoparticles for visual contrast, heat signals are collected for detection

detectors based on thermal imaging can achieve high-sensitivity detection of target objects and do not suffer from photobleaching or light instability like fluorescence or chemiluminescence methods. However, this operation requires expensive and bulky infrared image sensors. Small, inexpensive thermal imaging sensors have been introduced, and these technological advances, combined with the integration of photothermal tracers, may promote their application in the field of practical sensing. In addition, many excellent photothermal performance nanomaterials also provide more choices for the development of ex vivo sensing detection technology based on heat signals.

Under the irradiation of an external light source, the laser-induced heating of the target nanoparticles also causes rapid thermal expansion and contraction of the surrounding medium, thereby generating sound signals. These photoacoustic responses can be measured by acoustic sensors to quantify the target biomarkers in immunochromatography (Fig. 1.15). This method often uses metal nanoparticles with strong light absorption properties as photoacoustic agents. Compared with traditional detection strategies, lateral flow detectors based on photoacoustics have better detection performance. However, they require high-power laser sources to generate efficient photoacoustic signals, and for high-sensitivity detection, lateral flow test strips need to be enclosed in a soundproof room. These requirements make this technology cumbersome and costly to implement.

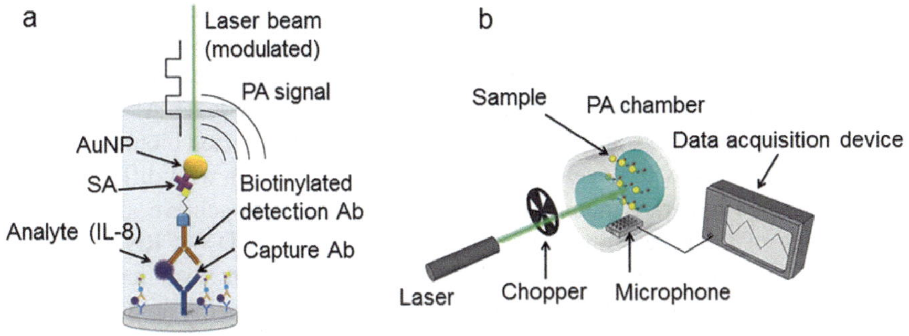

Fig. 1.15 (**a**) Photoacoustic immunoassay mode; (**b**) Device for measuring photoacoustic signals. The sample is placed in a sealed chamber, heated periodically, and therefore, expands and contracts periodically, generating pressure oscillations. The microphone detects these pressure oscillations and displays them as waveforms on the oscilloscope. The sandwich method is used for immunoassay, where the analyte is captured by immobilized antibodies and marked by detection antibodies, the detection antibodies are further labeled with coupled AuNPs, AuNPs generate photoacoustic signals in the above setup, and the signal strength reflects the concentration of the analyte

1.5 Trend Analysis

In the past 20 years, the demand for ultra-sensitive biosensors has been growing, stimulating a large amount of research into the design and development of functional nanoprobes. Micro-nano materials have been widely used in the development of new biosensors. A large number of detection methods have been explored, including fluorescence, SERS, electrochemistry and photochemistry, as well as magnetism and other technologies. Although many reported biosensors have shown advantages in convenience, selectivity, sensitivity, and multiplexing capabilities, there are still huge challenges in implementing these ultra-sensitive biosensors in practical applications. ELISA and PCR are still the gold standards for protein and nucleic acid detection in clinical diagnosis. In particular, the structure and performance stability of functional micro-nano probes have always been the main obstacles to their widespread application. Therefore, in addition to constantly striving to find new and better micro-nano materials to develop biosensing probes, we should also make greater efforts to face the practical challenges and needs brought about by hospital and other epidemic prevention and control medical conditions. Taking clinical applications as an example, new detection methods should meet the basic standards for diagnostic purposes, and show stable and reliable sensing detection performance for specific biomolecules under complex environments. Therefore, basic research on the interaction mechanism and interface properties of the bio/micro-nano material interface is crucial to the development of this field. Due to the high serum or whole blood background, the interface design of micro-nano probes is particularly important to combat strong non-specific adsorption. In some clinical trials, the determination time is another key parameter, for example, it is crucial to detect cardiac indicators in a few minutes rather than a few hours. Considering this, it is also very important to increase the binding affinity and efficiency of biomolecules, for example, by reducing the size of the interface or using self-propelled probe carriers to promote mass transfer mixing. Many diseases are complex and cannot rely on a single level of biomarkers to reflect the true level. Although DNA and protein microarrays are popular technologies, cross-species biomarker detection is still rare. How to develop micro-nano probe materials with different signal responses, and integrate them into a biosensing device, to achieve simultaneous monitoring of multiple biomedical test indicators, such as DNA, RNA, proteins, etc., in a miniaturized and user-friendly mode, will bring new technical platforms and hopes for the future application of micro-nano materials in the field of biomedical sensing detection.

Bibliography

1. ALI N, BILAL M, KHAN A, et al. Chapter 4 - Fabrication strategies for functionalized nanomaterials [M]//TAHIR M B, SAGIR M, ASIRI A M. Nanomaterials: Synthesis, Characterization, Hazards and Safety. Elsevier. 2021: 55-95.
2. FAROOQI Z U R, QADEER A, HUSSAIN M M, et al. Chapter 5 - Characterization and physicochemical properties of nanomaterials [M]//TAHIR M B, SAGIR M, ASIRI A M. Nanomaterials: Synthesis, Characterization, Hazards and Safety. Elsevier. 2021: 97-121.
3. Li Ling, Xiang Hang. Functional Materials and Nanotechnology [M], Chemical Industry Press, First Edition, July 2003.
4. AMEN R, MUKHTAR A, SAQIB S, et al. Chapter 1 - History and development of nanomaterials [M]//TAHIR M B, SAGIR M, ASIRI A M. Nanomaterials: Synthesis, Characterization, Hazards and Safety. Elsevier. 2021: 1-14.
5. Wang Xin, Liu Xiaoheng. Concise Course on Nanomaterials [M]. Beijing: Chemical Industry Press, 2010.
6. SIGEL R W, HU E, ROCO M C. Nanostructure science and technology-a worldwide study [C]. National Science and Technology Council (NSTC) Committee on Technology, 1999, 1-362.
7. BURDA C, CHEN X, NARAYANAN R, et al. Chemistry and Properties of Nanocrystals of Different Shapes [J]. ChemInform, 2005, 36(27).
8. WANG X, PENG Q, LI Y. Interface-Mediated Growth of Monodispersed Nanostructures [J]. Accounts of Chemical Research, 2007, 40(8): 635-43.
9. WILEY B, SUN Y, MAYERS B, et al. Shape-Controlled Synthesis of Metal Nanostructures: The Case of Silver [J]. Chemistry – A European Journal, 2005, 11(2): 454-63.
10. MURRAY C, KAGAN C, BAWENDI M G. Synthesis and Characterization of Monodisperse Nanocrystals and Close-Packed Nanocrystal Assemblies [J]. Annual Review of Materials Science - ANNU REV MATER SCI, 2000, 30: 545-610.
11. BAIG N, KAMMAKAKAM I, FALATH W. Nanomaterials: a review of synthesis methods, properties, recent progress, and challenges [J]. Materials Advances, 2021, 2(6): 1821-71.
12. DANG S, ZHU Q-L, XU Q. Nanomaterials derived from metal–organic frameworks [J]. Nature Reviews Materials, 2017, 3(1): 17075.
13. KOLAHALAM L A, KASI VISWANATH I V, DIWAKAR B S, et al. Review on nanomaterials: Synthesis and applications [J]. Materials Today: Proceedings, 2019, 18: 2182-90.
14. CHEN H, HU H, TAO C, et al. Self-Assembled Au@Fe Core/Satellite Magnetic Nanoparticles for Versatile Biomolecule Functionalization [J]. ACS Applied Materials & Interfaces, 2019, 11(27): 23858-69.

15. SHEN J, ZHU Y, YANG X, et al. Graphene quantum dots: emergent nanolights for bioimaging, sensors, catalysis and photovoltaic devices [J]. Chemical Communications, 2012, 48(31): 3686-99.
16. ENDO M, KIM Y A, HAYASHI T, et al. Structural characterization of cup-stacked-type nanofibers with an entirely hollow core [J]. Applied Physics Letters, 2002, 80(7): 1267-9.
17. LI D, XIA Y. Electrospinning of Nanofibers: Reinventing the Wheel? [J]. Advanced Materials, 2004, 16(14): 1151-70.
18. MEI J, LIAO T, SUN Z. Two-dimensional metal oxide nanosheets for rechargeable batteries [J]. Journal of Energy Chemistry, 2018, 27(1): 117-27.
19. LIU J, YANG Y, LYU P, et al. Few-Layer Silicene Nanosheets with Superior Lithium-Storage Properties [J]. Advanced Materials, 2018, 30(26): 1800838.
20. WANG L-B, WANG Y-C, HE R, et al. A New Nanobiocatalytic System Based on Allosteric Effect with Dramatically Enhanced Enzymatic Performance [J]. Journal of the American Chemical Society, 2013, 135(4): 1272-5.
21. LI G, MA P, HE Y, et al. Enzyme–Nanowire Mesocrystal Hybrid Materials with an Extremely High Biocatalytic Activity [J]. Nano Letters, 2018, 18(9): 5919-26.
22. AN J, LI G, ZHANG Y, et al. Recent Advances in Enzyme-Nanostructure Biocatalysts with Enhanced Activity [J]. Catalysts, 2020, 10(3).
23. KHAN I, SAEED K, KHAN I. Nanoparticles: Properties, applications and toxicities [J]. Arabian Journal of Chemistry, 2019, 12(7): 908-31.
24. FENG Cuiju. The peculiar properties of nanomaterials [J]. Modern Physics Knowledge, 2008, (04): 5-7.
25. MURTY B S, SHANKAR P, RAJ B, et al. Unique Properties of Nanomaterials [M]//MURTY B S, SHANKAR P, RAJ B, et al. Textbook of Nanoscience and Nanotechnology. Berlin, Heidelberg; Springer Berlin Heidelberg. 2013: 29-65.
26. ANDRIEVSKI R A, GLEZER A M. Size effects in properties of nanomaterials [J]. Scripta Materialia, 2001, 44(8): 1621-4.
27. SOSA I O, NOGUEZ C, BARRERA R G. Optical Properties of Metal Nanoparticles with Arbitrary Shapes [J]. The Journal of Physical Chemistry B, 2003, 107(26): 6269-75.
28. MCINTYRE R A. Common Nano-Materials and Their Use in Real World Applications [J]. Science Progress, 2012, 95(1): 1-22.
29. XIA T, CAO Y, OYLER N A, et al. Strong Microwave Absorption of Hydrogenated Wide Bandgap Semiconductor Nanoparticles [J]. ACS Applied Materials & Interfaces, 2015, 7(19): 10407-13.
30. REMES Z, NESLADEK M, HAENEN K, et al. The optical absorption and photoconductivity spectra of hexagonal boron nitride single crystals [J]. physica status solidi (a), 2005, 202(11): 2229-33.
31. CAMELIO S, TOUDERT J, BABONNEAU D, et al. Tailoring of the optical properties of Ag:Si3N4 nanocermets by changes of the cluster morphology [J]. Applied Physics B, 2005, 80(1): 89-96.
32. FENG M, CHEN Y, GU L, et al. CdS nanoparticles chemically modified PAN functional materials: Preparation and nonlinear optical properties [J]. European Polymer Journal, 2009, 45(4): 1058-64.
33. Zhang Guangping, Li Menglin, Wu Ximao, et al. Research progress on the effect of scale on the resistivity of metal materials [J]. Journal of Materials Research, 2014, 28(02): 81-7.
34. Che Yaping. Research on the improvement and mechanism of dielectric properties of polymer-based nanocomposites [D]; University of Science and Technology of China, 2020.
35. SAVAGE T, RAO A M. Thermal Properties of Nanomaterials and Nanocomposites [M]// TRITT T M. Thermal Conductivity: Theory, Properties, and Applications. Boston, MA; Springer US. 2004: 261-84.
36. YETTER R A, RISHA G A, SON S F. Metal particle combustion and nanotechnology [J]. Proceedings of the Combustion Institute, 2009, 32(2): 1819-38.

37. WRONSKI C R M. The size dependence of the melting point of small particles of tin [J]. British Journal of Applied Physics, 1967, 18(12): 1731-7.
38. Cheng Benpei, Kong Jie, Luo Ju, et al. The relationship between the structural stability and Debye temperature of nano TiO_2 and particle size [J]. Progress in Materials Science, 1993, (03): 240-3.
39. SOLER M A G, PATERNO L G. 6 - Magnetic Nanomaterials [M]//DA RóZ A L, FERREIRA M, DE LIMA LEITE F, et al. Nanostructures. William Andrew Publishing. 2017: 147-86.
40. SUN X-C. Microstructure characterization and magnetic properties of nanomaterials [J]. Molecular Physics, 2002, 100(19): 3059-63.
41. KIKUCHI K, TAKAKUSA H, NAGANO T. Recent advances in the design of small molecule-based FRET sensors for cell biology [J]. Trac-Trends in Analytical Chemistry, 2004, 23(6): 407-15.
42. CHEN H, ZHANG L, HU Y, et al. Nanomaterials as optical sensors for application in rapid detection of food contaminants, quality and authenticity [J]. Sensors and Actuators B-Chemical, 2021, 329.
43. LIU G, ZHANG R, HUANG X, et al. Visual and Colorimetric Sensing of Metsulfuron-Methyl by Exploiting Hydrogen Bond-Induced Anti-Aggregation of Gold Nanoparticles in the Presence of Melamine [J]. Sensors, 2018, 18(5).
44. REX M, HERNANDEZ F E, CAMPIGLIA A D. Pushing the limits of mercury sensors with gold nanorods [J]. Analytical Chemistry, 2006, 78(2): 445-51.
45. LANGER J, DE ABERASTURI D J, AIZPURUA J, et al. Present and Future of Surface-Enhanced Raman Scattering [J]. Acs Nano, 2020, 14(1): 28-117.
46. GAO L, ZHUANG J, NIE L, et al. Intrinsic peroxidase-like activity of ferromagnetic nanoparticles [J]. Nature Nanotechnology, 2007, 2(9): 577-83.
47. TOH R J, MAYORGA-MARTINEZ C C, SOFER Z, et al. MoSe2 Nanolabels for Electrochemical Immunoassays [J]. Analytical Chemistry, 2016, 88(24): 12204-9.
48. CUI Y, WEI Q Q, PARK H K, et al. Nanowire nanosensors for highly sensitive and selective detection of biological and chemical species [J]. Science, 2001, 293(5533): 1289-92.
49. LEE H, SUN E, HAM D, et al. Chip-NMR biosensor for detection and molecular analysis of cells [J]. Nature Medicine, 2008, 14(8): 869-74.
50. OSTERFELD S J, YU H, GASTER R S, et al. Multiplex protein assays based on real-time magnetic nanotag sensing [J]. Proceedings of the National Academy of Sciences of the United States of America, 2008, 105(52): 20637-40.
51. ZHANG S, HAN G, XING Z, et al. Multiplex DNA Assay Based on Nanoparticle Probes by Single Particle Inductively Coupled Plasma Mass Spectrometry [J]. Analytical Chemistry, 2014, 86(7): 3541-7.
52. LI S, ZHANG Y, WEN W, et al. A high-sensitivity thermal analysis immunochromatographic sensor based on au nanoparticle-enhanced two-dimensional black phosphorus photothermal-sensing materials [J]. Biosensors & Bioelectronics, 2019, 133: 223-9.
53. ZHAO Y, CAO M, MCCLELLAND J F, et al. A photoacoustic immunoassay for biomarker detection [J]. Biosensors & Bioelectronics, 2016, 85: 261-6.

Chapter 2
Surface Enhanced Raman Technology and SERS Sensor Detection Applications

2.1 Surface Enhanced Raman Technology

2.1.1 Introduction

Light is a natural phenomenon that can be seen everywhere in life. Scientific research defines light as an electromagnetic wave with a certain frequency. Its essence is an electromagnetic wave visible to the human eye. Light is also known as the visible spectrum, which is composed of tens of thousands of photons, possessing both particle and wave properties. Human research on light can be traced back to ancient Greece. At that time, humans had a certain understanding of the generation, propagation, and application of light. However, due to the combination of natural science and religious beliefs at that time, people's understanding of light stagnated until the nineteenth century when the research results of Scottish physicist James Clerk Maxwell appeared. Physicists gained a deeper understanding of the laws of light, and then began to further study the properties and applications of light. People's understanding of the essence of light has become clearer with these studies. Nowadays, the research on the properties and applications of light is still a hot direction in various scientific research fields, including aerospace, weapons development, material processing, medical equipment, ranging, analytical detection, photography, daily lighting, etc.

Light analysis and detection is the use of the interaction effect of light and matter to analyze the test object. The interaction process includes scattering, reflection, absorption, refraction, dispersion, radiation, etc. of light. Among them, light scattering refers to the phenomenon that the propagation direction of incident light in the medium changes when there are uneven physical properties such as magnetic field, electric field, frequency, etc. in the system. The light deviating from the original direction is called scattered light. There are many types of scattered light. From the perspective of frequency change, it can be divided into elastic scattering and

X. Ma et al., *Functional Micro/Nanomaterials for Intelligent Sensing Devices*, https://doi.org/10.1007/978-981-96-4990-7_2

inelastic scattering. Elastic scattering refers to the light not changing in frequency (wavelength) before and after scattering, that is, the frequency of incident light and scattered light is the same, such as Rayleigh scattering, Mie scattering, etc.; inelastic scattering refers to the change in wavelength (frequency) of light before and after scattering, that is, the frequency of incident light and scattered light is different, such as Raman scattering, Compton scattering, etc.

Physicists studied elastic scattering earlier. Faraday first discovered that when the particle size is relatively small compared to the wavelength of the incident light, small changes in the particle size will cause it to produce different colors. Later, Lord Rayleigh discovered in his research that when the particle size is much smaller than the wavelength of light, the intensity of scattered light is inversely proportional to the fourth power of the incident light wavelength. He also found the relationship between the intensity of scattered light and the angle distribution. These findings further helped people explain why the sky appears blue, among other things. As research deepened, humans began to study inelastic scattering of light. As early as 1923, A. Smekal and others theoretically predicted the existence of inelastic scattering phenomena. In 1928, C. V. Raman and K. S. Krishman first discovered spectral lines in the scattered light of carbon tetrachloride liquid that had different frequencies from the incident light and were extremely weak in intensity. This discovery was later called Raman scattering. By analyzing the Raman scattering spectrum, information about molecular rotation and vibration can be obtained, which is used in the study of molecular structure. C. V. Raman also made a significant contribution to the field of molecular detection with Raman spectroscopy, for which he received the Nobel Prize in Physics in 1930. However, due to poor experimental conditions at the time, the Raman signal produced was extremely weak. It wasn't until the 1960s that the advent of ruby lasers provided an ideal light source for Raman scattering experiments, providing a stable Raman signal and furthering research. Today, Raman spectroscopy can not only detect the composition, functional groups, stress, etc. of a substance, but also measure the symmetry, orientation, quality, etc. of a crystal. It can be said that Raman spectroscopy has become an indispensable technology in the field of molecular vibration and molecular structure characterization. However, in current applications, Raman spectroscopy still has limitations such as low sensitivity and weak signals, so a technology is needed to enhance the Raman signal. The emergence of surface-enhanced Raman scattering technology has solved this problem, and because of its high sensitivity, this technology can obtain internal molecular structure information that is not easily detected by conventional Raman spectroscopy.

2.1.2 Current Research Status of Surface-Enhanced Raman Technology

Surface-Enhanced Raman Scattering (SERS) refers to the phenomenon where the Raman scattering of molecules adsorbed on the surface of corrugated metals such as gold and silver nanoparticles is greatly enhanced. It was first observed in 1974 by

Fleischmann and his collaborators when measuring the surface Raman spectrum of pyridine adsorbed on an electrochemically roughened silver electrode. They found that the Raman signal measured by this method was unexpectedly large, but it was not recognized at the time as it was the first measurement of SERS. In 1977, Jeanmaire and Van Duyne further quantified the enhancement results of the Raman signal by organizing experiments and calculating data. They found that large chunks of pyridine provided the main Raman signal, which could be enhanced by 10^5–10^6 times. In the same year, Albrecht and Creighton also reported similar results, further indicating that the enhancement of Raman scattering signals is not accidental, but caused by the enhancement of Raman scattering efficiency. The discovery of SERS has laid a solid foundation for basic research and applications such as molecular research and material characterization, and has opened up a new field—surface-enhanced spectroscopy. SERS has been a hot direction for more than 40 years since its discovery, and tens of thousands of results in SERS-related fields have been published, including review papers and application articles, as well as related patents, and the number of research teams studying SERS mechanisms and their applications is also gradually increasing. Today, SERS technology is widely used in many fields such as biology, catalysis, medical detection, electrochemistry, etc., due to its advantage of being able to serve as a super-sensitive probe to determine the detailed structure and orientation of surface molecules.

2.1.3 Surface Enhanced Raman Technology Mechanism

Theoretical research on SERS is particularly important for its practical application. To date, the enhancement mechanism of SERS has not been thoroughly studied. Over the decades, people have found that the enhancement mechanism of SERS is not a single mechanism, but may be formed by the interaction of several mechanisms. The two mechanisms widely accepted by people are the physical electromagnetic enhancement mechanism and the charge transfer enhancement mechanism (chemical enhancement mechanism), which are completely different. The classical electromagnetic theory believes that the intensity of Raman scattering is proportional to the induced dipole moment of the molecule, and the induced dipole moment is closely related to the incident light field strength E and the molecular polarizability a. Nowadays, all theories and models about the SERS enhancement mechanism are based on the incident light field strength and molecular polarizability. The physical electromagnetic enhancement theory and the charge transfer enhancement theory correspond to enhancing the incident light field strength and enhancing the molecular polarizability, respectively. The electromagnetic enhancement theory mainly considers the enhancement of the local electric field on the metal surface formed by the interaction of the incident light and the electrons on the metal surface. The charge transfer enhancement (chemical enhancement) mainly considers the change in the molecular polarizability caused by the chemical interaction between the analyte molecule and the metal. It is generally believed that the electromagnetic enhancement effect is greater than the chemical enhancement effect, and most of the

SERS enhancement is essentially the mechanism of electromagnetic enhancement, but it will have some elements of the chemical enhancement mechanism.

2.1.3.1 Electromagnetic Enhancement Mechanism

The Electromagnetic Enhancement Mechanism (EM) believes that when incident light shines on a metal surface with appropriate roughness, the local electromagnetic field strength produced by this surface will greatly increase. Because the Raman scattering intensity is proportional to the square of the local electromagnetic field strength where the analyte molecule is located, the analyte molecule on this surface increases the probability of producing Raman scattering, thereby greatly enhancing the Raman signal of the surface-adsorbed molecule, forming an electromagnetic enhancement effect. The electromagnetic enhancement mechanism can use electrodynamics calculations to further quantify the enhancement effect. In SERS substrates, when plasmons are excited at frequency ω, an enhanced electric field amplitude E(ω) will appear. SERS enhancement is usually approximated by the average value of $|E(\omega)|^4/|E_0|^4$ on the irradiated molecule, where E_0 is the incident light field amplitude. Further subdividing, E(ω) can be divided into two parts, one part is the enhanced electric field amplitude $E(\omega_1)$ that appears when the aforementioned plasmon is excited.), and another part is the enhanced electric field amplitude $E(\omega_2)$ that appears when the Raman signal is emitted from the near field to the far field. It can also be said that it further considers the impact of Stokes displacement on electromagnetic enhancement. In this case, SERS enhancement can be approximately represented as $|E(\omega_1)|^2\ |E(\omega_2)|^2/|E_0|^4$. Studying the factors that affect electromagnetic field enhancement is also very important. These factors will indirectly affect the enhancement effect of the Raman signal. The main factors that cause electromagnetic field enhancement are as follows:

Surface Plasmon Resonance (SPR)

This factor is considered to be the main source of SERS enhancement. The electrons on the rough metal surface vibrate under the action of the incident laser, forming surface plasmon resonance at a specific frequency. After the surface plasmon resonance is generated, a very large local electric field is formed on the substrate surface, and the Raman signal of the measured molecules in this local electric field also increases significantly. It should be noted that the electric field intensity produced by surface plasmon resonance has a large distance effect and decreases exponentially with the increase of distance from the surface. Its maximum effect range can reach several nanometers, and this mechanism belongs to long-range effects. Different metals require different laser frequencies to excite surface plasmon resonance. The metals that can produce surface plasmon resonance under visible light excitation mainly include precious metals such as silver, gold, copper, and alkali metals and other free electron metals.

Lightning Rod Effect

The process of preparing the SERS substrate often produces some needle-shaped nanoparticles with a very small radius of curvature. These particles will generate a strong local surface electromagnetic field at the tip, and the surface field strength increases with the sharpness of the tip. Many metals can produce this enhancement effect. In group 11 metals, the electromagnetic field enhancement caused by the SPR effect is much higher, so the lightning rod effect only plays a secondary role in group 11 metals. In transition metals, Yang's theoretical calculation results believe that the lightning rod effect is the main source of electromagnetic field enhancement in the transition metal system and increases with the increase of the aspect ratio. However, many experimental results have observed the SPR effect on the surface of transition metals and believe it is the main reason for electromagnetic field enhancement.

Image Field Effect

The mirror field model assumes that the metal surface is an ideal mirror, the adsorbed molecule is a vibrating dipole, which generates a conjugate electric dipole within the metal, thereby forming a mirror photoelectric field on the surface. The incident light and the mirror photoelectric field both enhance the surface Raman signal of the adsorbed molecule, and coupled with the surface reflection causing a double local electric field enhancement, a cumulative 16-fold total enhancement effect can be obtained. The intensity of the mirror field is inversely proportional to the cube of the distance, that is, as the distance increases, the mirror field strength rapidly decreases, so it is a very short-range effect, which only explains the order of magnitude of surface enhancement within the atomic spacing range. After considering the orientation of molecular adsorption and different vibration modes, Moskovits and others believe that an enhancement effect of 5-6 times is more reasonable in the actual research system. In summary, this effect does not make a significant contribution to the overall enhancement effect. In addition, this model simplifies the multipole molecule into a dipole, and can only explain part of the surface enhancement factor.

2.1.3.2 Chemical Enhancement Mechanism

The electromagnetic field enhancement mechanism explains the occurrence of the SERS phenomenon from the perspective of photoelectric field enhancement, and also has certain guiding significance for the preparation of SERS substrates, but the electromagnetic field enhancement mechanism cannot explain all experimental phenomena, and needs to rely on the chemical enhancement mechanism. For example, N_2 and CO with the same Raman scattering cross-section are adsorbed on the same SERS active substrate, but their Raman enhancement effects are significantly different, with the latter being 200 times higher than the former. Even considering the difference in adsorption orientation on the surface, the peak intensities of the two cannot have such a large difference. Another major basis for the existence of chemical enhancement comes from the SERS research results of electrochemical systems. Experiments show that for many systems, the SERS intensity of molecules adsorbed on the electrode surface is often a function of the applied electrode potential. The SERS intensity of the probe molecules adsorbed on the surface varies with the change of the electrode potential and reaches a maximum value, and this maximum value often shifts with the change of the excitation light wavelength. The theoretical simulation results of single-molecule SERS enhancement also show that even under the best conditions, the electromagnetic field enhancement factor of SERS active sites can only reach 11–12 orders of magnitude, which is still two to three orders of magnitude different from the maximum 14 orders of magnitude enhancement factor obtained experimentally. It is reasonable to believe that the difference of two to three orders of magnitude is due to the contribution of the chemical enhancement mechanism. To explain these special phenomena, scientists have proposed many models to study the chemical enhancement mechanism from the perspective of the interaction and bonding between the adsorbate and the metal surface. The chemical enhancement mechanism mainly includes the following:

Molecule-Metal Bonding

The selective Raman enhancement caused by molecule-metal bonding occurs when there is a bonding interaction between the molecule and the surface metal axis, resulting in a modified polarizability. If the intensity of a characteristic peak of a vibration mode in the Raman spectrum is used as an internal reference, the relative intensities of different vibration peaks will be obtained, which is due to the selective Raman enhancement triggered by the chemical bonding of the adsorbate and the metal substrate. This theory believes that the formation of a molecule-metal bond is a prerequisite for Raman enhancement.

Photoinduced Charge Transfer

Currently, the most reported in the research of chemical enhancement mechanisms is photoinduced charge transfer (PICT). For example, Otto and others proposed a 4-step charge transfer process. Taking the charge transfer from metal to molecule as an example, it mainly goes through the following steps:

1. Electrons near the metal Fermi energy level are excited to a high energy level, creating a hole below the Fermi energy level, thus forming an electron-hole pair on the metal side;
2. The electron that has absorbed the photon energy moves to the electron affinity level of the adsorbed molecule;

3. After a short relaxation time, the electron migrates back to the metal, at which point the adsorbed molecule is in a vibrational excited state;
4. The returning electron recombines with the hole inside the metal and radiates a Raman photon. Light-induced charge transfer makes the transition process, which is allowed in the original molecular energy state but lacks direct transition energy, become a process with a higher probability of occurrence through the auxiliary role of the metal energy level. The system's polarization rate is significantly increased, resulting in a higher Raman intensity.

2.2 Raman Enhancement Substrate Materials

2.2.1 Metal Structure Nanoparticles SERS Substrate

Metal nanoparticles are one of the most commonly used SERS substrates because their synthesis method is simple and cheap. Eric and others use the mass coefficient to represent the SERS effect of metal nanoparticles. As shown in Fig. 2.1a, gold, silver, and copper show excellent performance as SERS substrates under visible light and near-infrared lasers. Due to the higher mass coefficient, gold and silver nanoparticles are more widely used. Metal nanoparticles can be prepared in wet chemical synthesis methods with controllable shapes and sizes. Therefore, a large number of researchers have synthesized metal nanoparticles of different shapes and structures for SERS sensing detection research. Gutes and others

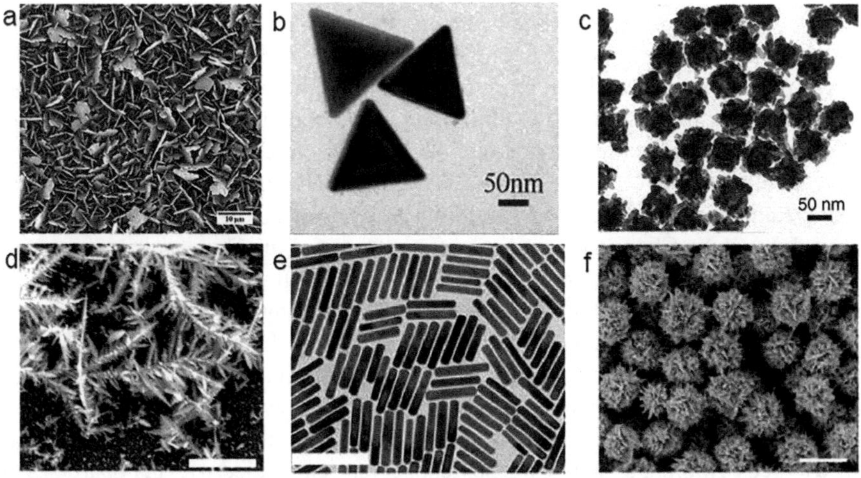

Fig. 2.1 Different structures of metal nanostructure SERS substrates. (**a**) Scanning electron microscope/transmission electron microscope images of silver rose nanostructures; (**b**) Triangular silver particles; (**c**) Rose-shaped nano silver structures; (**d**) Dendritic silver nanostructures; (**e**) Gold nanorod colloidal particles; (**f**) Sea urchin-shaped gold nanoparticles

synthesized desert rose-shaped silver nanostructures. Lombardi and others used tartrate and citrate as structure-directing agents to prepare triangular silver nanoparticles by light-driven methods (Fig. 2.1b). As shown in Fig. 2.1c, Jian Ye and others used gold nanoparticles as seeds to prepare petal-shaped gold nanoparticles. Sailor and others prepared a silver dendritic structure SERS substrate with random intervals on a porous silicon substrate (Fig. 2.1d). Murray and others used a binary surfactant mixture composed of cetyltrimethylammonium bromide (CTAB) and sodium oleate to synthesize one-dimensional gold colloidal nanorods (Fig. 2.1e). By controlling the secondary nucleation and growth process of nanoparticles, Hahn and others synthesized sea urchin-shaped gold nanoparticles with specific surface morphologies (Fig. 2.1f).

2.2.2 Thin Film Structure SERS Substrate Materials

Surface-enhanced Raman substrate materials based on thin-film structures have the characteristics of large surface area, light weight, and easy to carry. They can be used with portable Raman spectrometers and are suitable for large-scale production. As shown in Fig. 2.2a, Park et al. demonstrated a transparent and flexible SERS substrate with high sensitivity, which is based on a polydimethylsiloxane (PDMS) thin film embedded with gold nanostar (GNS) components. The flexible SERS substrate can be covered on any surface, and the optical transparency allows light to interact with the underlying contact surface, thereby enabling high-sensitivity detection of analytes adsorbed on any metal and dielectric surface, otherwise these surfaces will not provide any significant Raman signals of analytes. Figure 2.2b shows the gold thin film SERS substrate prepared by Gao et al. Based on the plasmonic gold film, it can utilize the high sensitivity of SERS for early detection and diagnosis of bacteria, and can be combined with flexible PDMS to become a bandage with sterilization effect. Guilin Wang et al. provided a method to study the Raman enhancement effect of non-metallic nanostructures. They attached a flexible polydimethylsiloxane film to flat and nanostructured surfaces, and they found an extra Raman enhancement up to 5 times on the nanoporous silicon surface (Fig. 2.2c). Yulin Chen et al. prepared different types of metal or semiconductor thin films to study the enhancement activity of SERS substrates by magnetron sputtering. By changing the conditions of magnetron sputtering, nanofilms with different surface morphologies can be obtained (Fig. 2.2d). As shown in Fig. 2.2e, Xuejiao Wang et al. reported an extremely simple but effective method, which achieves high-sensitivity SERS detection of analyte molecules (as low as 10^{-10} M) by using a thin film sandwich structure. The sandwich structure they proposed can be realized by reassembling a flexible transparent gel tape embedded with nanoparticles on a gold film decorated with analyte molecules. Xu et al. prepared a biodegradable flexible film by irreversibly longitudinally stretching a biocompatible poly (ε-caprolactone) film deposited with metal. This film substrate not only has good

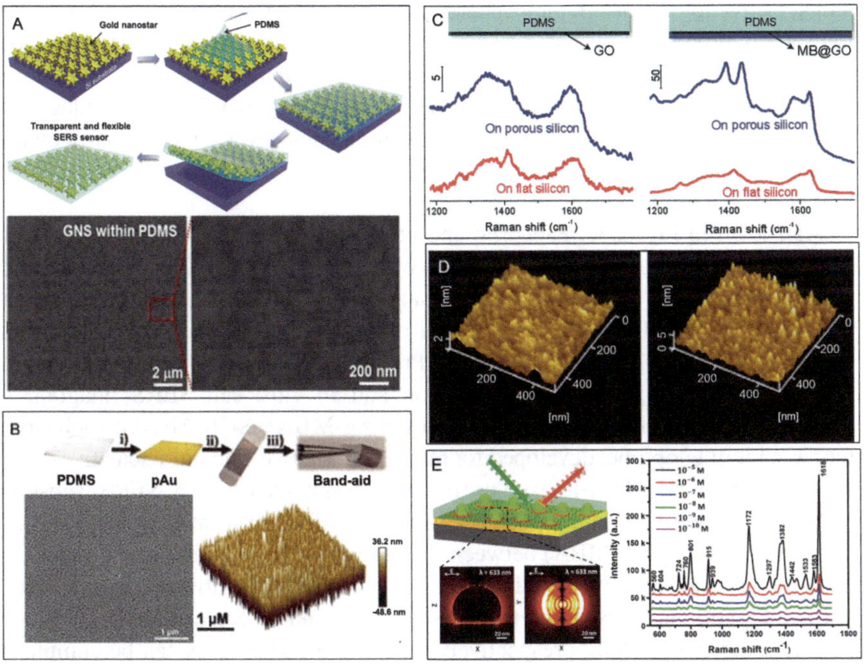

Fig. 2.2 (**a**) Schematic of the fabrication process of a flexible SERS sensor with a gold nanocolumn array. The self-assembled gold nanocolumn array is transferred from the silicon substrate to PDMS, SEM image of the self-assembled gold nanocolumn array embedded in PDMS; (**b**) Schematic of the preparation process of plasmonic gold film strips; (**c**) Raman spectra when PDMS thin films coated with GO and methylene blue@GO are attached to flat silicon (red curve) and nanoporous silicon (blue curve) surfaces, respectively. (**d**) AFM images of nanofilms: Si film (left) and Al film (right); (**e**) POF SERS substrate and its calculated electric field contours on the XZ plane and XY plane (left), Raman spectra of crystal violet detection (right)

flexibility and transparency, but can also be attached to any non-flat surface for in-situ SERS detection of various chemical molecules.

2.2.3 *Core-Shell Structure SERS Substrate Materials*

The core-shell structure is based on a particulate template core, which can be metal and other inorganic nanoparticles, covered by different shell materials. They can usually be prepared by chemical synthesis (such as seed-mediated wet chemical growth methods) or physical methods (such as ALD deposition) (Fig. 2.3a). Gia Chuong Phan-Quang and others have designed a 3D plasmonic structure through the self-assembly of Ag@MOF core-shell nanoparticles, which can actively adsorb and rapidly detect aerosols, gases, and volatile organic compounds, with a sensitivity reaching a few parts per billion (ppb level). This platform is highly responsive and sensitive to changes in atmospheric composition, as demonstrated by monitoring carbon dioxide gas over several cycles. They also demonstrated real-time remote and multiple quantitative analysis of polycyclic aromatic hydrocarbon mixtures under outdoor sunlight. The SERS sensing strategy they developed provides technology for highly sensitive monitoring of molecular-level air/gas environments, which has significant potential in environmental protection, disaster prevention, and homeland defense applications (Fig. 2.3b).

As shown in Fig. 2.3c, Pancorbo and others reported a new type of adjustable core-shell structure composite (Au-SiO_2-WO_3) nanoparticles. These nanoparticles show an easily improved $\approx 10^3$ SERS enhancement factor. WO_3 core nanoparticles with diameters of 20–50 nm are wrapped in a 10–60 nm silica intermediate layer, followed by a 5–20 nm thick gold shell. Preliminary results confirm their superior performance in single-molecule detection and in vitro cell SERS imaging. In another study, Diandian Deng and others used Ag-MBA@SiO_2-Ab core-shell structure as an immunoprobe, developed for rapid, quantitative, and ultra-sensitive detection of diclofenac (DCF) residues in water samples using a SERS substrate (Fig. 2.3d). The immunoprobe is obtained by placing the Raman signal molecule mercaptobenzoic acid (MBA) between the core and the shell, and fixing monoclonal antibody Ag@SiO_2 NPs on the surface of the anti-DCF antibody (mAb). Under optimal conditions, its SERS detection sensitivity and detection limit (LOD) are 9 pg mL^{-1} and 0.07 pg mL^{-1} respectively. The entire test process can be completed within 15 min, and the AgMBA@SiO_2-Ab core-shell structure immunoprobe remains stable for 4 months, with no significant reduction in the SERS probe signal intensity. The key to the success of this SERS probe is the use of Ag@SiO_2 NPs core-shell structure nanoparticles as the SERS substrate, which has excellent SERS enhancement ability, good biocompatibility, and structural component stability. In addition, Zhang and others proposed a lateral flow analysis (LFA) based on core-shell SERS nanotags, allowing for multiple and quantitative detection of cardiac biomarkers, and used for early diagnosis of diseases such as acute myocardial infarction (AMI) (Fig. 2.3e).

2.3 Current Status of Surface Enhanced Raman Sensing Detection

2.3.1 Detection of Biochemical Molecules

The establishment of a Raman scattering information database for biomolecules is very important for the identification and detection of living cells or pathogens, as well as for a deeper understanding of the structure of biological markers. This will help to advance the research of Raman enhanced detection technology in biological systems and develop potential biomedical application scenarios. At present, SERS detection technologies for various small biological molecules

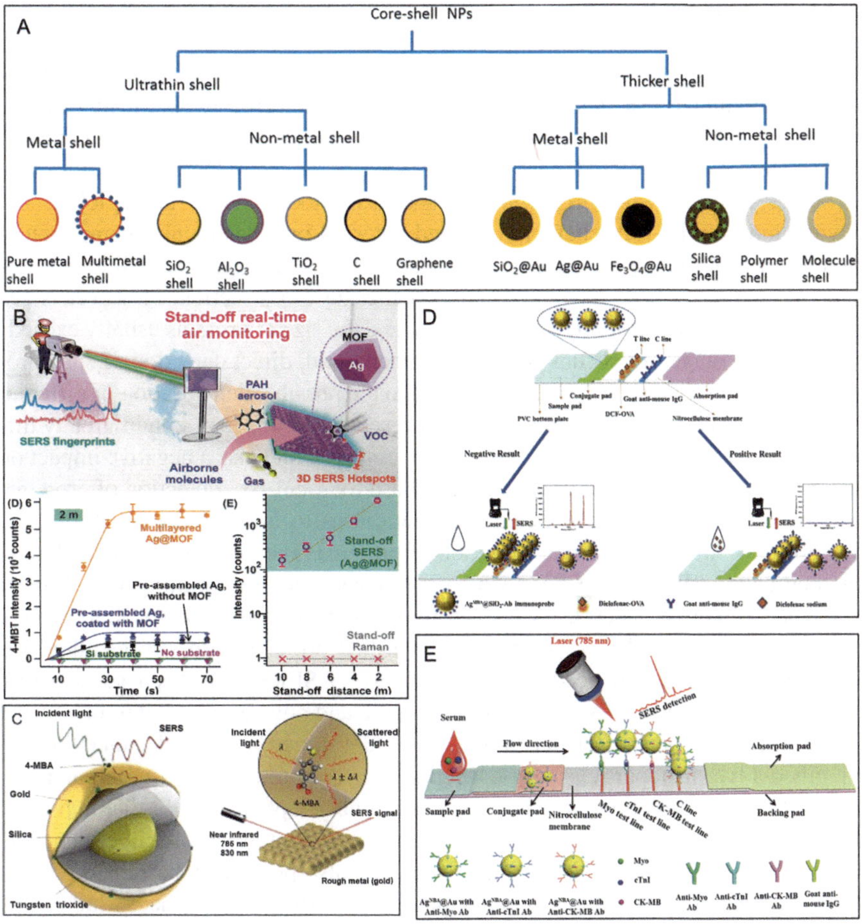

Fig. 2.3 (**a**) Classification of core-shell structure SERS substrate nanoparticles; (**b**) Schematic diagram of the working principle of the standing MOF-SERS platform for real-time air monitoring. Real-time 4-MBT intensity obtained in standing SERS detection using different platforms, and 4-MBT intensity obtained using Ag@MOF SERS platform and ordinary Raman detection at distances from 2 to 10 m; (**c**) Schematic diagram of the design of the core-shell structure composite SERS substrate structure and the effect of obtaining SERS signals on the gold nanoshell; (**d**) Self-assembly scheme of ICA test strip and use of AgMBA@SiO_2-Ab as a new SERS-ICA immune probe for rapid, quantitative, and ultra-sensitive detection of diclofenac residues in water samples; (**e**) Schematic diagram of the working principle of multiple LFIA based on core-shell structure SERS nano labels

(dopamine, folic acid, toxins, etc.) and large molecules (vascular endothelial growth factor, prostate-specific antigen, alpha-fetoprotein, DNA, etc.) have been reported. Among them, Raman enhanced detection of three important types of biochemical molecules has received particular attention: proteins, nucleic acids, and metabolites.

2.3.1.1 Proteins

Label-free detection of proteins and their structures is very important for life sciences, including drug screening, early disease diagnosis, clinical treatment, and proteomics. Label-free Raman enhanced detection can provide protein "fingerprint" information directly related to its structure in a simpler and more cost-effective way.

Some proteins with chromophores (such as cytochrome C, hemoglobin, and myoglobin) can be well characterized and detected by Raman enhancement technology, even at the single molecule level, because under resonance conditions, chromophores usually have a larger Raman scattering cross-section. However, since most proteins do not contain chromophores, and the size of proteins usually exceeds the effective range of the substrate enhancement field, direct detection of such proteins still poses challenges. In addition, due to the possible denaturation or different adsorption directions of proteins, different research groups may obtain greatly different Raman enhanced spectra of the same protein, which has a negative impact on the establishment of a Raman spectral database for label-free detection of proteins.

Han and others have conducted systematic tests to improve the reliability and versatility of label-free Raman detection of proteins. They developed an analytical method called "Western SERS", which combines protein imprinting with Raman enhanced detection. They used protein imprinting to purify and separate protein mixtures, then stained the separated and purified proteins with silver colloid for subsequent Raman enhanced detection. "Western SERS" has been successfully applied to the analysis of solutions of myoglobin and bovine serum albumin (BSA).

On the other hand, Ren and others have developed a method called iodide-modified silver nanoparticles (Ag@IMNPs) to avoid protein denaturation and balance adsorption direction, which can achieve highly repetitive and reliable protein Raman enhancement spectra (Fig. 2.4a). The iodide coating on the silver nanoparticles not only eliminates surface impurities, but also provides a barrier to prevent direct and intense interactions between proteins and the metal surface, which helps to maintain the natural structure of the protein. The Raman enhancement features of

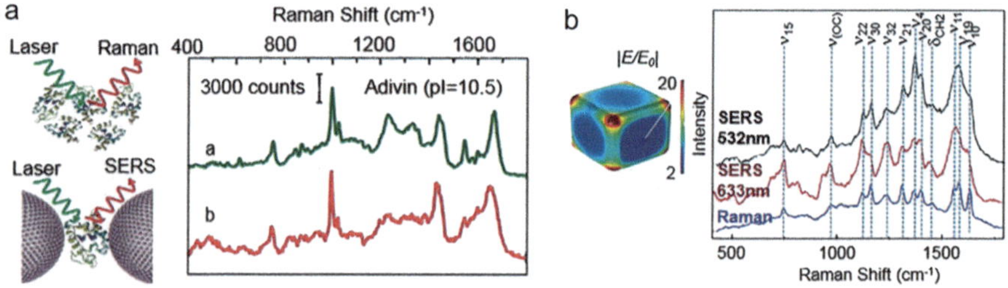

Fig. 2.4 Surface-enhanced Raman spectroscopy detection analysis of protein molecules. (**a**) Ordinary Raman spectra (green) and Raman enhancement spectra (red) of biotin protein in iodide-modified silver nanoparticle colloids; (**b**) Finite element simulation of the electric field distribution of nanocubes (50 nm), and Raman enhancement spectra of cytochrome C obtained on silver nanocubes using 532 nm and 638 nm excitation wavelengths. The concentration of cytochrome C is 0.1 nM, and the Raman signal of a 1 mM cytochrome C solution is used as a control (blue)

proteins without chromophores are almost the same as the corresponding normal Raman spectra, indicating that the natural structure of the protein is well maintained. This unique function allows the qualitative identification of proteins by simply calculating the intensity ratio of Raman peaks (tryptophan and phenylalanine residues), which is of great significance for reliably and accurately identifying protein molecules. In addition, this method can be used to accurately analyze protein mixtures based on known single Raman enhancement spectra. Matteini and others used single silver nanocubes to directly detect proteins through Raman enhancement scattering signals, as shown in Fig. 2.4b. The corner sites of the silver nanocubes serve as Raman enhancement scattering hotspots, providing 80% of the total Raman enhancement scattering signal. The polyvinylpyrrolidone (PVP) coating on the metal surface of the nanocubes can prevent proteins from directly contacting the surface, thereby promoting preferential interactions between proteins and PVP special sites. In this way, it can avoid signal fluctuations caused by variable adsorption configurations within different particle regions and cluster nanoparticles. This method is used to study the interaction between cytochrome C and the silver surface, and quantitative analysis of low concentration cytochrome C has been achieved.

2.3.1.2 Nucleic Acids

The technical difficulties of Raman spectroscopy for protein identification also exist in the field of nucleic acid Raman detection. Because nucleic acid molecules (such as DNA) are composed of four bases with different relative contents and sequences, as the DNA chain grows, the differences in composition will become very insignificant, especially when the contents of different bases are close. Bell and others used $MgSO_4$ as an aggregation reagent to obtain Raman enhancement spectra of DNA/RNA mononucleotides with silver colloidal particles. In addition, Halas and others have developed a gentle thermal cycling method that can relax DNA chains into extended conformations to obtain stable Raman enhancement signals (Fig. 2.5a). Another technical challenge in nucleic acid Raman enhancement detection is how to improve the repeatability of the detection signal, which is crucial for reliably distinguishing DNAs of different compositions through the slight differences in the relative intensity of Raman enhancement signals. To solve this problem, Ren and others redesigned and modified the iodide-modified silver nanoparticle method mentioned in the previous section (Fig. 2.5b). Adding $MgSO_4$ The neutralization of surface charge and enhancement of electrostatic interactions between DNA and iodide-modified silver nanoparticles allows for the detection of strong and repeatable Raman enhanced scattering signals. Studies have found that the vibrational band of the phosphate backbone can serve as an internal standard signal to calibrate the absolute signal of each base, thereby reliably determining the structure of DNA. This calibration scheme can determine the proportion of each base in oligonucleotides and obtain the absolute content of each base in the DNA sequence. Using this strategy, high-sensitivity detection of single bases in adenine/cytosine biopolymers in single-strand and double-strand DNA, as well as quantitative Raman enhanced detection analysis of base content, can be achieved. The Alvarez-Puebla research group synthesized positively charged silver nanoparticles (Ag NPs@Sp) with a spermine molecule coating and used it for DNA detection, as shown in Fig. 2.5c. Negatively charged DNA can bind to the surface of Ag NPs@Sp through electrostatic interactions, forming stable aggregates and obtaining repeatable and reliable Raman enhanced spectral signals. They further compared the Raman enhanced detection of DNA with positively charged spermine-coated Ag NPs and negatively charged halide anion-coated Ag NPs. The results showed that seemingly minor experimental changes can significantly alter the affinity of single-strand and double-strand DNA on silver nanoparticles, as well as the final Raman enhanced spectral data results.

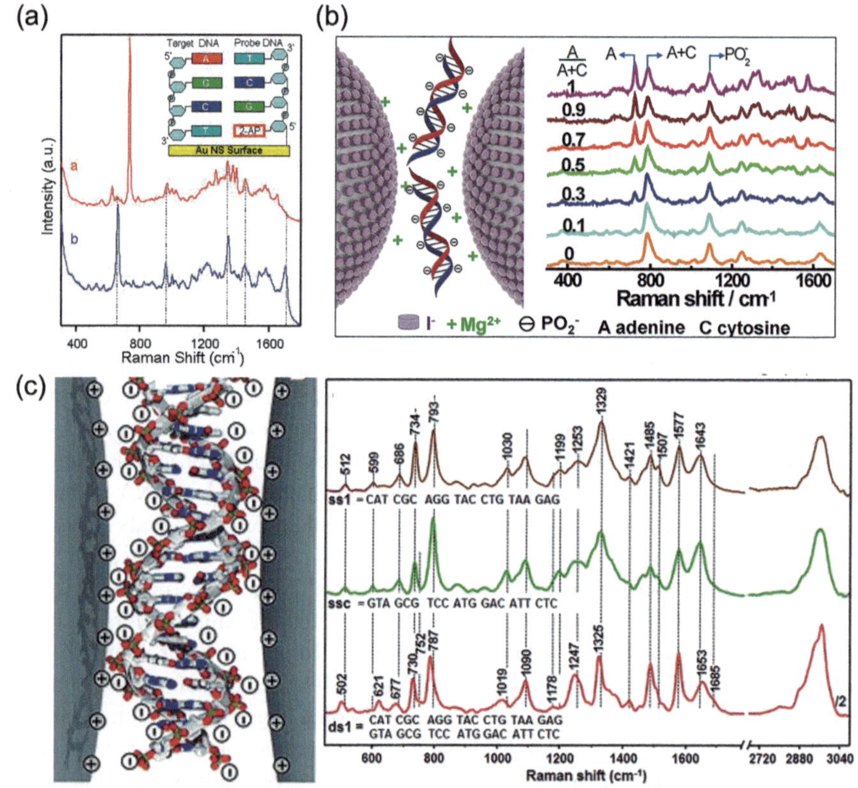

Fig. 2.5 Schematic diagram of surface Raman enhanced detection of DNA. (**a**) Raman enhanced detection of DNA by thermal annealing; (**b**) Iodide-modified silver nanoparticles used for DNA detection; (**c**) Spermine-modified silver nanoparticles used for DNA detection

2.3.1.3 Metabolites

Metabolites are usually referred to as small biological molecules produced in the process of life, playing a crucial role in basic life science research and disease diagnosis. Some common metabolites, such as glucose, uric acid, hypoxanthine, and lactic acid, can already be analyzed using Raman enhanced scattering spectroscopy. The most meaningful application background of metabolite detection is the analysis of body fluid samples. By systematically comparing the droplet deposition Raman spectra and Raman enhanced spectra used for detecting microliters of human tears, it is interesting to find that the main contribution to the tear droplet deposition Raman spectral signal comes from high-abundance proteins, while Raman enhanced spectra are dominated by uric acid and hypoxanthine, which are low in abundance but have strong interactions with silver nanoparticles. This result indicates that the Raman enhanced spectral signal is mainly dominated by biomolecules with high affinity for the Raman enhanced active metal surface. More importantly, these two Raman-based techniques are obviously complementary: droplet deposition Raman spectroscopy is beneficial for detecting high-abundance components such as proteins, while Raman enhanced spectroscopy can detect low-abundance components with high adsorption capacity for the substrate material. The combination of these two techniques provides multi-parameter information for the systematic analysis of clinical tears and can be further extended to the analysis of other body fluids. For the diagnosis and health management of diabetes, continuous glucose monitoring methods are greatly needed. Raman enhanced scattering spectroscopy has the potential to become an alternative glucose sensor for rapid and minimally invasive diabetes analysis. However, obtaining Raman enhanced signals from glucose is very challenging due to its small Raman scattering cross-section and weak affinity for metal surfaces. Van Duyne and his team designed a Raman-enhanced glucose sensor based on the silver film on nanospheres (AgFON) substrate they had previously developed. The substrate was functionalized with a mixed self-assembled monolayer (SAM) of decanethiol (DT) and mercaptohexanol (MH) to enhance the affinity between glucose and the SAM layer. For the first time, they achieved Raman-enhanced detection of glucose in vivo. By combining Raman-enhanced spectroscopy and spatially offset Raman spectroscopy (SORS), the sensor was implanted subcutaneously in rats and achieved transdermal monitoring of in vivo glucose concentration over a period of 17 days.

The Raman-enhanced spectroscopy database of biomolecules undoubtedly has significant implications for the future development of bioanalysis. However, establishing such a database requires collaboration from multiple parties. A prerequisite is that each research team can obtain similar Raman spectra of a certain biomolecule. Preventing the denaturation of biomolecules and controlling the adsorption/orientation of biomolecules on the surface of Raman-enhanced substrate materials are important ways to achieve this goal.

2.3.2 *Cell Detection*

With its molecular fingerprint information, Raman-enhanced scattering can reflect the molecular composition within living cells. Raman-enhanced scattering detection of living cells can be divided into static detection and dynamic detection. In static detection, Raman-enhanced scattering serves only as a diagnostic method to identify and classify cells, and can be used to identify different viruses and bacteria, as well as distinguish between cancer cells and normal cells, etc. In dynamic detection, the focus is on the spatiotemporal distribution and evolution of biomolecules during processes such as cell division, differentiation, and apoptosis. These time-series Raman-enhanced spectra or images can help researchers understand the life processes of cells and the pathogenesis of various diseases. The technical challenge of direct cell detection lies in the fact that Raman-enhanced scattering substrate materials can enhance any molecules near their surface, but the resulting Raman spectra are too complex to allow for effective dynamic analysis.

2.3.2.1 Detection of Cancer Cells

Raman-enhanced scattering spectroscopy technology is often combined with multivariate data analysis methods, enabling it to distinguish between cancer cells and normal cells. Due to differences in intracellular substances and components on the cell membrane surface (such as nucleic acids, proteins, sugars, and lipids), normal cells and cancer cells have different Raman-enhanced vibrational spectroscopic features. Choo and his team developed antibody-modified SERS labels based on gold nanorods and gold-silver bimetallic nanoparticles to monitor MCF7 cells overexpressing the breast cancer marker HER2, and HEK293 cells overexpressing phospholipase PLCγ1.

Research has reported an interesting and novel single-cell Raman enhancement detection label, known as nanocoral. As shown in Fig. 2.6, half of the surface of this metal nanocoral is coated with highly rough gold to increase its adsorption capacity and generate high-density Raman enhancement hotspots, while the blank polystyrene hemisphere can be targeted to specific cells through functionalization. By incubating the nanocoral suspension with anti-HER2 antibodies, specific targeting and Raman enhancement detection of breast cancer cells were achieved.

The detection of tumor cells in biological fluids is a hot topic in the field of cancer diagnosis, and rapid and accurate optical analysis techniques are one of the current important detection methods. However, the probe for tumor cells faces the challenge of low content of the test substance. The concentration of circulating tumor cells (CTCs) in the blood is extremely low, in stage IV cancer, the number of units in 8 milliliters of blood (the standard volume for analysis) is only a few dozen. Because the concentration of circulating tumor cells in the blood is extremely low, many methods need to be pre-enriched before the detection step, some of these enrichment strategies are based on the physical characteristics of cells, such as size, density or deformability, while others are based on the biological characteristics of

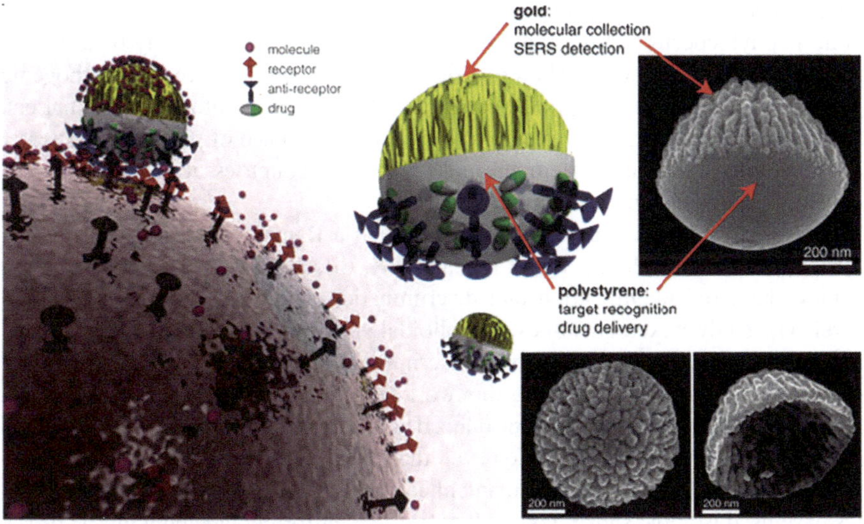

Fig. 2.6 Schematic diagram of the working principle and structure of metal nanocoral as a multifunctional nanoprobe for targeting, sensing, and drug delivery; the inset shows the scanning electron microscope image of the nanocoral probe

cells, such as the expression of characteristic proteins. So far, the most common strategy is to use antibody-modified magnetic particles, usually magnetite, functionalized with antibodies against epithelial cell adhesion molecules (EpCAM). The basic principle of this choice is that epithelial cells are not common in the blood, so epithelial cells in the blood are likely to be related to cancer. Epithelial cell adhesion molecule-modified magnetic particles are exposed to blood samples, bind to target cells after an appropriate time, and are finally enriched and extracted by an external magnetic field. And these systems usually use more than one antibody, based on coded plasmonic nanoparticles and Raman enhancement particles can perform multiplex quantification analysis of different circulating tumor cells. The advantages of this method include ultra-high analysis resolution and almost unlimited number of available markers. However, even the collection time of the more advanced Raman enhanced scattering spectrum usually exceeds 10 ms, which hinders the use of this technology in real samples. Therefore, by combining fluorescence signal acquisition technology with high response speed (as low as nanoseconds), the high-resolution advantage of Raman enhancement spectroscopy technology in cell detection applications can be brought into play.

2.3.2.2 Detection of Pathogenic Bacteria

The detection of specific pathogens of infectious diseases has also attracted great attention. The ultra-high sensitivity and label-free recognition characteristics of Raman enhancement spectroscopy technology make it very suitable for detecting low-abundance but highly harmful pathogens, which pose a huge risk to human health. The most common infections are caused by bacteria such as Escherichia coli, Staphylococcus aureus, and Salmonella typhi, which can usually be transmitted through food and water. Therefore, in order to prevent the spread of pathogenic bacteria, it is necessary to quickly identify and detect different pathogens that can infect humans. At the same time, due to the low prevalence of safe drinking water and common pollution problems in underdeveloped countries, it is also crucial to develop fast and cheap detection methods.

Jarvis and Goodacre, among others, reported a Raman enhancement detection technique that has reliable signal repeatability, directly detects bacteria, and can achieve bacterial identification and discrimination at the species and subspecies level. The study mixed bacteria with colloidal silver and dried them. After Raman enhancement detection, multivariate discriminant analysis was used for bacterial category identification. Following this work, Ziegler and his colleagues reported a similar analytical method, but introduced a recognition program that combined Raman enhancement barcodes. However, these studies did not attempt to determine the biochemical source of the Raman enhancement signal. In contrast, recent studies have used isotopic substrates, such as glucose uniformly labeled with ^{13}C or ^{15}N ammonium hydroxide during bacterial growth—a process known as Stable Isotope Probing (SIP). In this process, heavier isotopes are incorporated into bacterial biomass, causing changes in vibrational frequency (due to reduced mass in vibration), resulting in a shift of the Raman/Raman enhancement signal to lower frequencies. This process can clarify the origin of the vibration and, combined with the reference Raman enhancement spectra of natural and isotopic standards of key molecules, can achieve clear chemical identification. Premasiri and others' research shows that purine-related molecules are mainly detected through the Raman enhancement spectra of whole bacteria, and other studies using SERS-SIP also use this principle for bacterial identification analysis. This method is often used for imaging and identifying individual bacteria/cells and phenotypic functions in mixed communities. In summary, Surface Enhanced Raman Scattering (SERS) spectroscopy is becoming a powerful tool for characterizing bacterial pathogens, not only for assessing antibiotic sensitivity or resistance, but also for directly detecting bacteria or volatile organic compounds produced during bacterial culture.

Liz-Marzan and others do not directly detect bacteria, but measure their intercellular signal transmission process, namely quorum sensing (QS). Most bacteria in nature use biofilms as a medium for information transmission to achieve their quorum sensing, which is a mechanism of intercellular communication that allows bacteria to monitor cell density and environmental changes and respond to them. Since quorum sensing and biofilms are related to the ability of bacteria to cause diseases, it is very important to find the messenger molecules of the quorum sensing mechanism in natural bacterial populations and to analyze them non-invasively. Bodelon and others reported the preparation method of nanostructured plasmonic substrates to monitor the presence of pyocyanin using unlabelled Raman enhanced probes. Pyocyanin is a signal metabolite for Pseudomonas aeruginosa biofilms and bacterial communities to perform quorum sensing, and is a heterocyclic nitrogen-containing compound of the pyrazine family produced and excreted by Pseudomonas aeruginosa, playing an important role in the morphological changes of this microorganism's biofilm. This strategy is based on the use of mixed materials containing plasmonic components in a porous matrix, which only allows small molecules to diffuse. By using a micropatterned supercrystal gold nanorod array coated with mesoporous silica, Raman enhancement can detect quorum sensing behavior (i.e., pyocyanin expression) in the early stages of biofilm formation, and can image the pyrazines produced by small bacterial groups (average 25 μm^2), as shown in Fig. 2.7. Subsequent work has shown that this method can be used to study the interactions between different bacteria, including isolated bacterial colonies and mixed colonies.

2.3.3 *Environmental Pollutant Detection*

Surface-enhanced Raman scattering has been widely used to monitor and quantify toxins, chemicals, heavy metals, bacterial signal metabolites, etc. present in the environment. Various nanostructures including nanoparticle monomers and dimers

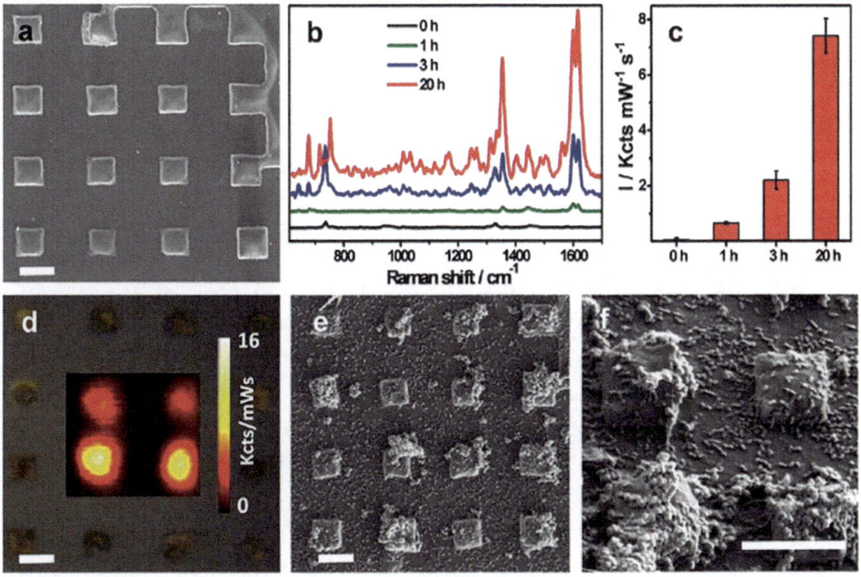

Fig. 2.7 In situ Raman detection and imaging of pyocyanin produced by Pseudomonas aeruginosa grown on a gold nanorod supercrystal substrate coated with a micro-silica array. (**a**) Scanning electron microscope image of the substrate; (**b**) Representative Raman enhancement spectra measured after 0, 1, 3, and 20 h of bacterial growth; (**c**) Relative Raman enhancement spectral feature peak intensities (1600 cm^{-1}) recorded after 0, 1, 3, and 20 h; (**d**) Substrate optical image and surface scanning image of pyocyanin (1600 cm^{-1}) Raman signal at 20 h of growth; (**e**, **f**) Scanning electron microscope images of supercrystals colonized by Pseudomonas aeruginosa (20 h) at different magnifications; Scale bar: 5 μm

(Fig. 2.8a), gold nanorods and nanoparticle chain structures (Fig. 2.8b), anisotropic structures (Fig. 2.8c), nano gold arrays, nanopillars, thin film structures, etc., have been used in the monitoring of environmental pollutants based on Raman enhanced spectroscopy. In addition, new nanomaterials including semiconductors, transition metal chalcogenides, and metal-organic frameworks have also been developed and used to enhance the performance of Raman-enhanced substrates.

Compared to sensing detection methods based on other spectral technologies, detection technology based on Raman enhanced spectroscopy has better sensitivity, even reaching the picomolar level. Compared with gold nanorod monomers and gold nanoparticle chains, the sharp tips on the surface of nanoparticles help to improve the performance of Raman enhanced detection. However, for the detection of environmental toxins or chemical substances, a chain of gold nanorods or a trimeric structure of nanoparticle-nanorod-nanoparticle formed in an "end-to-end" manner can provide a similar Raman enhancement effect and detection. Therefore, the ultra-sensitive detection strategy based on nanoparticle structure assembly and Raman enhancement technology can provide a detection scheme to overcome sensitivity limitations, but it is necessary to ensure the repeatability and stability of its Raman signal for the detection analysis of actual samples. Based on the fingerprint recognition characteristics of Raman spectroscopy, Raman enhancement technology can simultaneously detect

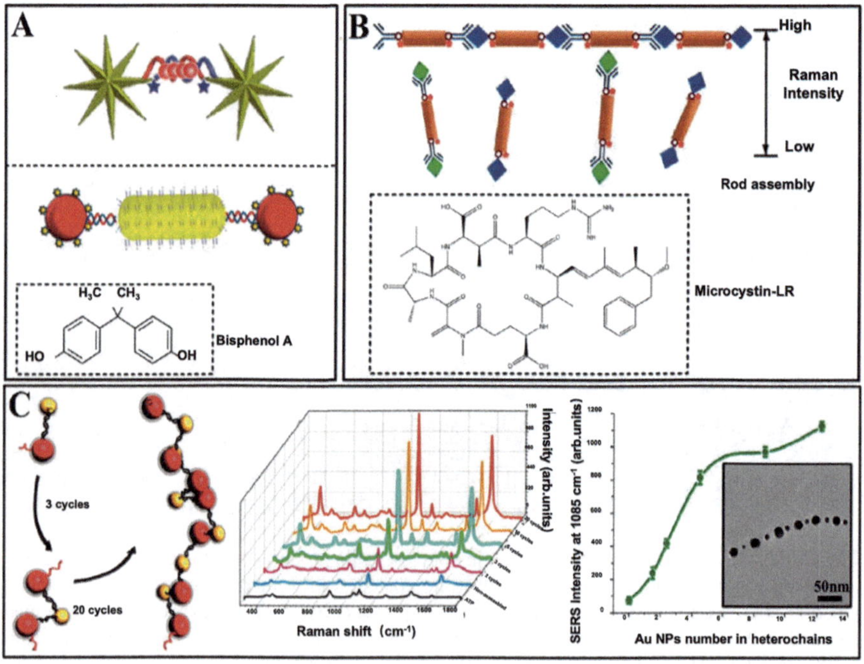

Fig. 2.8 (**a**) Raman-enhanced active gold nanostar dimers for mercury ion detection (top) and gold nanoparticle-nanorod assemblies for bisphenol A detection (bottom); (**b**) Raman-enhanced active gold nanorod polymers for toxin detection; (**c**) Gold nanoparticle heterochains and their Raman-enhanced detection performance

pollutants of various different molecular weights, and can even distinguish different organic chemical components with similar structures. Although multiple targets can be monitored simultaneously, the optimal parameters for Raman enhanced detection may vary for different molecules with similar structures, polarities, or molecular weights. In summary, Raman enhanced detection technology still faces many limitations and challenges in practical applications, such as non-specific aggregation of nanoparticles, conformational changes of surface chemical groups, and matrix effects of samples, all of which will affect detection performance (sensitivity, selectivity, and accuracy). Although single-molecule Raman enhanced detection technology can detect pollutant molecules in complex chemical and biological environments, the presence of impurities or the orientation of target molecules can cause spectral fluctuations, affecting the accuracy of detection. In summary, surface-enhanced Raman spectroscopy technology has great potential for establishing a fast, reliable, and practical sensing detection platform, but it is necessary to combine sample characteristics, select the appropriate Raman enhancement substrate, optimize the intensity of the detection excitation light source, signal detection time, and perform a large amount of sample statistical analysis to ensure the practicality and reliability of this technology.

2.4 Raman Detection Technology Based on Microfluidic Platform

With the continuous development of nanoscience technology, analytical techniques such as Surface Enhanced Raman Spectroscopy (SERS) can detect the structure and chemical information of various analyte molecules at ultra-low sample concentrations. Although this technology has made significant application progress in various fields, there are still huge challenges in the signal reproducibility of SERS sensing and the sensitivity to small molecule analytes. In this regard, the Surface Enhanced Raman Spectroscopy (MF-SERS) detection technology based on the microfluidic technology platform provides a feasible solution.

2.4.1 Introduction to Microfluidic Technology

Microfluidic technology manipulates fluids through fluid pumps in micron-sized channels. Microfluidic devices usually include microchannel chips, microfluidic pump systems, microvalve control systems, etc. Through these miniaturized microfluidic chip devices, a variety of tasks and functional applications can be completed in a highly integrated manner. Therefore, the microfluidic system is also known as a laboratory on a chip (Lab on Chip, LoC), for example, it can execute chemical micro-reactions and biological cell experimental processes with high precision through microfluidic automated driving.

When chemical processes occur in microchannels instead of flasks or test tubes, only very low amounts of reagents (microliters/nanoliters) are needed, and diffusion-controlled processes are greatly accelerated. At the same time, reagents or samples can be manipulated in different ways, such as pressure, optics, electricity, centrifugal force, acoustics, and magnetic fields. Microfluidic systems, with their unique advantages, have found wide applications in synthesis and analytical chemistry, biology, medical diagnostic research, drug screening discovery, and health testing. Especially in the past few decades, scientists have combined various analytical techniques (such as mass spectrometry, nuclear magnetic resonance, fluorescence spectroscopy, electrochemical techniques, SERS, and infrared spectroscopy) with microfluidic systems, developing a variety of microfluidic-based sensing detection technologies.

Among these sensing technologies, microfluidic sensing systems combined with SERS, due to their label-free recognition of Raman spectra and small volume sample analysis capabilities of microfluidics, have broad application potential and engineering practical value. Figure 2.9 shows the process of SERS detection analysis of analytes adsorbed on gold nanoparticles (AuNPs) in microfluidic channels.

This section will briefly outline the current status and advantages of microfluidic SERS technology, including the design of microfluidic SERS channels, the diversified preparation of SERS substrates in microfluidic devices, and their current applications in different sensing detection fields. Finally, the challenges and future development trends of microfluidic SERS in the field of intelligent sensing are discussed.

2.4.2 *Advantages of Microfluidic SERS*

2.4.2.1 SERS Quantitative Detection

Due to the uncertainty of metal nanostructures, quantitative detection is a very important but challenging topic for SERS technology. In this regard, microfluidic SERS technology can provide a reliable and stable SERS substrate in the

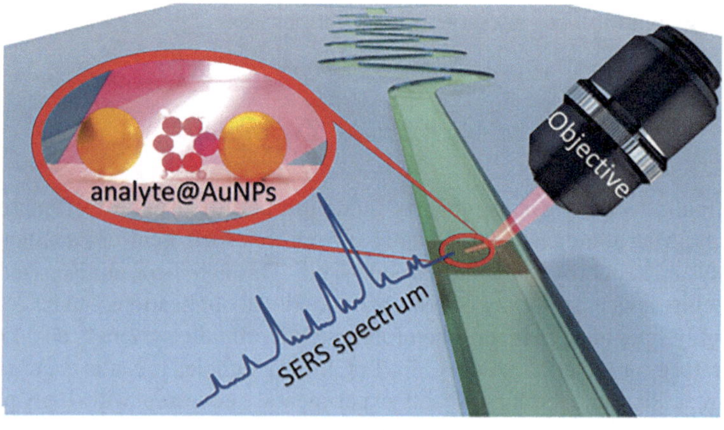

Fig. 2.9 Schematic diagram of SERS detection analysis of the test substance in the microfluidic channel

microchannel, ensuring the stable reproduction of SERS signals, and achieving quantitative detection of analytes. For example, ultra-low concentration drugs can be quantitatively detected online using a microfluidic SERS system.

2.4.2.2 Cell and Droplet Classification

Whether it is basic laboratory research or clinical medical detection applications, effective and reliable classification of cells and droplets using microfluidic systems is very meaningful. For example, the principle of high-throughput screening of single cells in the microfluidic platform can also be successfully applied to Raman activated droplet sorting (RADS). In this microfluidic sorting application, Raman detection technology has unique advantages in cell sorting, and the cell sorting system based on microfluidic Raman (RACS) can even surpass traditional fluorescence activated cell sorting technology in some indicators.

2.4.2.3 In-situ Synthesis of Nano SERS Substrate Materials

In SERS sensing applications, it is best to use SERS substrate materials with controllable morphological structures to achieve repeatable SERS signal measurements, but the preparation of highly uniform nanostructures is a challenge in the field of micro and nano material processing and manufacturing. Based on microfluidic systems, metal nanocrystals can be continuously and controllably prepared using droplet microreactors, which can effectively control the morphology of nanocrystals and concentrate local content. At the same time, several methods have been proposed for preparing silver nanostructures using microreactors, all of which demonstrate the important potential of microfluidic systems in preparing SERS active substrate nanostructures with controllable morphology.

2.4.2.4 Reproducibility and Reliability of SERS Signals

In SERS detection application research, the metal nanostructures are easily affected by air oxidation and environmental pollutants, which can reduce the sensitivity and reproducibility of SERS sensing. However, using a closed microchannel system for SERS detection can effectively solve these problems, such as slowing down the oxidation process of the SERS substrate material by air. In addition, the continuous flow of microfluids in the microchannels can effectively reduce the heating effect of the excitation light source on the sample, thereby avoiding unnecessary photochemical reactions or damage to the sample to be tested. Although the sensitivity of SERS in microfluidic systems is relatively lower compared to traditional SERS detection, due to the continuous flow of analyte molecules, the reproducibility and reliability of SERS sensing detection in microfluidic systems are relatively higher.

2.4.2.5 System Portability

Nowadays, there are several handheld Raman spectrometers available for SERS measurements, which have obvious advantages for the portability of Raman spectroscopy technology for on-site analysis. The advantage of microfluidic devices is that they consume a small amount of sample, and the microfluidic chip also has the advantage of miniaturization. Therefore, the combination of microfluidic SERS technology and portable Raman spectrometers is very suitable for on-site analysis and detection applications. For example, a portable Raman spectrometer with fiber coupling combined with a SERS sensor in a microfluidic system can already be used to detect biomarkers of Alzheimer's disease, such as amyloid beta protein.

2.4.2.6 Analysis of Biological Samples in Solution Environments

SERS technology can provide the "fingerprint peak" spectral data of biomolecules, and the microfluidic system only needs a microliter level of sample volume to capture, classify or detect and identify SERS signals. In this regard, we can perform rapid concentration of solution samples and rapid detection and analysis of targets (such as bacteria) in sewage samples by using disposable microfluidic chips combined with SERS detection technology.

2.4.2.7 Real-time Detection and Universality

Another key advantage of microfluidic SERS technology is that it can be used for online dynamic continuous monitoring of chemical/biological reaction processes. For example, Kneipp and others proposed a plug-in multifunctional microfluidic platform, the central part of which is a plug-in SERS probe. This plug-in design can be easily integrated into existing Raman microanalysis systems. This platform can perform sensitive optical characterization of chemical reaction processes, can also detect reactions under controlled gas atmospheres, and is not affected by surface adsorption or memory effects of nanoparticles, demonstrating strong universality.

2.4.3 SERS Detection Microfluidic Channel Design

For SERS detection application requirements, effective mixing between metal nanocolloid particles and analytes in the microfluidic channel is very important. Passive mixing channels, due to their simplicity and operational applicability, have been most widely used in laminar mixing. Figure 2.10 shows several different passive channel designs that can be used for rapid and effective mixing of two different converging flows.

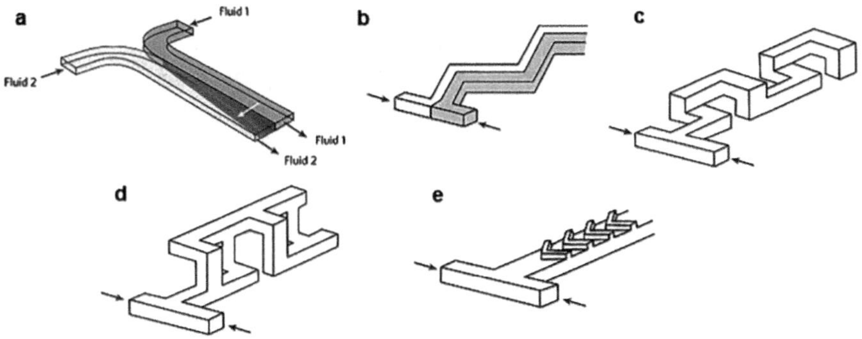

Fig. 2.10 Microfluidic channels for mixing continuous flow (**a**) Simple diffusion mixing on the fluid interface; (**b**) Serrated channel for chaotic mixing at high Reynolds numbers; (**c**) Three-dimensional serpentine channel for chaotic flow at medium and low Reynolds numbers; (**d**) Three-dimensional serpentine channel for chaotic flow at medium and low Reynolds numbers; (**e**) Microchannel for chaotic mixing at low Reynolds numbers

Compared to other chaotic channels, the crocodile tooth-shaped microfluidic channel shows a higher mixing efficiency, as vertical and lateral diffusion generate strong chaotic advection during the confluence. Analyte molecules effectively adsorb onto the surface of silver nanoparticles through the sawtooth-shaped microfluidic channel, achieving the detection of SERS signals. Generally speaking, the results produced by on-chip Raman detection in microfluidic channels have better repeatability than those under static conditions. In the microfluidic channel, the nanocolloid and analyte are introduced into a mutually mixed channel. When the mixed sample passes through the laser beam, Raman signals are accumulated from the flowing fluid; at the laser detection point, SERS signals from different aggregates are accumulated and collected, thereby significantly improving the reproducibility of the detection signal. Through effective mixing channel design and optimal flow rate settings, quantitative analysis of analytes can be performed based on changes in SERS signal intensity.

Currently, microdroplet microfluidic channels have been used for serial high-throughput microanalysis, and microscopic SERS technology has also been applied to the detection and quantitative analysis of analytes in liquid/liquid segmented microfluidic systems. This technology can overcome the "memory effect" caused by the light deposition of analyte/colloid conjugates on the channel wall.

2.4.4 Diversity of SERS Substrates in Microfluidic Devices and Their Detection Applications

The signal enhancement performance of SERS depends on several aspects, such as the size, shape, material properties, and colloidal particle spacing of the active nanostructures of the SERS substrate; the chemical structure of the analyte and its interaction with the SERS substrate; Raman instrument parameters, including laser wavelength, intensity, and exposure time, etc. One of the keys to high-sensitivity SERS signal collection is the rational selection and optimization of SERS substrates. In microfluidic SERS detection applications, they can be roughly divided into two categories: colloidal SERS substrates and fixed SERS substrates. The development of SERS substrates is one of the main research directions in the SERS field. In the past few decades, people have been committed to developing plasmonic nanostructures for microfluidic SERS applications. The following text introduces the diversity of SERS substrates in microfluidic devices and their detection applications in different fields, using colloidal SERS substrates and fixed SERS substrates as examples.

2.4.4.1 Colloidal SERS Substrates

Chemically synthesized silver and gold colloids are the most important substrate materials in SERS detection applications. In most cases, due to their simplicity and reusability in microfluidic systems, metal colloids have obvious advantages. They are mainly prepared by chemical methods, such as the Lee Meisel method, the Lendl method, and the Creighton method.

Gold Colloid

Figure 2.11a shows different types of metal colloids and their applications in microfluidic platforms. Gold nanostructures, due to their biocompatibility and chemical stability, have been widely used in microfluidic SERS research. As shown in Fig. 2.11a(ii), Rodríguez-Lorenzo and others have functionalized Au nanostars with a monoclonal antibody for foodborne pathogen detection. Due to their sharp edges, these nanostars can exhibit higher SERS enhancement performance. As shown in Fig. 2.11a(i), Choo and others have developed a magnetic composite immunosorbent SERS substrate based on gold nanoparticles, combined with a microfluidic system, successfully used for early diagnosis of prostate cancer. In addition, to obtain repeatable Raman enhancement signals, it is necessary to assemble plasmonic nanostructures uniformly. As shown in Fig. 2.11a(iv), Jing and others reported a method for preparing double emulsion droplets (water-oil-water), with high concentrations of gold nanoparticles assembled inside the droplets, used for high-sensitivity SERS detection.

Silver Colloid

As shown in Fig. 2.11a(iii), Andreou and others used Ag colloid and aggregating agent, based on the microfluidic system, to detect drug molecules in saliva within a few minutes. Compared with traditional SERS detection, this work improved the interaction between the analyte and nanoparticles, controlled the degree of

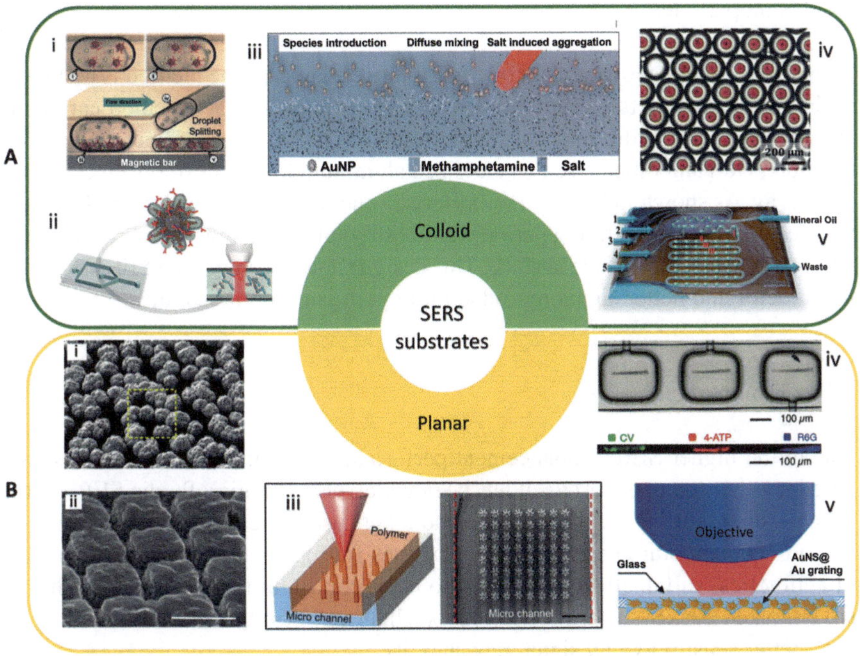

Fig. 2.11 Different types of metal colloids and their applications in microfluidic platforms. (**a**) Colloidal SERS substrates for microfluidic SERS detection: (i) Schematic of magnetic composite nanoparticles; (ii) Schematic of gold nanostars with SERS tags; (iii) Microfluidic channel with AgNPs; (iv) Double emulsion droplets containing gold nanoparticles prepared under hypertonic conditions; (v) With Ag colloidal droplet microfluidic chip. (**b**) Fixed SERS substrates for microfluidic SERS detection: (i) Nanopillar SERS substrates; (ii) Structured quartz chips covered with a 40 nm silver layer; (iii) Plasmonic nano-gap structures manufactured using a switchable self-assembly method; (iv) Pseudo-color Raman maps of CV, 4-ATP, and R6G on each segmented Ag line; (v) Hybrid microfluidic device based on surface plasmon polaritons and localized surface plasmon coupling

nanoparticle aggregation in the microfluidic channel, thereby achieving regulation of SERS detection performance. However, in some cases, analyte molecules can induce nanoparticle aggregation, leading to stronger SERS signals. As shown in Fig. 2.11a(v), Xidi and others observed this analyte-induced nanoparticle aggregation behavior in a droplet microfluidic device using levofloxacin samples, used for detecting levofloxacin in human urine samples.

2.4.4.2 Fixed SERS Substrate

The main advantage of the fixed SERS substrate is that it can achieve higher precision than the morphology of metal nanostructures, therefore, it can further improve the sensitivity and repeatability of SERS detection. Figure 2.11b shows different fixed SERS substrates built inside the microfluidic chip.

Gold Nanoparticles

To integrate gold nanostructures into microfluidic devices, one method is to directly build gold nanostructures on the microfluidic channel, another method can be the integration of gold film and gold nanostructures in the microfluidic channel. For example, Huh and others used an anodized aluminum film to develop a metal nanotube array. As shown in Fig. 2.11b(v), Burtsev and others improved the SERS detection sensitivity by integrating periodic metal structures (gratings) and branched gold nanoparticles on the grating surface. This design showed a very high SERS enhancement factor and better signal reproducibility in the microfluidic chip.

Silver Nanoparticles

Due to the higher SERS enhancement performance of nano-silver structures compared to other metal materials, it has been widely used in microfluidic SERS detection. As shown in Fig. 2.11b(ii), Popp and others used electron beam lithography to prepare nanostructured silver SERS substrates for high-sensitivity SERS detection of antibiotics (sulfamethoxazole) at nanomolar concentrations. They found that the silver-coated SERS substrate showed excellent uniformity on a millimeter scale, with better sensitivity and signal repeatability.

3D Micro-Nano Structures

Three-dimensional (3D) SERS substrates with uniform hot spots (gap <10 nm), can produce significant SERS enhancement effects due to the effective concentration of the electromagnetic field. Figure 2.11b(i) shows a nanopillar SERS substrate manufactured by a plasmonic stripping technique, which showed good signal reproducibility in the microfluidic chip, confirming the superior performance of microfluidic devices combined with SERS detection. At the same time, the continuous flow of analyte molecules in the microchannel significantly reduced the relative error in SERS measurements. In addition, as shown in Fig. 2.11b(iii), Lao and others proposed a switchable self-assembly method for detecting anticancer drugs on 3D plasmonic nanostructures in the microchannel.

2.4.5 Challenges and Development Trends of Microfluidic SERS Technology

Microfluidic devices with multi-layer SERS substrates perform excellently in terms of sensitivity and reproducibility, but it is difficult to detect molecules that do not interact with the SERS substrate. In this case, the functionalization of metal nanostructures has opened up new avenues for the indirect detection of such analytes. Some studies have found that metal colloids can block microchannels, thereby affecting the reproducibility of SERS detection signals in microchannels. Therefore, core-shell nanostructures and shell-separated nanoparticles are often used for microfluidic SERS detection, avoiding precipitation or blockage in microchannels through surface property regulation.

Other SERS-enhanced alternative nanomaterials, such as graphene, metal oxides, carbon nanowires, and hybrid nanomaterials, can also be used for microfluidic SERS detection due to their good chemical stability and recyclability. With the rapid development of nanochemistry, more and more new materials and processing synthesis techniques are applied to microfluidic SERS technology, which can improve the overall performance of microfluidic SERS detection methods.

The signal reproducibility of SERS detection is a major challenge currently faced by this technology. To overcome this problem, microfluidic SERS detection is considered an ideal platform. Generally, the reproducibility of SERS detection signals can be affected by factors such as irregular plasmonic nanostructures, oxidation and contamination of nanostructures, and sample heating effects. However, microfluidic SERS detection technology can improve signal reproducibility to some extent. Firstly, more uniform metal nanostructures can be prepared and assembled in droplet-based microfluidic systems. Secondly, the continuous flow of analyte molecules in microchannels can avoid sample heating and unnecessary photochemical reactions. Thirdly, the enclosed microchannels can slow down the oxidation and contamination of SERS substrate materials. Over the past twenty years, some researchers have demonstrated that high reproducibility SERS detection can be performed using microfluidic platforms, even for quantitative analysis of sample analytes.

Challenges faced by microfluidic SERS include short interaction time between metal nanostructures and analytes, long exposure time, Raman signal interference from the material of the microfluidic device itself, and reduced signal sensitivity due to the presence of liquid phase. Generally, spectroscopic techniques require longer collection times and optimal optical paths to achieve high sensitivity detection. However, it is difficult to achieve or guarantee this in microchannels due to continuous flow. For example, molecules containing thiol, such as mercaptobenzoic acid, can interact strongly with metal nanostructures within a few seconds, so the flow profile will not significantly affect the SERS intensity. However, if there is no strong interaction between the SERS substrate and the analyte, it is necessary to reduce the flow rate to improve their interaction.

Although the application and market related to microfluidic detection are growing rapidly, factors such as the integration of SERS substrate materials in microfluidic chips and their manufacturing costs still restrict the widespread promotion and use of microfluidic SERS. Therefore, the development of low-cost and simple microfluidic chip systems and SERS enhancement substrates is still crucial. For example, polymer materials and paper microfluidic chips can be used to reduce device costs, reduce operation time, and manufacturing steps, etc. In general, in the past decade, the application of microfluidic SERS in analytical chemistry is only a small part, and there are still many problems and technical challenges, such as how do we use SERS signals to detect tiny droplets or molecules to be tested in microchannels in a few seconds? How to reuse SERS substrate materials multiple times without processing the device? How to achieve highly sensitive SERS detection for analyte molecules with low Raman cross-sections? Can we effectively combine other analytical techniques such as chip chromatography, mass spectrometry, electrophoresis, isoelectric focusing, and electroanalysis with microfluidic SERS? These issues all require future researchers to invest more scientific research efforts to develop intelligent detection analysis technologies that are oriented towards practical applications and solve core issues in the sensing field.

Bibliography

1. FLEISCHMANN M P, HENDRA P J, MCQUILLAN A J. Raman spectra of pyridine adsorbed at a silver electrode [J]. Chemical Physics Letters, 1974, 26(2):163-166.
2. Cheng Guangxu. Raman Brillouin Scattering: Principles and Applications [M]. Science Press, 2001.
3. FARADAY M. Experimental relations of gold (and other metals) to light [J]. Philosophical Transactions of the Royal Society of London Series I, 1857, 147: 145-181.
4. RAYLEIGH L. On the scattering of light by small particles [J]. Philosophical Magazine, 1871, 41(275): 447-454.
5. ERIC C L R, PABLO G E, et al. Quantifying SERS enhancements [J]. MRS bulletin, 2013, 38: 631-640.
6. RAMAN C V, KRISHNAN K S. A New Type of Secondary Radiation [J]. Nature, 1928, 121: 501-502.
7. ZHAO L, JENSEN L, SCHATZ G. Pyridine Ag 20 Cluster: A model system for studying surface-enhanced Raman scattering [J]. Journal of the American Chemical Society, 2006, 128: 2911-2919.
8. KNEIPP K, YANG W, KNEIPP H, et al. Single molecule detection using surface-enhanced Raman scattering (SERS) [J]. Physical Review Letters, 1997, 78: 1667-1670.
9. ZHU W, ESTEBAN R, BORISOV A G, et al. Quantum mechanical effects in plasmonic structures with subnanometre gaps [J]. Nature Communications, 2016, 7: 1-14.
10. ALESSANDRI I, LOMBARDI J R. Enhanced Raman Scattering with Dielectrics [J]. Chemical Reviews, 2016, 116: 14921-14981.
11. YEO B S, STADLER J, SCHMID T, et al. Tip-enhanced Raman spectroscopy – Its status, challenges and future directions [J]. Chemical Physics Letters, 2009, 472: 1-13.
12. EMILY, L, KELLER, et al. Ultrafast surface-enhanced Raman spectroscopy [J]. Analyst, 2015, 140: 4922-4931.
13. FAN M, ANDRADE G F S, BROLO A G. A review on the fabrication of substrates for surface enhanced Raman spectroscopy and their applications in analytical chemistry [J]. Analytica Chimica Acta, 2011, 693: 7-25.
14. LIN X M, CUI Y, XU Y H, et al. Surface-enhanced Raman spectroscopy: substrate-related issues [J]. Analytical and Bioanalytical Chemistry, 2009, 394: 1729-1745.
15. RAMON A, ALVAREZ-PUEBLA, LUIS M, et al. Traps and cages for universal SERS detection [J]. Chemical Society reviews,2017, 46: 3866-3885.
16. ZHU W, BANAEE M G, WANG D, et al. Lithographically fabricated optical antennas with gaps well below 10 nm [J]. Small, 2011, 7(13): 1761-1766.
17. RODRIGUEZ-LLORENTE R A. Surface-enhanced vibrational spectroscopy [J]. Journal of Molecular Structure, 1997, 408: 17-22.

18. ZHANG Y, ZHAO S, ZHENG J, HE L. Surface-enhanced Raman spectroscopy (SERS) combined techniques for high-performance detection and characterization [J]. Trends Anal. Chem. 2017, 90: 1-13.
19. SHEN W, LIN X, JIANG C, et al. Reliable quantitative SERS analysis facilitated by core–shell nanoparticles with embedded internal standards [J]. Angewandte Chemie International Edition, 2015, 54: 7308-7312.
20. LOFRUMENTO C, RICCI M, PLATANIA E, et al. SERS detection of red organic dyes in Ag-agar gel [J]. Journal of Raman Spectroscopy Jrs, 2018, 49: 997-1005.
21. DUYNE R V, JEANMAIRE D L, SHRIVER D F. Mode-locked laser Raman spectroscopy. New technique for the rejection of interfering background luminescence signals [J]. Analytical Chemistry, 1974, 46(2): 213-222.
22. WECK P F, GORDON M E, GREATHOUSE J A, et al. Infrared and Raman spectroscopy of α-ZrW_2O_8 : A comprehensive density functional perturbation theory and experimental study [J]. Journal of Raman Spectroscopy, 2018, 49: 954-981.
23. CIALLA-MAY D, ZHENG X S, WEBER K, et al. Recent progress in surface-enhanced Raman spectroscopy for biological and biomedical applications: from cells to clinics [J]. Chemical Society Reviews, 2017, 46, 3857-4112.
24. PAHLOW S, MEISEL S, CIALLA-MAY D, et al. Isolation and identification of bacteria by means of Raman spectroscopy [J]. Adv Drug Deliv Rev, 2011, 11: 1013-1021.
25. WALTER A, MARZ A, SCHUMACHER W, et al. Towards a fast, high specific and reliable discrimination of bacteria on strain level by means of SERS in a microfluidic device [J]. Lab on A Chip, 2011, 11: 1013-1021.
26. GUTES A, CARRARO C, MABOUDIAN R. Silver nanodesert rose as a substrate for surface-enhanced Raman spectroscopy [J]. Acs Applied Materials & Interfaces, 2009, 1(11): 2551-2555.
27. ZHOU J, AN J, TANG B, et al. Growth of tetrahedral silver nanocrystals in aqueous solution and their SERS enhancement [J]. Langmuir the ACS Journal of Surfaces & Colloids, 2008, 24(18): 10407-10413.
28. ZHANG Y, GU Y, HE J, et al. Ultrabright gap-enhanced Raman tags for high-speed bioimaging [J]. Nature Communications, 2019, 10: 3905.
29. LIN H, MOCK J, SMITH D, et al. Surface-enhanced Raman scattering from silver-plated porous silicon [J]. Journal of Physical Chemistry B, 2004, 108: 11654-11659.
30. YE X, ZHENG C, CHEN J, et al. Using binary surfactant mixtures to simultaneously improve the dimensional tunability and monodispersity in the seeded growth of gold nanorods [J]. Nano Letters, 2013, 13: 765-771.
31. FANG J, DU S, LEBEDKIN S, et al. Gold mesostructures with tailored surface topography and their self-assembly arrays for surface-enhanced Raman spectroscopy [J]. Nano Letters, 2010, 10: 5006-5013.
32. PARK S, LEE J, KO H. Transparent and flexible surface-enhanced Raman scattering (SERS) sensors based on gold nanostar arrays embedded in silicon rubber film [J]. ACS Applied Materials & Interfaces, 2017, 9: 44088-44095.
33. GAO X, WU H, HAO Z, et al. A multifunctional plasmonic chip for bacteria capture, imaging, detection, and in situ elimination for wound therapy [J]. Nanoscale, 2020, 12: 6489-6497.
34. WANG G, YI R, ZHAI X, et al. A flexible SERS-active film for studying the effect of non-metallic nanostructures on Raman enhancement [J]. Nanoscale, 2018, 10: 16895-16901.
35. CHEN Y, FANG Y. Surface enhanced Raman scattering (SERS) activity studies of Si, Fe, Ti, Al and Ag films prepared by magnetron sputtering [J]. Spectrochimica Acta Part A: Molecular and Biomolecular Spectroscopy, 2008, 69: 733-737.
36. WANG X, ZHU X, CHEN Y, et al. Sensitive surface-enhanced Raman scattering detection using on-demand postassembled particle-on-film structure [J]. Acs Applied Materials & Interfaces, 2017, 9: 31102-31110.

37. XU K, RONG J, HONG M. Uniaxially stretched flexible surface plasmon resonance film for versatile surface enhanced Raman scattering diagnostics [J]. ACS Applied Materials & Interfaces, 2017, 9 (31): 26341-26349.
38. LI J F, ZHANG Y J, DING S Y, et al. Core-shell nanoparticle-enhanced Raman spectroscopy [J]. Chem. Reviews, 2017, 117: 5002-5069.
39. PHAN QUANG G C, YANG N, LEE H K, et al. Tracking airborne molecules from afar: three-dimensional metal-organic-framework - surface-enhanced Raman scattering (MOF-SERS) platform for stand-off and real-time atmospheric monitoring [J]. ACS Nano, 2019, 13: 12090-12099.
40. PANCORBO P M, THUMMAVICHAI K, CLARK L, et al. Novel Au–SiO_2^- WO_3 core–shell composite nanoparticles for surface-enhanced Raman spectroscopy with potential application in cancer cell imaging [J]. Adv. Funct. Mater, 2019, 29(46): 1903549.
41. DENG D D, YANG H, LIU C, et al. Ultrasensitive detection of diclofenac in water samples by a novel surface-enhanced Raman scattering (SERS)-based immunochromatographic assay using AgMBA@SiO2-Ab as immunoprobe [J]. Sensors and Actuators B: Chemical, 2019, 283: 563-570.
42. ZHANG D, HUANG L, LIU B, et al. Quantitative and ultrasensitive detection of multiplex cardiac biomarkers in lateral flow assay with core-shell SERS nanotags [J]. Biosens Bioelectron, 2018, 106: 204-211.
43. TANG L J, LI S, HAN F, et al. SERS-active Au@Ag nanorod dimers for ultrasensitive dopamine detection [J]. Biosensors & Bioelectronics, 2015, 71: 7-12.
44. GAO F L, LIU L Q, CUI G, et al. Regioselective plasmonic nano-assemblies for bimodal subfemtomolar dopamine detection [J]. Nanoscale, 2017, 9(1): 223-229.
45. WU X L, CHEN X, GAN F L, et al. Pt@AuNPs integrated quantitative capillary-based biosensors for point-of-care testing application [J]. Biosensors & Bioelectronics, 2016, 75: 55-68.
46. ZHU Y, KUANG H, XU L, et al. Gold nanorod assembly based approach to toxin detection by SERS [J]. Journal of Materials Chemistry, 2012, 22(6): 2387-2391.
47. MA W, YIN H, XU L, et al. Ultrasensitive aptamer-based SERS detection of PSAs by heterogeneous satellite nanoassemblies [J]. Chemical Communications, 2014, 50(68): 9737-9740.
48. ZHAO S, MA W, XU L, et al. Ultrasensitive SERS detection of VEGF based on a self-assembled Ag ornamented–AU pyramid superstructure [J]. Biosensors & Bioelectronics, 2015, 68: 593-597.
49. FENG F L, XIAO H X, LI Y F, et al. Influence of Acarbose on Plasma Glucose Fluctuations in Insulin-Treated Patients with Type 2 Diabetes: A Pilot Study [J]. International Journal of Endocrinology, 2015, 2015(2015): 1-5.
50. ZHAO Y, LIU L, KUANG H, et al. SERS-active Ag@Au core–shell NP assemblies for DNA detection [J]. Rsc Advances, 2014, 4(99): 56052-56056.
51. HAN X X, HUANG G G, ZHAO B, et al. Label-free highly sensitive detection of proteins in aqueous solutions using surface-enhanced Raman scattering [J]. Analytical Chemistry, 2009, 81(9): 3329-3233.
52. XU L J, ZONG C, ZHENG X S, et al. Label-free detection of native proteins by surface-enhanced Raman spectroscopy using iodide-modified nanoparticles [J]. Analytical Chemistry, 2014, 86(4): 2238-2245.
53. MATTEINI P, COTTAT M, TAVANTI F, et al. Site-selective surface-enhanced Raman detection of proteins [J]. ACS Nano, 2017, 11(1): 918-926.
54. BELL S, SIRIMUTHU N. Surface-enhanced Raman spectroscopy (SERS) for sub-micromolar detection of DNA/RNA mononucleotides [J]. Journal of the American Chemical Society, 2006, 128(49): 15580-16691.
55. BARHOUMI A, ZHANG D, TAM F, et al. Surface-enhanced Raman spectroscopy of DNA [J]. Journal of the American Chemical Society, 2008, 130(16): 5523-5529.
56. XU L J, LEI C Z, LI J X, et al. Label-free surface-enhanced Raman spectroscopy detection of DNA with single-base sensitivity [J]. Journal of the American Chemical Society, 2015, 137(15): 5149-5154.

57. GUERRINI L, KRPETI E, LIEROP D V, et al. Direct surface-enhanced Raman scattering analysis of DNA duplexes [J]. Angewandte Chemie International Edition, 2015, 127(4): 1160-1164.
58. STUART D A, YUEN J M, SHAH N, et al. In vivo glucose measurement by surface-enhanced Raman spectroscopy [J]. Analytical Chemistry, 2006, 78(20): 7211-7215.
59. YUEN J M, SHAH N C, WALSH JT JR, et al. Transcutaneous glucose sensing by surface-enhanced spatially offset Raman spectroscopy in a rat model [J]. Analytical Chemistry, 2010, 82(20): 8382-8325.
60. PRAK H, LEE S, CHEN L, et al. SERS imaging of HER2-overexpressed MCF7 cells using antibody-conjugated gold nanorods [J]. Physical chemistry chemical physics, 2009, 11(34): 7444-7449.
61. HA S H, KANG K H, CHOO J, et al. Biological imaging of HEK293 cells expressing PLC gamma 1 using surface-enhanced raman microscopy [J]. Anal. Chem, 2007, 79(3): 916-922.
62. WU L Y, ROSS B M, HONG S, et al. Bioinspired nanocorals with decoupled cellular targeting and sensing functionality [J]. Small, 2010, 6(4): 503-507.
63. PALLORO A, HOONEJANI M R, BRAUN G B, et al. Rapid identification by surface-enhanced Raman spectroscopy of cancer cells at low concentrations flowing in a microfluidic channel [J]. ACS nano, 2015, 9(4): 4328-4336.
64. WU X X, LUO L Q, YANG S G, et al. Improved SERS nanoparticles for direct detection of circulating tumor cells in the blood [J]. ACS Applied Materials & Interfaces, 2015, 7(18): 9965-9971.
65. JARVIS R M, GOODACRE R. Discrimination of bacteria using surface-enhanced Raman spectroscopy [J]. Analytical chemistry, 2004, 76(1): 40-47.
66. PATEL I S, PREMASIRI W R, MOIR D T, et al. Barcoding bacterial cells: A SERS-based methodology for pathogen identification [J]. Journal of Raman Spectroscopy, 2008, 39(11): 1660-1672.
67. MUHAMADALI H, CHISANGA M, SUBAIHI A, et al. Combining Raman and FT-IR spectroscopy with quantitative isotopic labeling for differentiation of E. coli cells at community and single cell levels [J]. Analytical Chemistry, 2015, 87(8): 4578-4586.
68. PREMASIRI W R, LEE J C, SAUER-BUDGE A, et al. The biochemical origins of the surface-enhanced Raman spectra of bacteria: a metabolomics profiling by SERS [J]. Analytical and Bioanalytical Chemistry, 2016, 408(17): 4631-4647.
69. BODELON G, MONTES-GARCIA V, LOPEZ-PUENTE V, et al. Detection and imaging of quorum sensing in Pseudomonas aeruginosa biofilm communities by surface-enhanced resonance Raman scattering [J]. Nature Materials, 2016, 15(11): 1203-1211.
70. FENG J J, XU L G, Cui G, et al. Building SERS-active heteroassemblies for ultrasensitive Bisphenol A detection [J]. Biosensors & Bioelectronics, 2016, 81: 138-142.
71. MA W, SUN M, XU L, et al. A SERS active gold nanostar dimer for mercury ion detection [J]. Chemical Communications, 2013, 49(44): 4989-4891.
72. ZHAO Y, XU L G, LIZ-MARZAN L M, et al. Alternating plasmonic nanoparticle heterochains made by polymerase chain reaction and their optical properties [J]. Journal of Physical Chemistry Letters, 2013, 4(4): 641-647.
73. LIN E C, FANG J, PARK S C, et al. Effective collection and detection of airborne species using SERS-Based detection and localized electrodynamic precipitation [J]. Advanced Materials, 2013, 25(26): 3554-3559.
74. HUNG Z L, MENG G W, HUANG Q, et al. Improved SERS performance from Au nanopillar arrays by abridging the pillar tip spacing by Ag sputtering [J]. Advanced Materials, 2010, 22(37): 4136-4139.

Chapter 3
Lateral Flow Immunoassay Probe Materials and Sensing Detection Applications

3.1 Introduction to Immunoassay Technology

Lateral Flow Immunoassay (LFIA) technology emerged in the late 1960s as a low-cost, fast, and convenient immunodetection method. Its basic principle is to couple detectable probes to antigens, antibodies, or aptamers, and achieve macroscopic visualization of the analyte through signal aggregation amplification pathways. Compared with traditional immunodetection methods, LFIA does not require professional operation, has flexible usage conditions, and has broad application prospects in the field of point-of-care testing (POCT).

3.1.1 Overview of Immunoassay

In 1976, the first LFIA strip based on radioimmunoassay for detecting human chorionic gonadotropin (hCG) in urine was launched in the United States, but its high price made it unaffordable for most ordinary families at the time. In the 1990s, Osikowicz and others used colored colloidal selenium to replace radioactive probes for hCG determination in urine, greatly reducing manufacturing costs. Since then, various new types of immunoassay strips have flooded the domestic and foreign medical markets.

In recent years, immunoassay technology based on different micro and nano probes has developed rapidly and has been widely used in the POCT field. Micro and nano probe technology refers to the use of materials with good biocompatibility at the micro and nano scale as the medium for signal transmission. By coupling them with target protein antibodies or aptamers, after binding to the analyte through a specific immune response, the physical signal emitted by the probe material is used for qualitative and quantitative sensing detection of the analyte in the sample.

X. Ma et al., *Functional Micro/Nanomaterials for Intelligent Sensing Devices*,
https://doi.org/10.1007/978-981-96-4990-7_3

Traditional colloidal gold immunoassay technology based on colloidal gold nanoparticle probes has the advantages of low cost and simple operation, and is therefore widely used in medical diagnosis, drug detection, and food safety. However, the detection of colloidal gold probe signals is mostly done by colorimetric qualitative analysis, that is, using the naked eye to distinguish the strength of colors, so there are problems such as low sensitivity and difficulty in achieving precise quantification. With the increasing demand for accuracy and sensitivity in in vitro testing in the medical industry, the research and development of new nano probes is gradually becoming a research hotspot. At the same time, immunoassay instruments characterized by speed, convenience, high sensitivity, and networking have emerged to meet the integrated needs of today's digital medical development.

3.1.2 Structure of Lateral Flow Immunoassay Strip

The LFIA strip consists of a sample pad, a conjugate pad, a nitrocellulose membrane (NC membrane), and an absorption pad. The sample pad is the starting point of the immunoassay strip, used to carry the added sample solution; the conjugate pad is fixed after the sample pad, used to carry the solution of the probe-labeled protein complex; the NC membrane is the core area where the immune reaction occurs, with a test line (T line) and a quality control line (C line) fixed on it for signal reading and strip quality control; the absorption pad is located at the end, providing capillary driving force for the entire strip and absorbing excess sample solution.

3.1.3 Principle of Immunoassay

The detection principle of LFIA is based on the specific binding of antigen and antibody. The nanoprobes coupled with antibodies are pre-fixed on the conjugate pad and can specifically react with the antigens in the sample solution. When the sample solution is dropped onto the sample pad, the liquid will flow forward through capillary action, passing through the conjugate pad, NC membrane, and finally reaching the absorbent pad at the end of the test strip. During the chromatography process, the T line and C line will capture a certain amount of probes according to the different detection forms and the concentration of the analyte. We can calculate and quantify the content of the analyte in the sample solution by reading the intensity of the probe signals on the T and C lines.

Immunoassay detection can be divided into sandwich method and competitive method. The sandwich method is suitable for large biomolecules with two or more immune binding sites. Specifically, as shown in Fig. 3.1 (left), the T line and C line on the NC membrane are fixed with primary antibody (Ab1) and secondary antibody (Ab2) respectively; when the sample solution contains the target antigen (Ag),

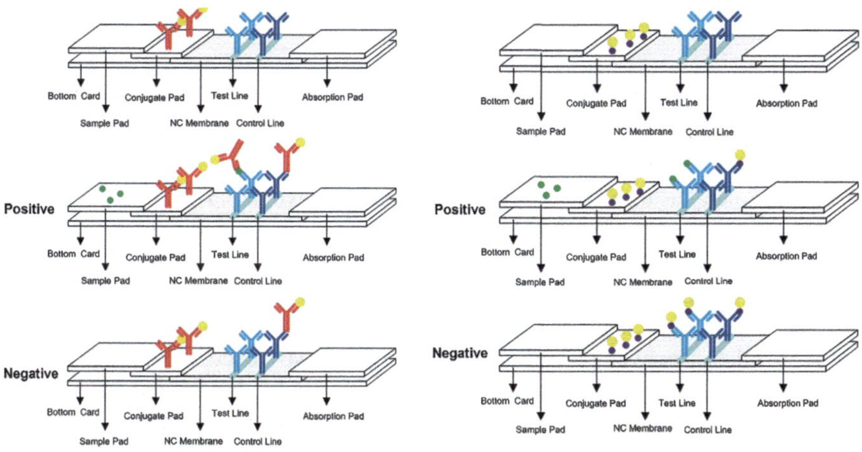

Fig. 3.1 Diagram of the sandwich method immunoassay principle (left) and competitive method immunoassay principle (right)

Ag specifically binds with the labeled antibody (Ab) on the conjugate pad to form an Ab-Ag complex, which is then captured by Ab1 on the T line, forming a sandwich form of Ab-Ag-Ab1, and the uncaught complex is then captured by Ab2 on the C line. Due to the presence of nanoprobes, we can analyze the test results by measuring the signal intensity on the T and C lines. Since the Ab2 fixed on the C line can specifically capture Ab to form an Ab-Ab2 complex, a signal will appear on the C line regardless of whether the sample contains an antigen, serving as a quality control indicator for the effectiveness of the chromatography process.

The competitive detection method is suitable for small molecule analytes with only one binding site. Generally, the competitive method can be divided into two types: the first type of process can be briefly described as the antibody (Ab) labeled with a probe is fixed on the conjugate pad, while the semi-antigen (Ag2) is fixed on the T line, and the secondary antibody (Ab2) is fixed on the C line. When the sample solution contains an antigen (Ag), the Ag2 on the T line will compete with Ag to form an Ab-Ag or Ab-Ag2 complex. Therefore, the signal on the T line will decrease as the concentration of the test sample increases. In the absence of the test antigen in the sample, all labeled Abs will be captured by the Ag2 on the T line, resulting in a strong signal on thc T line; for the second type of competitive method, as shown in Fig. 3.1 (right), the antibody Ab is fixed on the T line to capture the semi-antigen (Ag2) labeled with a probe on the binding pad. When the sample solution contains the test antigen (Ag), Ag and Ag2 will compete with the Ab on the T line. Therefore, if the antigen concentration in the sample is too high, the signal intensity on the T line will decrease or even be 0, and if there is a high intensity signal on the T line, it can be determined that the immunoassay is a negative test result.

3.2 Immunoassay Nanoprobe Materials and Sensors

3.2.1 Colloidal Gold Nanoprobes and Immunoassay Sensors

Colloidal Gold Immunochromatography (CG-LFIAs) emerged in the early 1980s, with colorimetry as its main detection method. As shown in Fig. 3.2, CG-LFIAs use gold nanoparticles (AuNPs) as probes, which are usually prepared by the reduction of chloroauric acid. AuNPs can be coupled with large biological molecules such as immunoglobulins, toxins, antibiotics, hormones, etc., through electrostatic action, without affecting the activity of the biological molecules. CG-LFIAs, due to their simple synthesis, convenient detection, and low cost, have been widely used in medical diagnosis, drug detection, food safety, and other fields. Sahoo and others found that the particle size of AuNPs mainly depends on the concentration of sodium citrate added during the preparation process. In addition, the concentration of chloroauric acid solution, pH, and temperature also affect the particle size. To improve the sensitivity of CG-LFIAs, Parolo and others proposed using enzyme-modified AuNPs as probes to amplify the detection signal, increasing the detection sensitivity by an order of magnitude. Surface modification of AuNPs with a core-shell structure can effectively improve the detection stability of CG-LFIAs. Lu and others proposed using highly stable silica to coat AuNPs, and successfully completed the qualitative detection of alpha-fetoprotein (AFP) and vanillin with detection limits of 300 pg/mL and 100 ng/g. Lin and others proposed using polyethylene glycol (PEG)

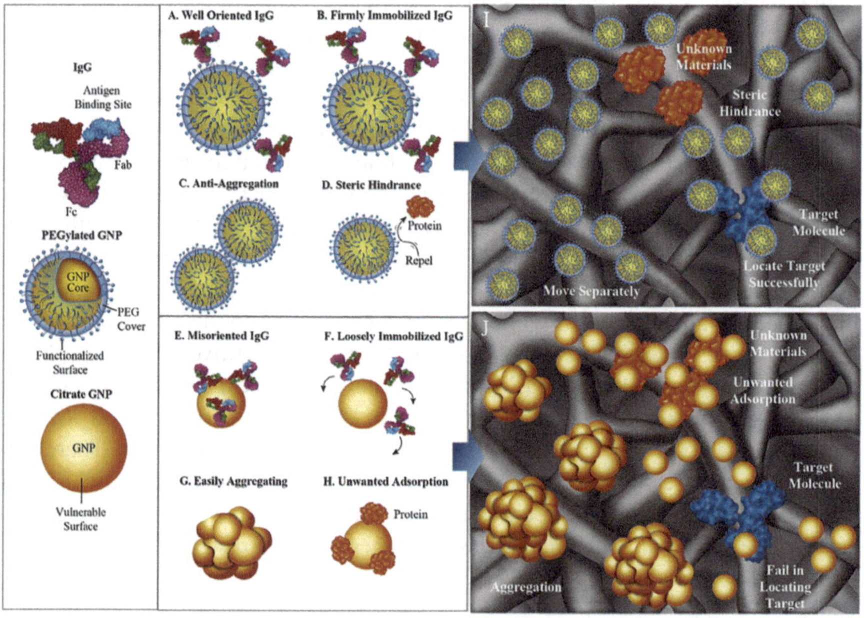

Fig. 3.2 Preparation of colloidal gold nanoprobe and its chromatographic detection application

to modify the surface of AuNPs. Since PEG is amphiphilic, nanoparticles wrapped in PEG can maintain good stability in different environments, making the entire coupling and detection process stable and reliable. Ultimately, this system achieved a visual detection limit of 0.8 ng/mL for bisphenol A (BPA). Because colorimetry is difficult to accurately judge the color signal intensity on the test strip, achieving quantitative detection of CG-LFIAs is a challenging problem. Liu and others first applied the reflection light signal of AuNPs to quantitatively determine the concentration of chromium ions in water and serum samples. Although its quantitative range is relatively small and the detection gradient is not accurate enough, this work has opened the door for quantitative detection of CG-LFIAs. To improve the sensitivity and detection range of quantitative detection, Yu and others introduced an electron microscope and a scattering analyzer to achieve quantitative detection of ciprofloxacin (CIP) in animal muscle, which has broad application prospects in the field of instant detection of food safety. To make quantitative CG-LFIAs more portable, researchers have developed an immunodetection system based on smartphones. Yu and others proposed using a smartphone camera to quantitatively detect alkaline phosphatase (ALP) in milk samples; Hou and others developed a dual-mode imaging system for quantitative detection of hCG and cancer embryonic cells. This system can not only detect the color signal of AuNPs but also read the fluorescence signal of quantum dots. Its high sensitivity and portability provide a reliable method for POCT diagnosis.

3.2.2 *Fluorescent Nanoprobes and Immunochromatographic Sensors Fluorescent*

Nanoprobes can emit strong fluorescence signals under the excitation of specific wavelength light sources. LFIA based on fluorescent nanoprobes has the advantages of good signal stability, high sensitivity, and quantitative detection. The mainstream fluorescent nanoprobes used for immunochromatographic detection can be divided into quantum dot nanomaterials, time-resolved fluorescence nanomaterials, and upconversion fluorescence nanomaterials.

3.2.2.1 Quantum Dot Nanoprobes

Quantum dots (QDs) are a type of semiconductor nanoparticles composed of III–V and II–VI group elements. Compared with traditional fluorescent dyes, QDs have advantages such as adjustable emission wavelength, wide absorption cross-section, strong fluorescence emission intensity, and high quantum yield. Since 2010, QDs-LFIA has been extensively developed for qualitative or quantitative in vitro instant detection. Studies have shown that compared with colloidal gold probes, the immunochromatographic system with quantum dot probes has a lower detection limit. To further improve the sensitivity of QDs-LFIA, as shown in Fig. 3.3a, it has been proposed that the fluorescence quantum yield can be increased by synthesizing quantum dot particles with a core-shell structure. This method has achieved quantitative detection of human hepatitis B surface antigen with a detection limit as low as 0.05 ng/mL. However, there are still some limitations to the application of QDs. Firstly, the quantum dots with surface modification have a relatively large particle size, which can easily cause aggregation or blockage when binding with biomolecules or during immunochromatographic detection, thereby reducing the detection stability and sensitivity. In addition, the heavy metal elements contained in quantum dots give them potential toxicity, and finding safe and green synthesis methods will become the next research goal.

3.2.2.2 Time-Resolved Fluorescence Nanoprobes

As shown in Fig. 3.3b, the fluorescence decay time of time-resolved fluorescence probes is about 10^3–10^6 times that of conventional fluorescence, making delayed detection possible. Secondly, the main raw material for synthesizing time-resolved fluorescence probes—lanthanide chelates, have a large fluorescence Stokes shift and a wide excitation spectrum, which can greatly reduce background interference and thus improve detection sensitivity. In recent years, scientists have paid great attention to the application of time-resolved fluorescence immunosensing in clinical detection. Some studies have proposed using time-resolved fluorescence

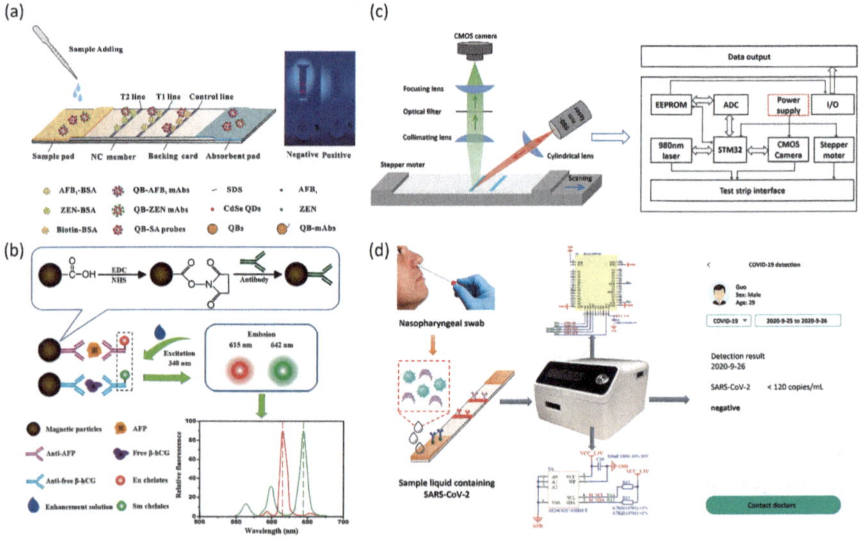

Fig. 3.3 (**a**) Schematic diagram of quantum dot immunochromatography; (**b**) Schematic diagram of time-resolved fluorescence probe; (**c**) Principle of upconversion immunochromatographic detection; (**d**) Portable immunochromatographic sensor device for COVID-19 antigen detection

immunosensing technology to simultaneously detect β2-microglobulin, ferritin, and cystatin-C in human serum, which are indicators of kidney function. Other studies have proposed using time-resolved fluorescence probes to double-label anti-Mycoplasma pneumoniae IgM and IgG for early diagnosis of pneumonia infection. When these experimental results were compared with existing commercial detection methods, they all showed good consistency and reliability. However, the time-resolved fluorescence immunosensing system has high requirements for the optical performance of its detection equipment. At present, most equipment is expensive and complicated to operate, which poses a significant obstacle to market application and promotion in digital medicine.

3.2.2.3 Upconversion Fluorescence Nanoprobes

Upconversion fluorescence nanoprobes (UCNPs) are a type of fluorescent nanomaterial doped with rare earth elements. Under near-infrared light (such as 980 Under the excitation of a laser (with a wavelength of nm), UCNPs can emit short-wavelength light signals such as blue and green light. Because their luminescence characteristics can be summarized as using low-energy photons to excite the emission of high-energy photons, this luminescence method is called "upconversion luminescence". Compared with quantum dot probes and time-resolved fluorescence, UCNPs have a great advantage in resisting endogenous fluorescence interference due to their unique luminescence characteristics. In addition, UCNPs are safe and non-toxic to synthesize, gradually becoming one of the most ideal fluorescent probes in LFIA. However, UCNPs chelated with rare earth elements have poor water solubility and dispersibility, so it is necessary to chemically modify their surface and coat them with a shell. A common modification method is to coat their surface with a silica shell, giving them good hydrophilicity and biocompatibility. Since 2018, UCNPs-LFIA has been reported in the detection of inflammatory factors, myocardial marker diagnosis, and drug detection. After the outbreak of the COVID-19 epidemic in 2019, as shown in Fig. 3.3d, a study proposed an LFIA system based on mesoporous silica-coated upconversion nanoprobe for rapid quantitative detection of COVID-19 antigens. The study also developed a portable upconversion fluorescence detection device enabled by the medical Internet of Things, which is palm-sized and can upload and analyze detection results in real time by connecting to edge devices (such as mobile phones, computers), contributing to the realization of digital medicine.

3.2.3 Carbon Nanoprobe and Immunochromatographic Sensor

Carbon nanomaterials, such as carbon nanotubes, have excellent spatial topology, good chemical stability, high specific surface area, and high absorbance, making them ideal probe materials for constructing biosensors. Compared with colloidal gold probes or fluorescent nanoprobe, carbon nanoprobe has the advantages of environmental protection, easy preparation, high stability, and has been applied in the construction of immunochromatographic systems. As shown in Fig. 3.4, Qiu et al. proposed an immunochromatographic sensor based on multi-walled carbon nanotubes (MWCNTs), using a sandwich structure to achieve rapid DNA sequence detection. The detection principle is to observe the characteristic black bands of carbon nanotubes and use software for auxiliary analysis and quantitative detection. Compared with single-walled carbon nanotubes, carbon nanoprobe based on composite materials can generate more modal signals, which helps to improve detection stability and sensitivity. Huang et al. proposed a new LFIA using magnetized carbon nanotubes (MCNTs) as probes. This work modified the surface of carbon nanotubes with magnetite by deposition method, and then captured the tumor marker in the blood, CA19–9 as the target sample through surface biochemistry and coupling

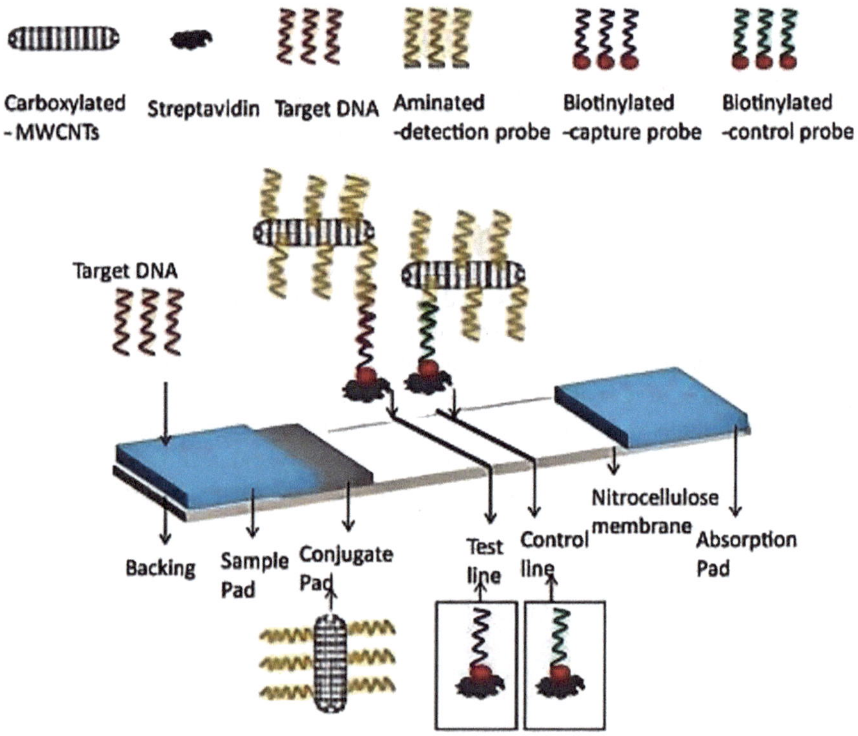

Fig. 3.4 Principle of immunochromatographic detection based on carbon nanotube probes

technology. Since MCNTs can generate magnetic signals, the detection results can be observed by the naked eye and quantitatively analyzed by a portable magnetic signal reader. Due to the excellent conductivity of carbon nanomaterials, carbon nanoprobe is also used as a signal enhancement method in electrochemical LFIA (ELFIA). Du et al. developed an ELFIA based on nanotubes for the quantitative detection of organophosphorus (OP) pesticides and nerve agents. The signal enhancement effect provided by the carbon nanoprobe makes the detection more sensitive at low potentials. Zhu et al. embedded carbon nanotubes into the conductive paper of ELFIA, Used for detecting DNA damage. This work uses carbon nanotubes as the working electrode, which is connected to an electrochemical analyzer, ultimately achieving quantitative detection of 8-OHdG in urine.

3.2.4 Magnetic Nanoprobe and Immunochromatographic Sensor

Uniform magnetic nanoparticles at the nanoscale are one of the new types of nanoprobe used in immunochromatographic detection in recent years. Compared with traditional fluorescent probes, magnetic nanoparticles are easy to prepare and preserve. At the same time, the Magnetic Lateral Flow Immunoassay (MLFIA) system based on magnetic nanoprobe suffers less background interference, and its magnetic signal detection can be achieved with a portable magnetic detector (MAR), eliminating the dependence of fluorescent immunochromatography on expensive detection equipment. In addition, because magnetic nanoparticles exhibit paramagnetism, biomolecules labeled with magnetic probes can undergo controlled directional movement under the drive of an external magnetic field, achieving the localization, enrichment, and separation of the detection sample.

As shown in Fig. 3.5, conventional magnetic nano-immunosensors use permanent magnets and variable magnetic resistance to analyze the voltage changes produced after the magnetic field lines of the magnetic nanoprobe are cut, thereby determining the quantity of antigens or antibodies loaded on the probe. In recent years, research has proposed using MLFIA to quantitatively detect the drug content (methamphetamine, morphine, ketamine) in human hair, with a detection time of less than 10 min and a sensitivity reaching the nanogram level. Compared with single-type magnetic nanoprobe, composite magnetic nanoprobe has multimodal signal output, which can apply immunodetection to more medical application scenarios. Recently, researchers have prepared a composite nanoprobe of gold-coated magnetism, that is, a layer of gold nanoshell is coated on the surface of the magnetic nano by in-situ growth method, so that the probe has a color signal and surface-enhanced Raman signal in addition to the original magnetic signal. Its color signal can be used for convenient qualitative detection, suitable for home or outdoor detection environments; while the Raman enhanced signal can be used for ultra-high

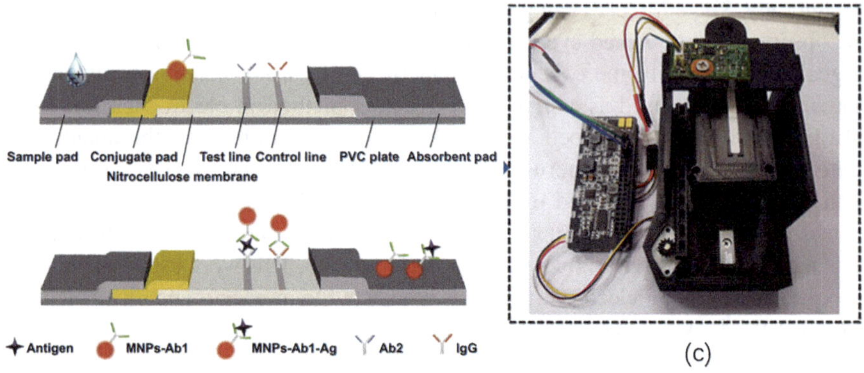

Fig. 3.5 Immunochromatography based on magnetic nanoprobe (left) and matching portable magnetic reader (right)

sensitivity quantitative detection, which is an important technical means for trace analysis in the laboratory.

3.3 POCT Application Based on Immunochromatography

LFIA as one of the most representative analysis platforms for POCT so far, compared with traditional laboratory analysis methods, its greatest advantage lies in extremely low detection costs and simple operation procedures. Because of its convenience, speed, and accuracy, LFIA can make clinical diagnosis, treatment, and nursing decisions rapid and efficient; at the same time, it also makes personal health monitoring, home health management, and even real-time diagnosis on the battlefield possible. In recent years, LFIA has been widely used in the detection fields of infectious disease antigens, antibodies, tumor biomarkers, bacteria, viruses, hormones, drug molecules, etc.

3.3.1 Multi-Index Antigen/Antibody (Human Immune Index) Detection

Based on the principle of antigen-antibody specific binding, LFIA is widely used in the detection of various human immune indicators. In the medical market, LFIA was first used to assist in diagnosing early pregnancy. Its detection principle is to use colloidal nano probes to qualitatively detect human chorionic gonadotropin (hCG) in blood or urine. Users can determine whether they are pregnant within a few minutes by observing the coloration of the T-line and C-line on the test strip. Early pregnancy test strips are now extremely common in both hospital and home use.

Today, in addition to this successful application of hCG detection, LFIA has been widely used in the detection of various hormones and immunoglobulins. For example, in 2017, Choi and others used gold nanoparticles (AuNPs) as probes and surface-enhanced Raman scattering (SERS) detection to develop LFIA test paper for thyroid-stimulating hormone (TSH) in serum. In 2019, Li Sihui and others used fluorescent microspheres as biomarkers to develop anti-Müllerian hormone LFIA test strips based on the principle of double antibody sandwich. In addition, there are reports on the application of LFIA in the detection of parathyroid hormone-like hormone, estrogen, cortisol, insulin, progesterone, prolactin, saliva E3, etc., and some have been commercialized. Figure 3.6 shows some of the commercially available hormone LFIA products.

Cancer is one of the diseases with the highest incidence and mortality rates in the world, and early detection and diagnosis of cancer are crucial for the treatment and care of cancer patients. Traditional immunological detection methods, including enzyme-linked immunosorbent assay, radioimmunoassay, chemiluminescence, fluorescence immunoassay, etc., usually require complex operations, long detection times, and expensive costs, which are not conducive to cancer prevention, early screening, monitoring diagnosis, and prognosis treatment. LFIA technology can complement traditional testing methods in terms of flexibility and timeliness, especially in recent years, various signal enhancement strategies and the development and application of new nano probes have made it widely used in clinical tumor biomarker detection. For example, in the diagnosis of liver cancer, colloidal gold immunochromatography can be used to detect alpha-fetoprotein (AFP) content. In 2020, Laura and others labeled inorganic nanoparticles to the detection antibody, and after the immunological recognition step, they controlled the gold deposition on the surface of the nanoparticles to achieve high sensitivity of the sensor. This

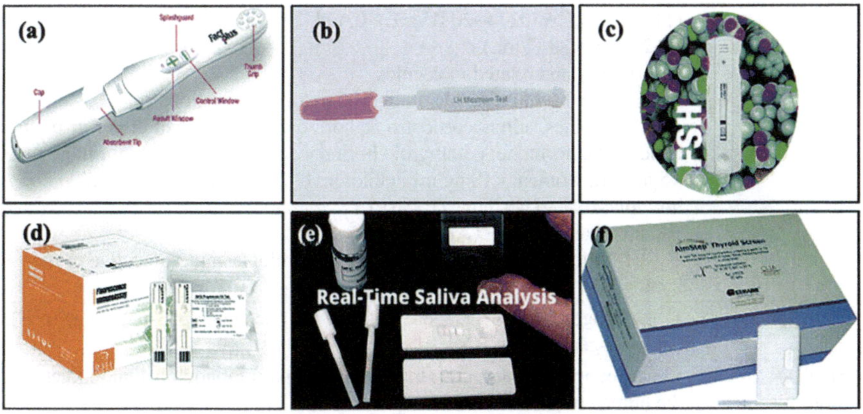

Fig. 3.6 Marketed hormone LFIA products: (**a**) FactPlus hCG pregnancy test strip; (**b**) BTNX LH midstream test strip; (**c**) Alfa Scientific Designs Instant-View FSH urine test strip; (**d**) CTK Biotech prolactin, progesterone, testosterone, T3 and T4 LFIA test strips; (**e**) SOMA Bioscience cortisol lateral flow device; (**f**) Germaine Laboratories rapid test kit AimStep thyroid screening TSH test

sandwich-type optical immunosensor can be used for ultra-sensitive detection of prostate-specific antigen (PSA), with a PSA detection limit of 1.1 fg/mL. In addition, biomarkers for gastric cancer screening such as pepsinogen I (PG I) and pepsinogen II (PG II), carcinoembryonic antigen, biomarkers for monitoring epithelial ovarian cancer (EOC) such as cancer antigen 125 (CA125), etc., have all developed corresponding LFIA tests. Recently, Tohid Mahmoudi and others summarized the tumor biomarkers detected by LFIA, as shown in Table 3.1.

In addition to the above applications, LFIA is also commonly used to detect biomarkers related to cardiovascular diseases. According to a report by the WHO, ischemic heart disease is the world's biggest killer, estimated to claim the lives of 17.9 million people each year, accounting for 16% of the total number of deaths worldwide. The development of highly sensitive instant detection devices is of great significance for the detection of cardiovascular diseases. In 2018, Du Xun and others used mouse-derived anti-cTnI monoclonal antibodies as colloidal gold-labeled antibodies, and with the help of colloidal gold immunochromatography technology and the principle of double-antibody sandwich method, they achieved the detection

Table 3.1 Common Tumor Biomarkers

Type of biomarker	Biomarkers (threshold value)
Enzyme	• Neuron specific enolase (NSE) (12.5 μg/mL); alkaline phosphatase (ALP) (44–147 mIU/mL); alpha-glutathione S-transferase (GST) (<8 ng/ml)
Protein	• Glycoproteins – Carcinoembryonic antigen (CEA) (3 ng/mL); alpha-fetoprotein (AFP) (10 ng/mL);human chorionic gonadotropin (hCG) (5.0 mIU/mL); prostate specific antigen (PSA) (4 ng/mL);Squamous cell carcinoma antigen (SCCA) (>20 ng/mL), platelet-derived growth factor (PDGF) • Mucin glycoproteins – CA125 (35 IU/mL); CA15–3 (25 IU/mL); CA19–9 (37 IU/mL); CA549 (11 IU/mL);CA50 (14–20 IU/mL); CA24–2 (20 IU/Ml); CA72–4 (6 IU/mL); CA27–29 (<38 U/mL) • Hormones and related molecules – Ectopic hormones; calcitonin (100 pg/mL); thyroglobulin; catecholamines; adrenocorticotropic hormone; hormones of digestive tract; antidiuretic hormone; parathyroid hormone; prolactin; growth hormone (men: <5 ng/mL, women: <10 ng/mL, children: 0–20 ng/mL); parathyroid hormone-related hormone (PTHrP) • Acute phase reactant (APR) – a1-antitrypsin; a1-antichymotrypsin; ceruloplasmin; c-reactive protein; haptoglobin;fibrinogen
DNA/RNA	• ras-oncogene; myc- oncogene; erbB-2/neu oncogene; p53-oncogene (p53); bcl- oncogene; micro RNA 21; let-7a
Small molecules	• Fucose; hormone receptor; transferrin receptor; monoclonal antibodies to the human tumor; immune complexes
Circulating tumor cell	• Epithelial cell adhesion molecules (EpCAM, 5CTCs); vimentin; Main dome protein
Extracellular vesicle	• Including exosomes, microvesicles, and apoptotic bodies

of cardiac troponin I within 10 min, with the lowest detectable mass concentration being 1.0 ng/mL. In recent years, there have been many reports on the related LFIA research of myoglobin, creatine kinase isoenzyme, creatine kinase, high-sensitivity C-reactive protein (hs-CRP), and lipoprotein-associated phospholipase A2 (Lp-PLA2). Nowadays, there are many immunodetection products developed based on LFIA technology on the market. In terms of human immune index detection, immunochromatography technology has greatly improved the detection rate compared with previous chemiluminescence or other detection methods, which is of great significance for the prevention, timely diagnosis, and treatment recovery of major diseases.

3.3.2 Bacterial Detection

Pathogens are ubiquitous in air, water, soil, and food, mainly spreading through physical contact. Once bacteria enter the body, they can potentially cause food poisoning and systemic infections, such as endophthalmitis, sepsis, pneumonia, diarrhea, and vomiting. If not treated in time, it can pose a serious threat to human life and even lead to death. In food microbiological testing, the main cause of food poisoning is foodborne pathogens. A survey report by the United Nations Food and Agriculture Organization pointed out that every year, 600 million people worldwide fall ill from eating food contaminated by bacteria, viruses, toxins, or chemicals, 11 million people die from unhealthy diets, and 420,000 people die from eating unsafe food. Therefore, due to the convenience and efficiency of LFIA, it is widely used in the detection of foodborne bacteria. At present, LFIA test strips have been developed for Salmonella, Listeria, E. coli, Brucella, Shigella, Staphylococcus aureus, Clostridium botulinum, Vibrio cholerae, Vibrio parahaemolyticus, Vibrio anguillarum, and Vibrio alginolyticus. Figure 3.7 shows a schematic diagram of the application of lateral flow immunoassay in pathogen detection. In 2020, a study used rabbit polyclonal antibodies as capture antibodies and a monoclonal antibody against O157:H7 as the detection antibody to construct a double-antibody sandwich chemiluminescent enzyme-linked immunodetection method for E. coli O157:H7, which was used to detect enterohemorrhagic E. coli in milk. The method has a minimum detection limit of 2.5×10^{-4} CFU/mL, which is a highly accurate and specific detection method. Ju Y and others developed a colloidal gold test strip for detecting Streptococcus suis type 2. The detection sensitivity of this method is as high as 106 CFU/mL, and there is no cross-reaction with other serotypes of Streptococcus (except SS1/2 type, because SS1/2 and type 2 Streptococcus have the same antigenic determinant cluster). Sohrabi and others have recently summarized the research on pathogen detection LFIA, including the main principles, challenges, and future prospects of LFIA in sensing pathogens.

In addition, LFIA is also widely used in the detection of foodborne parasitic diseases. Many parasitic diseases are highly prevalent and infectious, so there is a great need for simple, fast, efficient, and inexpensive diagnostic methods for

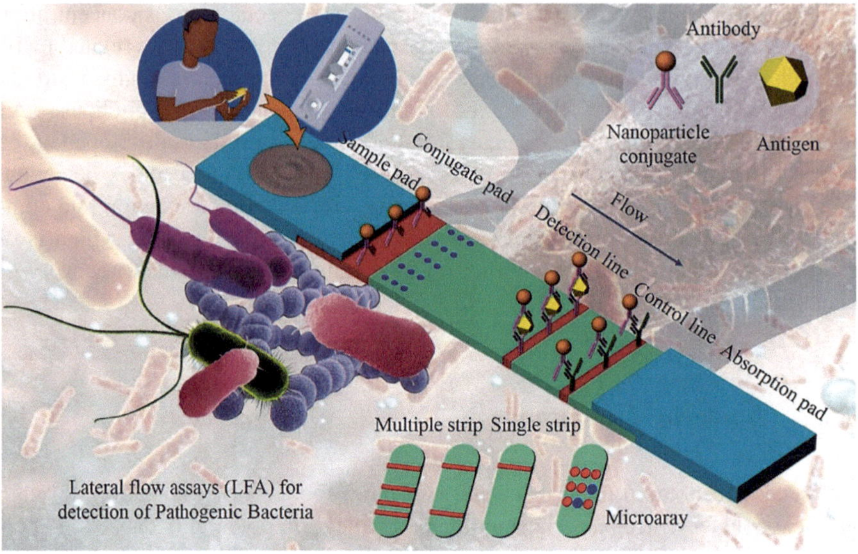

Fig. 3.7 Application of lateral flow immunoassay in pathogen detection

large-scale use. At present, immunoassay diagnostic methods for malaria, schistosomiasis, leishmaniasis, amoebiasis, filariasis, trypanosomiasis, and cryptosporidiosis have been established. For example, in 2018, Shi Feng and others used the prepared adult soluble antigen as the coating antigen, and colloidal gold-labeled G protein as the detection probe, to prepare an immunoassay strip with high sensitivity and specificity for detecting hookworm-specific antibodies. In 2017, Wang Zhaozhe and others successfully expressed and purified the recombinant protein of the SAG antigen epitope, and developed a colloidal gold immunoassay strip with this multi-antigen epitope recombinant protein as the inspection antigen, which can quickly detect Toxoplasma positive serum, laying the foundation for the rapid diagnosis of grassroots toxoplasmosis.

3.3.3 Virus Detection

LFIA immunoassay also plays a significant role in virus detection. Since ancient times, viruses have always been "invisible killers" threatening human health. According to the World Health Organization, the Marburg virus, known as the deadliest virus in history, has had a fatality rate of over 80% in multiple outbreaks; the HIV virus, known as the "nemesis" of the immune system, has killed 36 million people since its discovery. Although powerful antiviral drugs can allow infected people to live for several years, the infection rate of this disease can reach 95% in many low- and middle-income countries; the recent global outbreak of the novel coronavirus has led to more than 500 million confirmed cases and over six million unfortunate deaths in just a few years. However, there is currently a lack of specific drugs for most viral infections, so improving early screening methods for artificial immunity is the most effective way to prevent viral infections. For example, during the global outbreak of the novel coronavirus, the development of convenient methods for detecting the new coronavirus was urgent. Under resource-limited conditions, LFIA detection methods played a crucial role, enabling various immunoassay techniques for detecting SARS-CoV-2 infection based on colloidal gold nanoparticles, fluorescent latex, and up-conversion fluorescence to develop rapidly in a short time. Some companies have developed and marketed LFIA test kits for the detection of new coronavirus antibodies and antigens, which can complete the test within 20 min, and some of the kits have sensitivity and specificity close to 100%, and are widely used in the screening of COVID-19. In addition, LFIA can also be used to detect HIV, hepatitis virus, rabies virus, avian influenza virus, dengue fever virus, yellow fever (YF) virus, respiratory viruses, etc.

3.3.4 Molecular Detection (Drugs, Food Toxins, etc.)

According to reports, in 2021, approximately 275 million people worldwide used drugs, spanning over 200 countries and regions, with 200,000 people dying from drug use each year. Therefore, actively carrying out the detection of drugs and toxins is an important task in current clinical and forensic identification. Traditional laboratory-based detection methods, despite their centralized and high-precision advantages, are limited by their long detection time cycle, expensive detection equipment, and the need for specialized technicians to operate, making them difficult to use for extensive and in-depth drug detection work, especially for special work needs such as drug enforcement. LFIA technology, due to its simple and fast characteristics, is considered one of the most ideal on-site rapid detection platforms for drugs, drug molecules, etc. In 2020, Li and others developed an immunochromatographic assay (ICA) based on Surface Enhanced Raman Scattering (SERS) to detect morphine. The antibodies in ICA modified gold nanoparticles with Raman reporter labels, using indirect competitive immunoabsorption, achieved excellent results with a detection limit for morphine of 2.4×10^{-4} ng/ml, which is four orders of magnitude lower than the detection limit of visual methods. In addition, LFIA is also widely used for pesticide detection, and the application of LFIA in pesticide residue detection is crucial for environmental protection work. It is estimated in the literature that there are about 385 million cases of acute pesticide poisoning worldwide each year. Based on the global agricultural population of about 860 million, this means that about 44% of farmers are poisoned by pesticides each year. In 2019, Lan and others developed a colloidal gold bar assay based on a broadly specific monoclonal antibody (mAb) for simultaneous determination of CBF and 3-OH-CBF. The limit of detection (LOD) for both carbofuran and 3-OH-CBF was 7^{-10} ng/mL, and it can also be used for rapid simultaneous detection of CBF and

3-OH-CBF in water samples, as well as for determining illegal additions of CBF in other pesticide formulations.

3.3.5 *Heavy Metal Detection*

Heavy metal pollution (AMD/ARD) is one of the biggest environmental problems faced by the global mining industry. The impact of heavy metal pollution (AMD/ARD) on the global economy is estimated to exceed 10 billion Australian dollars, and its effects can last for more than 2000 years. Various environmental and health agencies have set maximum allowable limits for metals in water, air, and food. Analytical techniques for detecting heavy metals include Atomic Absorption Spectroscopy (AAS), Inductively Coupled Plasma Mass Spectrometry (ICP-MS), and Inductively Coupled Plasma Optical Emission Spectroscopy (ICP-OEC). However, due to the large size of the equipment, the need for expert personnel to operate, and the complexity of sample preparation, these techniques are not suitable for on-site detection applications. Therefore, LFIA is expected to become a powerful alternative for on-site heavy metal detection. For example, in 2021, Xu and others developed a rapid Latex Microsphere Immunoassay (LMIA) based on a monoclonal antibody (mAb) with high specificity and high affinity (affinity constant Ka = 1.83×10^{10} L/mol) for the detection of Cd^{2+} in asparagus. The 50% inhibitory concentration (IC50) of the test strip was determined to be 0.2 ng/mL, and the qualitative (LOD, visual) and quantitative detection (LOQ, data simulation) limits (IC10) of the test were 2 ng/mL and 0.054 ng/mL, respectively. In 2018, Yang and others used fluorescent microsphere-labeled monoclonal antibodies. (McAb) A one-step fluorescent immunoassay (FICA) probe has been developed for the rapid detection of copper ions in herbal plants and soil samples. Under optimal conditions, FICA can effectively detect copper ions within 15 min, with a detection limit of 0.2 μg/L. In addition, there are many reports and applications of LFIA technology for the detection of heavy metals such as chromium ions, lead ions, and mercury ions.

3.4 Development Trends Introduction

Lateral Flow Immunoassay (LFIA) technology has been successfully applied to the diagnosis of many diseases and conditions over the past 60 years, becoming an indispensable tool in medical diagnosis. The prevalence of these LFIA diagnostic platforms in medical institutions is increasing, especially in those with limited budgets and time. It is also widely used for personal health monitoring at home. In addition, LFIA has a wide range of applications in drug testing and environmental protection, making it one of the most popular rapid detection techniques. With the development of new probe material technology and the shift in health management concepts, the demand for personalized medical models is increasing, leading to a gradual shift from hospital-centered, symptomatic treatment models to community and home-based daily health care models. China has a large number of patients with chronic diseases such as diabetes and cardiovascular diseases. As the national self-care awareness continues to increase, the simplification of medical testing methods and the miniaturization of testing equipment are major future needs and development trends, which will further promote the steady growth of LFIA. At the same time, the requirements for the sensitivity, reproducibility, and multiplexing feasibility of LFIA will become higher and higher, but this is somewhat contradictory to the portability of the LFIA detection system platform. Therefore, how to further develop a portable rapid detection system that meets modern high requirements is still a highly challenging task.

Bibliography

1. J. Guo, S. Chen, J. Guo, X. Ma, "Nanomaterial Labels in Lateral Flow Immunoassays for Point-of-Care-Testing [J]. Journal of Materials Science & Technology, 2021, 60:90-104.
2. Wang Yinbiao, Liu Xiao, Li Qingmei, et al. Progress in immunochromatographic paper detection technology [J]. Journal of Henan University of Science and Technology: Medical Edition, 2017, 35(3):5
3. Huang L, Tian S , Zhao W , et al. Multiplexed detection of biomarkers in lateral-flow immunoassays[J]. Analyst, 2020, 145.
4. Li-Kai Lin, Uzunoglu A, Stanciu L A. Aminolated and Thiolated PEG-Covered Gold Nanoparticles with High Stability and Antiaggregation for Lateral Flow Detection of Bisphenol A[J]. Small, 2018, 14(10).
5. Y. Shao, H. Duan, L. Guo, Y. Leng, W. Lai, Y. Xiong, Quantum dot nanobead-based multiplexed immunochromatographic assay for simultaneous detection of aflatoxin B1 and zearalenone[J]. Analytica Chimica Acta, 2018, 1025:163-171.
6. J. Y. Hou, T. C. Liu, Z. Q. Ren, M.J. Chen, G.F. Lin, Y.S. Wu, Magnetic particle-based time-resolved fluoroimmunoassay for the simultaneous determination of α-fetoprotein and the free β-subunit of human chorionic gonadotropin[J]. Analyst, 2013, 138(13):3697-3704.
7. J. Guo et al., 5G-Enabled Ultra-Sensitive Fluorescence Sensor for Proactive Prognosis of COVID-19[J]. Biosensors and Bioelectronics, 2021, 181, 113160.
8. Yanjun Xie, Ying Yang, Weijun Kong, et al. Application of Lateral Flow Immunoassay Technology Based on Different Nanomaterials in Fungal Toxin Detection[J]. Analytical Chemistry, 2015, 043(004):618-628.
9. J. Guo, J. Zhang, S. Tian, K. Liu and J. Guo, An Up Conversion Optical System Based on Mesoporous Silica Encapsulated Up-converting Nanoparticles Labeled Lateral Flow Immunoassay for Procalcitonin Quantification in Plasma[J]. IEEE Journal of Selected Topics in Quantum Electronics, 2021, 27(5): 1-7.
10. W. Qiu, H. Xu, S. Takalkar, A.S. Gurung, B. Liu, Y. Zheng, Z. Guo, M. Baloda, K. Baryeh, G. Liu, Carbon nanotube-based lateral flow biosensor for sensitive and rapid detection of DNA sequence[J]. Biosensors & Bioelectronics, 2015, 64:367-372.
11. Z. Chu *et al.*, Magnetic Resistance Sensory System for the Quantitative Measurement of Morphine[J]. IEEE Transactions on Biomedical Circuits and Systems, 2021, PP(99):1-1.
12. Farka Z , Juík, Tomá, D Ková, et al. Nanoparticle-Based Immunochemical Biosensors and Assays: Recent Advances and Challenges[J]. Chemical Reviews, 2017:9973.

13. Bahadr E B , MK Sezgintürk. Lateral flow assays: Principles, designs and labels[J]. Trac Trends in Analytical Chemistry, 2016, 82:286-306.
14. Zhonghua Xiao, Haibing Huang. Progress in the Application of Lateral Flow Immunoassay in Clinical Instant Testing[J]. Chongqing Medicine, 2016, 045(028):4008-4011.
15. Ha Y, Ko S, Kim I , et al. Recent Advances Incorporating Superparamagnetic Nanoparticles into Immunoassays[J]. Acs Applied Nano Materials, 2018:acsanm.7b00025.
16. Andryukov B G. Six decades of lateral flow immunoassay: from determining metabolic markers to diagnosing COVID-19[J]. AIMS Microbiology, 2020, 6(3):280-304.
17. Khelifa L, Hu Y, Jiang N, et al. Lateral flow assays for hormone detection. 2022.
18. Mahmoudi T, Baradaran B , Guardia M . Lateral Flow Assays towards Point-of-Care Cancer Detection: A Review of Current Progress and Future Trends[J]. TrAC Trends in Analytical Chemistry, 2020.
19. Syedmoradi L, Norton M L , Omidfar K . Point-of-care cancer diagnostic devices: From academic research to clinical translation[J]. Talanta, 2020, 225:122002.
20. Cardiovascular diseases [Z]. https://www.who.int/health-topics/cardiovascular-diseases#tab=tab_1
21. Regan B, O'Kennedy R, Collins D. Advances in point-of-care testing for cardiovascular diseases[J]. Advances in Clinical Chemistry, 2021, 104: 1-70.
22. White H D, Chew D P . Acute myocardial infarction: The Lancet[J]. Lancet.
23. Sohrabi H, Majidi M R, Fakhraei M, et al. Lateral flow assays (LFA) for detection of pathogenic bacteria: A small point-of-care platform for diagnosis of human infectious diseases[J]. Talanta, 2022: 123330.
24. Boedeker W, Watts M , Clausing P , et al. The global distribution of acute unintentional pesticide poisoning: estimations based on a systematic review[J]. BMC Public Health, 2020, 20(1).
25. Calabria D, Calabretta M M, Zangheri M, et al. Recent advancements in enzyme-based lateral flow immunoassays[J]. Sensors, 2021, 21(10): 3358.
26. Di Nardo F, Chiarello M, Cavalera S, et al. Ten years of lateral flow immunoassay technique applications: Trends, challenges and future perspectives[J]. Sensors, 2021, 21(15): 518

Chapter 4
Optical Nanoprobes and Their Biomedical Detection Applications

4.1 Luminescence Mechanism of Optical Nanoprobes

An optical nanoprobe is a nanoscale probe that detects chemical or biological samples through optical signals. Its main components are usually inorganic nanoparticles, organic supramolecular assemblies, and organic-inorganic composites, etc. Currently, optical nanoprobes have been widely used in environmental monitoring, biomedicine, food safety, and other fields. This section mainly summarizes the luminescence mechanisms and biomedical detection applications of several common optical probe materials, including fluorescence, phosphorescence, upconversion luminescence, luminescence resonance energy transfer, and chemiluminescence, etc.

4.1.1 Fluorescence and Phosphorescence

There are many ways for optical probes to emit light, but they can be divided into two categories according to the length of the luminescence lifetime: fluorescence and phosphorescence. Both fluorescence and phosphorescence are photoluminescence, which are cold light phenomena. As shown in Fig. 4.1, if a molecule in a certain excited state emits a photon and transitions to a lower electronic state with the same multiplicity as the excited state, usually from the singlet excited state (S_1) to the singlet ground state (S_0), the light emitted at this time is called fluorescence. The luminescence lifetime of fluorescence is very short ($\leq 10^{-8}$ s), and the fluorescence almost disappears immediately after the excitation light is turned off. If a molecule in a certain excited state emits a photon and transitions to a lower electronic state with a different multiplicity from the excited state, usually from the triplet excited state (T_1) to the singlet ground state (S_0), the light emitted at this time

X. Ma et al., *Functional Micro/Nanomaterials for Intelligent Sensing Devices*,
https://doi.org/10.1007/978-981-96-4990-7_4

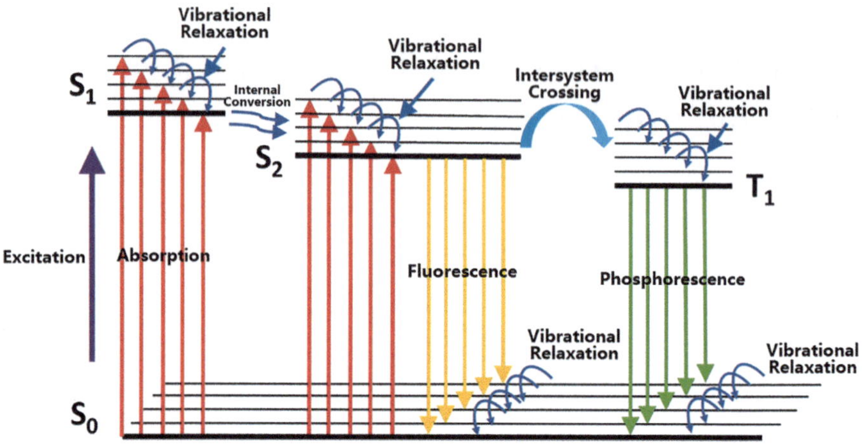

Fig. 4.1 Schematic diagram of the luminescence mechanism of fluorescence and phosphorescence materials

is called phosphorescence. Phosphorescence is a slow afterglow phenomenon. The de-excitation process of phosphorescence is forbidden by the transition selection rules of quantum mechanics, so the luminescence phenomenon still exists after the excitation light is turned off, and the luminescence lifetime of phosphorescence can last for milliseconds, seconds, or even several days. The non-radiative transition from the singlet excited state (S_1) to the triplet excited state (T_1) is called intersystem crossing, during which the electron of the excited molecule undergoes spin reversal. When the lower vibrational energy level of the singlet excited state (S_1) overlaps with the higher energy level of the triplet excited state (T_1), intersystem crossing will occur.

4.1.2 Fluorescence Resonance Energy Transfer

Fluorescence Resonance Energy Transfer (FRET), FRET (Fluorescence Resonance Energy Transfer) refers to the process where the excitation energy is transferred from the donor to the acceptor through dipole-dipole coupling in a non-radiative manner. This theory was first proposed by Förster in 1948; hence it is also known as the Förster principle. As shown in Fig. 4.2, when the emission spectrum of the donor group overlaps with the absorption spectrum of the acceptor group and the distance between the two fluorescent groups is appropriate (1–10 nm), the excited donor fluorescent group transfers energy to the nearby acceptor fluorescent group through dipole-dipole interaction in a non-radiative manner, exciting the acceptor and causing the quenching of the donor fluorescence and the emission of the acceptor fluorescence. The transfer process does not involve the emission and reabsorption of photons, so it is non-radiative.

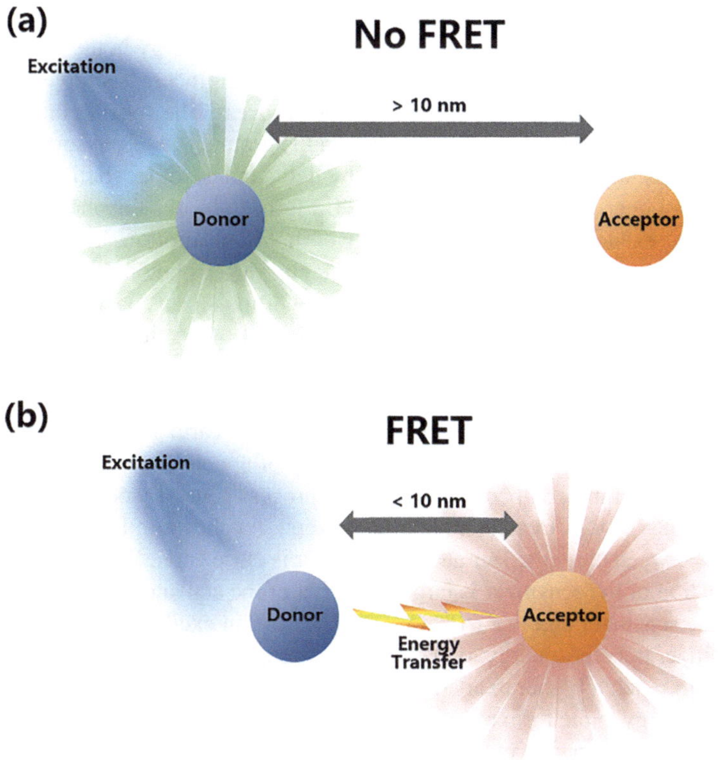

Fig. 4.2 Schematic diagram of the mechanism of fluorescence resonance energy transfer

4.1.3 Upconversion Luminescence

Stokes' law suggests that when a material is excited by high-energy light, it can only emit lower-energy light, i.e., when excited by high-frequency, short-wavelength light, it can only emit low-frequency, long-wavelength light. In 1960, Auzel, Ovsyankin, and Feofilov first proposed the concept of upconversion. Upconversion luminescence is an anti-Stokes phenomenon, which refers to the emission of high-energy photons when a material is excited by low-energy photons. Current research on upconversion luminescent materials mainly focuses on rare earth-based nanomaterials doped with lanthanide elements. Due to the rich energy level structure of rare earth elements, which can absorb multiple long-wave energy radiations, it is very conducive to the upconversion process. Based on different upconversion luminescence mechanisms, Excited State Absorption (ESA), Energy Transfer Upconversion (ETU), Photon Avalanche (PA), and Coordinative Energy Transfer (CET) have been proposed.

In 1959, Bloembergen et al. first proposed the process of excited state absorption, which refers to the same electron continuously absorbing two or more photons, transitioning from the ground state to the excited state and then to a higher excited state, and finally returning to the ground state, releasing energy in the form of light. Because the mechanism of the excited state absorption process is simple and direct, it is the most basic upconversion luminescence process. The specific luminescence process is shown in Fig. 4.3a. First, the ion in the ground state E_0 absorbs the pump photon with energy ω_1 after being excited by light, and the outer shell electron transitions to the intermediate metastable state E_1. It then absorbs the photon energy ω_2 to reach the excited state E_2, forming a two-photon absorption. The excited state electron returns to the ground state E_0, while emitting a photon with frequency ω. Although the ESA mechanism is simple, the conditions for its occurrence are stringent. It requires that the absorption cross-section of the excited state ion must be sufficient to absorb the second pump photon, which leads to its low absorption efficiency. Therefore, the ESA process only occurs at lower doping concentrations of rare earth ions, while high concentrations of rare earth ion doping will cause cross-relaxation, thereby affecting the efficiency of upconversion luminescence.

The Energy Transfer Upconversion (ETU) process generally occurs between two different types of ions (activators and sensitizers). As shown in Fig. 4.3b, when the distance between the sensitizer and the activator is close enough, and the absorption cross-section of the sensitizer is larger than that of the activator, the excitation light causes the sensitizer and the activator to each reach an excited state. Subsequently, the sensitizer transfers energy to the activator through energy resonance, causing the activator to reach a higher excited state and complete the upconversion process by releasing energy in the form of light. In this process, the concentration of doped ions determines the distance of energy transfer, which can greatly affect the efficiency of upconversion luminescence. Energy transfer upconversion is so far the most effective upconversion luminescence mechanism.

In 1979, Chivian and others first proposed a new upconversion luminescence mechanism in the study of infrared quantum counters based on Pr^{3+} ions, namely the process of photon avalanche (PA). As shown in Fig. 4.3c, in this process, the difference in energy levels between the ground state and the intermediate state of the

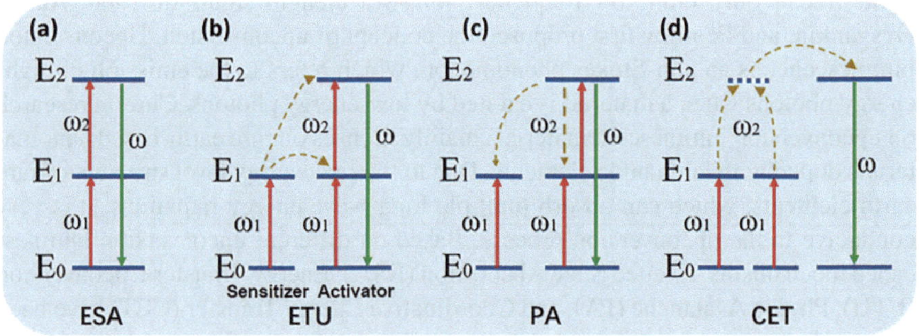

Fig. 4.3 Schematic diagram of different upconversion luminescence mechanisms: (**a**) Excited state absorption; (**b**) Energy transfer upconversion; (**c**) Photon avalanche; (**d**) Coordinative energy transfer

luminescent ion does not match the energy of the pump photon. In this case, the luminescent ion will first undergo an ESA process to reach a higher energy level (E_2), and then cross-relaxation occurs between it and the nearby ion in the ground state, causing both ions to reach an intermediate energy level (E_1). The ion at the E_1 energy level can absorb the energy of the excited state and reach the E_2 energy level. This process repeats, and finally, the number of electrons at the E_2 energy level increases exponentially over time like an avalanche. When the electron at the E_2 energy level transitions to the ground state, it emits a photon to complete the upconversion process. Photon avalanche is a relatively inefficient upconversion luminescence process.

As shown in Fig. 4.3d, the Cooperative Energy Transfer (CET) process is similar to ETU, and both the activator and the sensitizer are indispensable for the CET process. The main difference between the two luminescence mechanisms is that there is no real long-lived intermediate energy level in the activator during the CET process. Two sensitizers cooperate and are simultaneously excited to a virtual excited state, and then transfer energy to a nearby activator. The CET process mostly occurs in bulk materials and polymers, and there are relatively few reports of it occurring in nanomaterial systems.

4.1.4 Chemiluminescence

Chemiluminescence is a type of light radiation phenomenon that accompanies the process of chemical reactions in substances. The principle is that some substances in the reaction system absorb the chemical energy produced by the reaction, transition from the ground state to the excited state, and then return to the ground state, accompanied by the process of energy release in the form of light radiation. It is usually divided into direct chemiluminescence and indirect chemiluminescence. As shown in Fig. 4.4, direct chemiluminescence is the light radiation produced by the product C generated after the reactant undergoes a chemical reaction, and the energy

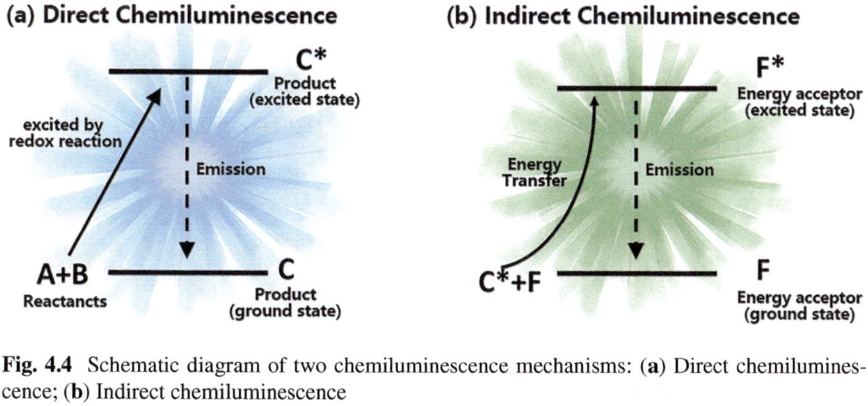

Fig. 4.4 Schematic diagram of two chemiluminescence mechanisms: (**a**) Direct chemiluminescence; (**b**) Indirect chemiluminescence

released by the reaction system is absorbed by the C molecule, causing the C molecule to transition to the excited state C*. Since C is the product of the reaction, this type of chemiluminescence is called direct chemiluminescence. As shown in Fig. 4.5, the chemiluminescence of luminol belongs to direct chemiluminescence. Indirect chemiluminescence is the light radiation produced when the excited state product C* (energy donor) generated after the reactant undergoes a chemical reaction releases energy to the luminescent substance F (energy acceptor) to excite it to the excited state F*, and then returns to the ground state.

Although traditional chemiluminescence nanoprobes have strong anti-interference and high sensitivity, most chemiluminescence nanoprobe materials have a limited range of applicable test objects, because they usually require the test object to be the reactant of the chemiluminescence reaction. The Amplified Luminescent Proximity Homogeneous Assay Linked Immunosorbent Assay (AlphaLISA) nanoprobe can avoid this disadvantage and take advantage of its strengths. The AlphaLISA nanoprobe produces chemiluminescence by exciting the photosensitizer to react with the luminescent agent to produce singlet oxygen. The role of the test object is to bring the photosensitizer and the luminescent agent closer to the diffusion distance of singlet oxygen through the specific immune results such as antibodies. The test object itself does not participate in the chemiluminescence reaction, thus solving the problem of the small range of applicable test objects for other chemiluminescence nanoprobes. However, the chemiluminescence efficiency induced by singlet oxygen is high and the background signal is low. The emission wavelength of the AlphaLISA nanoprobe material is shorter than the excitation wavelength, and it has the same advantages as upconversion nanoparticles. Compared with traditional methods such as Enzyme Linked Immunosorbent Assay

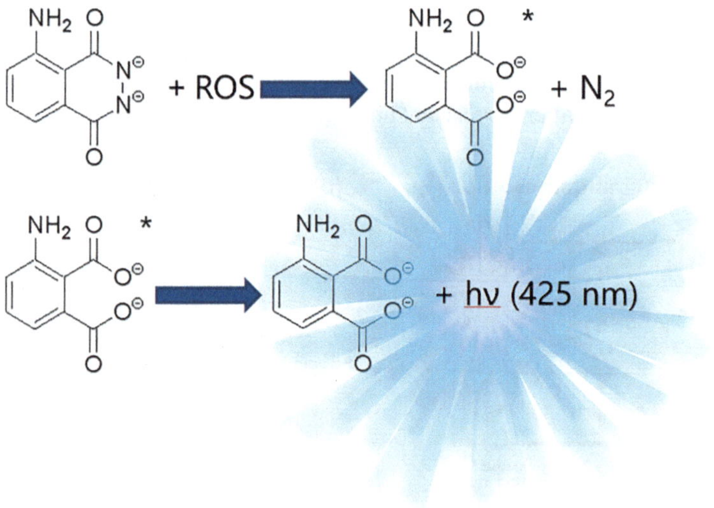

Fig. 4.5 Schematic diagram of the mechanism of luminol being oxidized by ROS

(ELISA) and fluorescence immunoassay, the photoexcited chemiluminescence immunoassay has obvious advantages. One is homogeneous reaction, eliminating the washing step; the second is that the emission wavelength is shorter than the excitation wavelength, avoiding the interference of background fluorescence.

As shown in Fig. 4.6, the nanoprobe materials used in the photoexcited chemiluminescence immunoassay include donor beads and acceptor beads, both of which have a particle size of about 200 nm. The donor beads carry the photosensitizer, which produces singlet oxygen under laser excitation; the acceptor beads carry the chemiluminescent agent and the fluorescent agent. When the acceptor beads come into contact with singlet oxygen, the chemiluminescent agent on the acceptor beads is oxidized by singlet oxygen, and the excited state product produced by Chemiluminescence Resonance Energy Transfer, The CRET (Chemiluminescence Resonance Energy Transfer) stimulates the fluorophore to emit light, with the emitted light having a shorter wavelength than the excitation light. Two different monoclonal antibodies of the antigen to be tested are modified on the receptor and donor spheres respectively. When the sample containing the antigen to be tested is incubated with the donor and receptor spheres in the sample pool, a "donor sphere-antibody1-antigen-antibody2-receptor sphere" sandwich complex is formed. In the sandwich complex, the singlet oxygen produced by the donor sphere diffuses to the receptor sphere and emits light upon interaction. Since the diffusion distance of singlet oxygen in water is only about 200 nm, the nanospheres that have not formed the sandwich complex do not participate in the luminescence. Therefore, the luminescence intensity of the sample pool is directly proportional to the total amount of the sandwich complex, which can be used to measure the content of the antigen to be tested, eliminating the need for a washing step. In summary, this section summarizes the luminescence mechanisms of different optical probes, including fluorescence, phosphorescence, upconversion luminescence, chemiluminescence, and fluorescence resonance energy transfer. Based on these luminescence mechanisms,

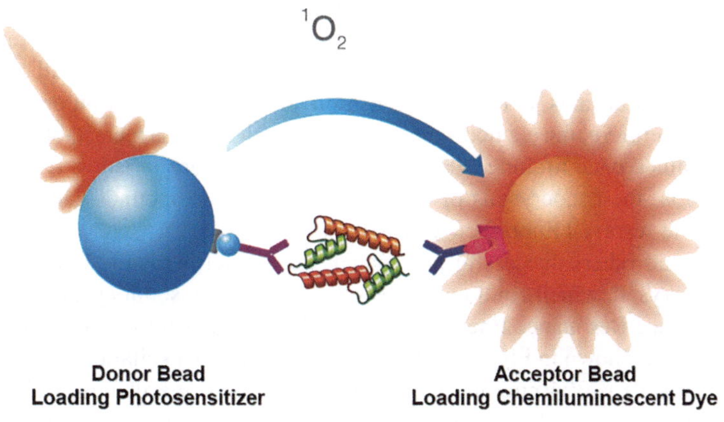

Fig. 4.6 Schematic diagram of the principle of photoexcited chemiluminescence immunoassay

a variety of optical nanoprobes have been developed and widely used in the field of biomedical detection.

4.2 Classification of Optical Nanoprobe Materials

4.2.1 Time-Resolved Fluorescence Nanoprobes

The fluorescence lifetime of the dye loaded on the time-resolved fluorescence nanoprobe is longer than the background fluorescence. The detector detects the light signal after a delay following the removal of the excitation light. At this time, the scattered light of the excitation light has disappeared, and the background fluorescence has also decayed to a very low level, while the fluorescence of the dye is still strong, thereby reducing the interference of scattered light and background fluorescence. Therefore, the luminescence lifetime of the time-resolved fluorescence nanoprobe must be significantly longer than the background fluorescence of the biological matrix. The background fluorescence lifetime of biological samples is generally only at the nanosecond level, while the luminescence lifetime of time-resolved fluorescence nanoprobes is mostly at the microsecond to millisecond level. Commonly used time-resolved fluorescence nanoprobes mainly include rare earth inorganic nanocrystals, rare earth chelates, and quantum dots.

4.2.1.1 Rare Earth Inorganic Nanocrystals

Rare earth compounds are a treasure trove of luminescent materials. Rare earth ions have rich 4f electron energy levels. Since 4f is a shielded inner shell layer, it is less affected by the crystal field or ligand field, so the emission spectrum of rare earth ions is close to the atomic spectrum, showing sharp peaks, rather than the broad peaks of organic fluorescent molecules, which is beneficial for optical signal detection applications. The 4f → 4f transition of free trivalent rare earth ions is forbidden due to the same parity. Although the ligand field and crystal field will affect the parity of the 4f orbitals, making the transition allowed, the transition probability is very low. Therefore, the excited state of the rare earth ions is a metastable state with a long luminescence lifetime, showing time-resolved characteristics. After surface modification, the nanocrystals of rare earth inorganic compounds can be directly used as time-resolved fluorescence probes.

Zheng and others successfully prepared a core-shell structured CaF_2:Ce,Tb@ CaF_2 nanocrystal, which continuously excites fluorescein isothiocyanate (FITC) as a Fluorescence Resonance Energy Transfer (FRET) donor for time-resolved imaging of cells. As shown in Fig. 4.7, CaF_2:Ce, The maximum excitation wavelength of Tb nanocrystals is 304 nm, and the maximum emission wavelength is 541.2 nm; CaF_2:Ce,Tb nanocrystals are spherical in shape, with an average particle size of

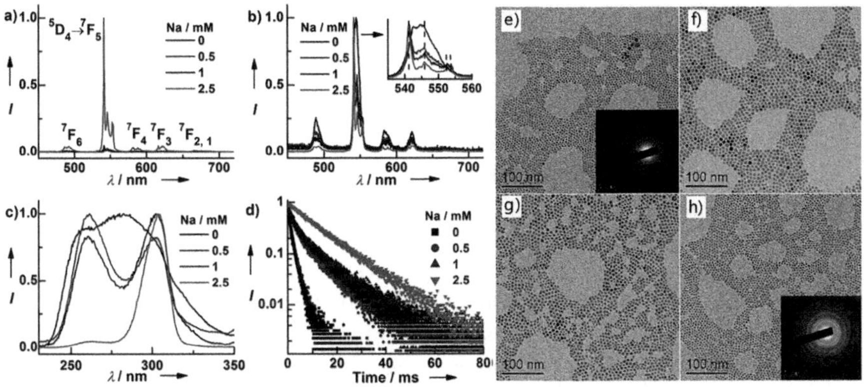

Fig. 4.7 (**a**) Emission spectra of CaF_2:Ce,Tb nanocrystals with different amounts of Na^+ input, normalized to 1 at the peak of the brightest sample; (**b**) Emission spectra of CaF_2:Ce,Tb nanocrystals with different amounts of Na^+ input, normalized to 1 at the peak of each sample; (**c**) Absorption spectra of CaF_2:Ce,Tb nanocrystals with different amounts of Na^+ input; (**d**) Fluorescence decay of CaF_2:Ce,Tb nanocrystals with different amounts of Na^+ input; (**e**) TEM image of CaF_2:Ce,Tb nanocrystals with a Na^+ concentration of 0 at the time of reaction; (**f**) TEM image of CaF_2:Ce,Tb nanocrystals with a Na^+ concentration of 0.5 at the time of reaction; (**g**) TEM image of CaF_2:Ce,Tb nanocrystals with a Na^+ concentration of 1 at the time of reaction; (**h**) TEM image of CaF_2:Ce,Tb nanocrystals with a Na^+ concentration of 2.5 at the time of reaction

about 3.8 nm, and a fluorescence lifetime of up to 12.5 ms. The particle size of CaF_2:Ce,Tb@CaF_2 nanocrystals depends on the number of shell coating layers and the concentration of Na^+, all of which are less than 10 nm, and are spherical in shape.

Wang et al. successfully synthesized Ce^{3+} and Tb^{3+} co-doped $LiYF_4$ nanocrystals using the thermal decomposition method. These nanocrystals have a fluorescence lifetime of 5.7 ms, carry a positive charge on their surface, and adsorb negatively charged fluorescent dyes. They act as donors for fluorescence resonance energy transfer (FRET), continuously exciting these dyes. When glyphosate is present in the sample, it replaces the adsorbed dye, causing the fluorescence of the dye to weaken, thereby detecting glyphosate. As shown in Fig. 4.8, $LiYF_4$:Ce,Tb

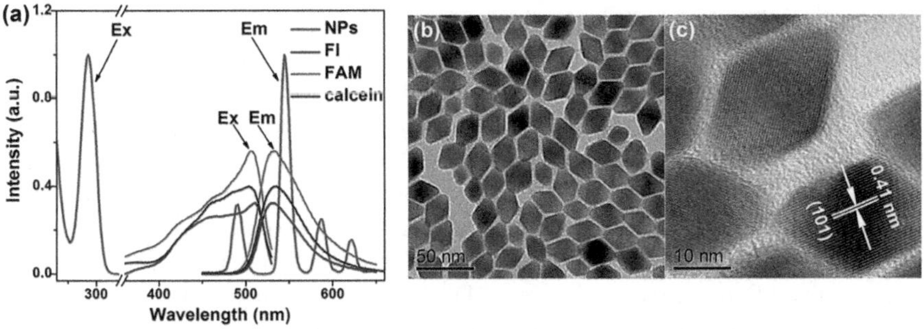

Fig. 4.8 Wang et al.'s $LiYF_4$:Ce/Tb(5%/5%) time-resolved fluorescence nanoprobes (**a**) excitation and emission spectra (green line in the figure), (**b**) TEM image, (**c**) HRTEM image

nanocrystals are rhombic, with an average length of about 21 nm; $LiYF_4$:Ce,Tb nanocrystals have a maximum excitation wavelength of 280 nm, and there are four stronger peaks in the emission spectrum at 489, 542, 585, 622 nm, with 542 nm being the strongest.

Rare earth metal ions are doped into inorganic matrices to form substitutional solid solutions. The difference in charge and radius between the doped cations and the matrix cations will cause lattice distortion, leading to a decrease in luminescence efficiency. Therefore, it is necessary to introduce ions that have the opposite effect to the doped ions to alleviate the lattice distortion and thus enhance the luminescence efficiency. For example, the CaF_2 nanocrystals co-doped with Ce^{3+} and Tb^{3+} prepared by Zheng et al., as shown in Fig. 4.7a, significantly improved the luminescence efficiency after doping with Na^+. In summary, the fluorescence lifetime of rare earth inorganic nanocrystals is significantly longer than the background fluorescence of the biological matrix, with good time resolution, strong emission, sharp emission lines, and is not easily disturbed. Moreover, these nanoprobes, as inorganic materials, have the advantages of simple preparation methods, stable chemical properties, and high resistance to photobleaching. However, the wide distribution of structural dimensions of these optical nanoprobe materials and their poor monodispersity in solution to some extent limit their application scope in biomedicine.

4.2.1.2 Rare Earth Chelates

Rare earth metal chelates also have time-resolved fluorescence properties and can be used as time-resolved fluorescence dyes. Although the 4f → 4f transition of trivalent rare earth ions becomes allowed under the influence of the crystal field or ligand field, its absorbance is still not high. The introduction of antenna ligands to chelate with it can enhance the absorption of the rare earth ions to the excitation light, thereby improving the luminescence ability. In such dye molecules, the absorbance of the ligand to the excitation light is much greater than that of the metal ion itself, so it is called an antenna ligand. The antenna ligand absorbs the excitation light and then transfers the energy to the metal ion to make it luminescent. If the antenna ligand is chosen properly, the luminescence efficiency of the chelate will be very high. Rare earth chelates, as small molecule compounds, need to be loaded into a nanomatrix and the surface of the nanomatrix modified with specific groups or molecules to make time-resolved fluorescence nanoprobes. The main functions of the matrix are: improving the dispersibility of the time-resolved fluorescence probe in water; facilitating surface modification so that the probe can specifically bind to the analyte; reducing the concentration of the dye to avoid self-quenching. Chen et al. prepared SiO_2 nanospheres loaded with Tb (III) chelate Cs124-DTPA-Tb for time-resolved fluorescence detection of DNA. The research group first synthesized the ligand Cs124-DTPA, which consists of two parts: one part is 7-amino-4-methyl-2-quinolone (Cs124), which acts as an antenna ligand responsible for absorbing the excitation light, and the other part is diethylenetriaminepentaacetic acid (DTPA), which acts as a multidentate ligand responsible for chelating with Tb^{3+} ions, and finally loaded onto SiO_2 nanospheres. As shown in Fig. 4.9, the nanoprobe has a spherical morphology, an average particle size of about 50 nm, a maximum excitation wavelength of 328 nm, a maximum emission wavelength of 546 nm, and a fluorescence lifetime of about 1.5 ms.

Yang et al. prepared a time-resolved fluorescence nanoprobes Eu@SMA using co-precipitation method, with a fluorescence lifetime of about 620 μs. The probe was successfully used for time-resolved fluorescence imaging of tumors in mice after being modified with the tumor-targeting agent transferrin (Transferrin, Tf) and Arg-Gly-Asp tripeptide. Eu@SMA uses styrene and maleic acid copolymer SMA nanospheres as the matrix, and the carboxyl groups on the surface of the matrix are used to connect with the amino groups of Tf or Arg-Gly-Asp tripeptide. The rare earth chelate loaded on Eu@SMA is $Eu(tta)_3bpt$. As shown in Fig. 4.10, the maximum excitation wavelength of this nanoprobe is 410 nm, the maximum emission wavelength is 618 nm, it is spherical in shape, with an average particle size of about 15 nm, and this nanoprobe also has the ability to emit upconversion luminescence under two-photon excitation.

Rare earth chelate nanoprobes and rare earth inorganic nanocrystals both have the advantages of good time-resolved effect, strong luminescence, and sharp emission lines. In addition, by choosing the appropriate matrix, rare earth chelate nanoprobes are easier to achieve surface functionalization modification than rare earth inorganic nanocrystals, and the water solubility and monodispersity of the probe

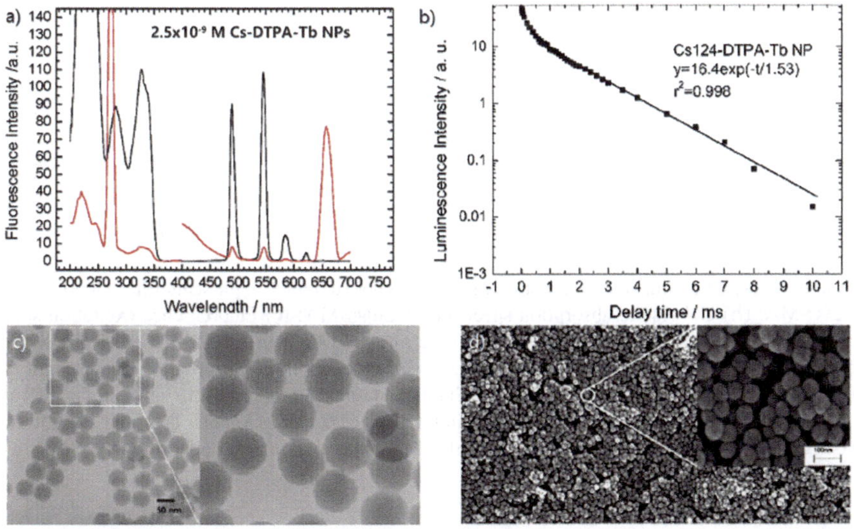

Fig. 4.9 (**a**) Real-time fluorescence (red line) and time-resolved fluorescence (black line) of SiO_2 nanospheres loaded with Cs124-DTPA-Tb; (**b**) Fluorescence decay of SiO_2 nanospheres loaded with Cs124-DTPA-Tb; (**c**) TEM image of SiO_2 nanospheres loaded with Cs124-DTPA-Tb; (**d**) SEM image of SiO_2 nanospheres loaded with Cs124-DTPA-Tb

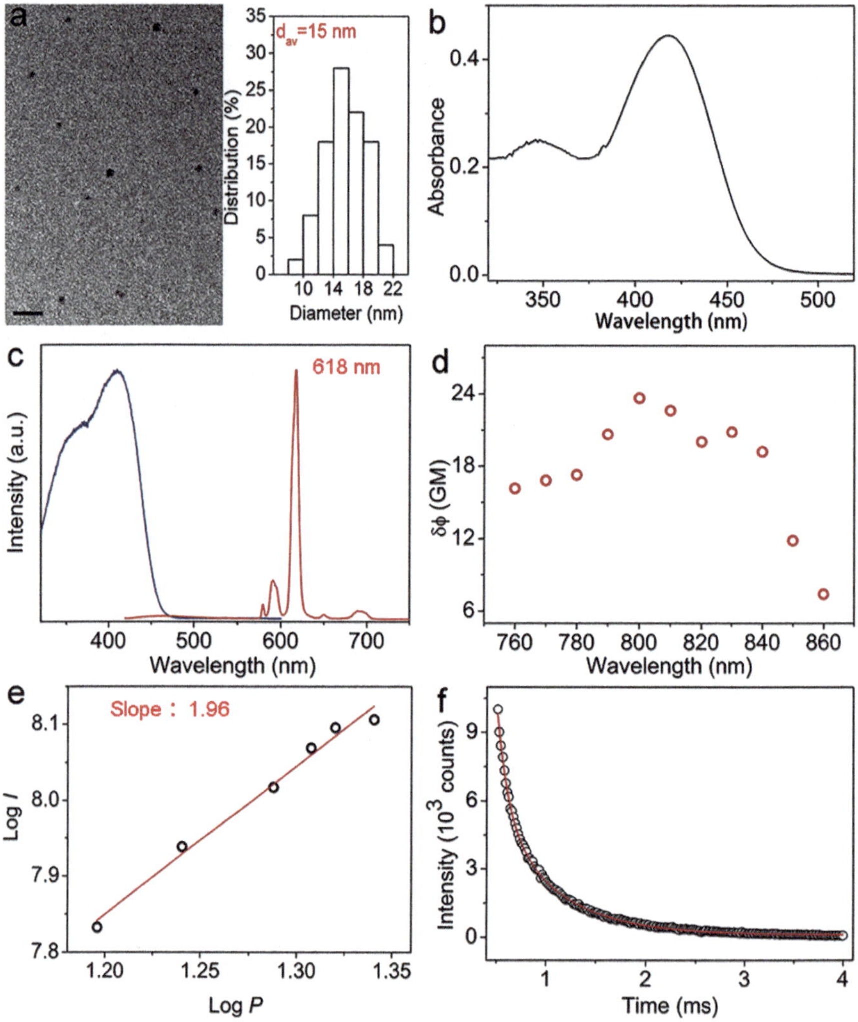

Fig. 4.10 (**a**) TEM image and dynamic light scattering (dynamic light scattering, DLS) data of Eu@SMA; (**b**) UV-visible absorption spectrum of Eu@SMA; (**c**) Fluorescence excitation spectrum (blue line) and fluorescence emission spectrum (red line) of Eu@SMA; (**d**) Relationship between the two-photon excitation absorption cross-section and excitation wavelength of $Eu(tta)_3bpt$ in Eu@SMA under near-infrared light excitation; (**e**) Relationship between the fluorescence at 615 ± 12.5 nm and the excitation light intensity of Eu@SMA under 800 nm near-infrared laser excitation; (**f**) Fluorescence decay of Eu@SMA under 410 nm excitation light

material are also better. However, as an organic material, the chemical stability and photobleaching resistance of rare earth chelate nanoprobes are worse than those of rare earth inorganic nanocrystals.

4.2.1.3 Quantum Dots

The fluorescence lifetime of quantum dots ranges from tens of nanoseconds to microseconds, while the fluorescence lifetime of organic fluorescent dyes and the background fluorescence of biological samples are both at the nanosecond level. Therefore, quantum dots can be used as time-resolved fluorescence probes to reduce the interference of background fluorescence on the detection signal. Quantum dots are nanoscale semiconductor crystals. Compared with bulk semiconductor materials, the carriers of quantum dots are size-limited in all three dimensions, resulting in quantum confinement effects, making the originally quasi-continuous energy bands become discrete energy levels, and the band gap widens. The width of the band gap depends on the particle size; the smaller the particle size, the wider the band gap, and the excitation spectrum and emission spectrum shift to the blue. Therefore, quantum dots have the advantage of adjusting the spectrum by controlling the size. The excitation wavelength range of quantum dots is wide, as long as the photon energy is higher than the band gap, it can be excited, which can overcome the limitation of organic fluorescent dyes requiring specific excitation wavelengths. The same excitation wavelength can excite multiple quantum dots with different emission wavelengths, so it can be used for simultaneous detection of multiple markers. Due to the quantum confinement effect, the energy levels are discrete, so the emission peak is narrow, and it has strong anti-interference ability when used as a fluorescence probe. Chen and others prepared a core-shell structure of CdTe@CdS:Cu near-infrared quantum dots with a lifetime of about 1 μm, which was successfully used for two-dimensional coding. The average particle size of CdTe@CdS:Cu near-infrared quantum dots is less than 10 nm, regulated by reaction time, the excitation wavelength is uniformly 490 nm, and the maximum emission wavelength is between 700–910 nm, regulated by its particle size. Hydrophilic quantum dots can be used as nanoprobes directly after surface modification, but the luminescence efficiency of hydrophilic quantum dots is lower than that of hydrophobic quantum dots. Hydrophobic quantum dots have poor dispersion in water, which can be improved by loading them into hydrophilic polymer nanomatrixes. Chen and others successfully prepared hydrophilic polystyrene nanospheres loaded with hydrophobic CdSe@ZnS quantum dots, CdSe@ZnS@PS-COOH, by the swelling method. The morphology of CdSe@ZnS@PS-COOH nanospheres is spherical, with an average particle size of about 200 nm, and the maximum emission wavelength is 605 nm. Quantum dots are mostly inorganic materials, and their chemical stability and resistance to photobleaching are far superior to organic fluorescent dyes. However, commonly used quantum dots are mainly zinc-oxygen compounds, especially CdSe is the most commonly used. Cd is a highly toxic heavy metal, which poses a huge environmental hazard and can accumulate in plants and humans, causing long-term harm, such as chronic Cd poisoning can cause "Itai-itai disease". Quantum dots containing S, Se, and Te are dangerous chemicals that produce highly toxic and flammable and explosive H_2S, H_2Se, and H_2Te gases when they encounter acid. Therefore, the toxicity and safety of quantum dot probe materials are one of the important considerations for their application in the field of biomedical sensing.

4.2.2 Long Afterglow Phosphorescent Nanoprobes

4.2.2.1 Inorganic Long Afterglow Phosphorescent Nanoprobes

The phosphorescence of inorganic long afterglow phosphorescent materials can last for tens of seconds or even several days, which is visible to the naked eye. Its afterglow lifetime is much longer than the fluorescence lifetime, and it can also achieve time-resolved effects. Chen and others used the hydrothermal method to prepare near-infrared inorganic long afterglow materials $Zn_{1.1}Ga_{1.8}Ge_{0.1}O_4:Cr^{3+}$ nanospheres, with a lifespan of about 20 s, have been successfully used for tumor imaging in mice. These nanoprobe are spherical in shape, with an average particle size of 56.8 nm, a maximum excitation wavelength of 650 nm, and a maximum phosphorescence emission wavelength of 695 nm. The luminescence lifetime of inorganic long afterglow phosphorescent materials is not only longer than that of background fluorescence, avoiding the interference of background fluorescence, but it can also be directly observed with cheap equipment such as mobile phone cameras for afterglow observation. When used for in vitro detection, expensive instruments are not required, making it particularly suitable for the POCT field. When used for in vivo imaging, it can be pre-excited before injection into the body, and imaging can be performed without the need for continuous excitation after injection, which can solve the problem that most excitation light is difficult to penetrate human tissues during photodiagnosis and treatment, and can also avoid to some extent the damage of excitation light to the human body.

4.2.2.2 Organic Long Afterglow Phosphorescent Nanoprobe

Ultralong Organic Phosphorescence (UOP) mostly has a phosphorescence lifetime of several hundred milliseconds to several seconds, which is much longer than the luminescence lifetime of background fluorescence. The key points in the preparation of UOP materials are: (1) to construct a rigid environment to restrict the thermal motion of luminescent molecules, thereby weakening non-radiative transitions, increasing luminescence efficiency, and extending phosphorescence lifetime; (2) to enhance intersystem crossing (ISC). As shown in Fig. 4.11, Shi and others successfully prepared carbazole derivatives 1,4-dicarbazolylbenzene (PDCz) and 1,4-dibromo-2,5-dicarbazolylbenzene (PDBCz), where PDBCz uses the steric hindrance of Br atoms to restrict molecular vibration and rotation in the crystal, thereby

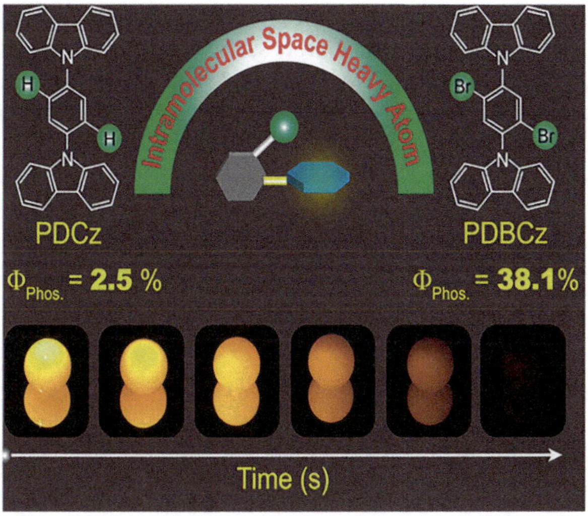

Fig. 4.11 Structure of carbazole derivatives PDCz and PDBCz and phosphorescence of PDBCz

weakening non-radiative transitions. At the same time, Br atoms enhance ISC, and the phosphorescence lifetime of PDBCz is as long as 217.3 ms, with an efficiency as high as 38.1%. This compound has two strong phosphorescence peaks at 546 nm and 595 nm in the crystal state.

UOP can be loaded into a nano matrix, and the surface of the nano matrix can be modified with specific groups or molecules to produce time-resolved fluorescence nanoprobe. You and others loaded 5 kinds of carbazole derivative UOPs into hollow mesoporous SiO_2 nanospheres, obtained hydrophilic phosphorescent probe water dispersion, the longest phosphorescence lifetime is 1.066 s, and it has been successfully used for HeLa cell imaging and mouse in vivo imaging. This UOP nanoprobe has high dispersion in water, is spherical in shape, has highly uniform particle size, and the excitation spectrum, emission spectrum, and fluorescence lifetime are very similar to those of the corresponding pure UOP crystal. The team further used a combination of precipitation and polymer self-assembly to prepare low-toxicity, water-soluble, high-brightness UOP nanocrystals, which were successfully used for in vivo imaging of zebrafish.

Some UOPs have aggregation-induced luminescence capabilities. These types of UOPs do not require a matrix and can be made into nanoscale molecular crystals. After surface modification, they can be used directly as nanoprobe. Nicol and others successfully prepared 1-(diphenyl[b,d]furan-2-yl)phenylmethanone UOP nanocrystals, which were successfully used as probes for imaging HeLa cells. This nanoprobe is rod-shaped, with lengths between 100 nm and 1 μm, an average hydrated particle size of 476.71 nm, and a maximum absorption wavelength of 350 nm, with a maximum phosphorescence emission wavelength of 520 nm, a phosphorescence lifetime of up to 232 ms, and a quantum yield of up to 34%.

In summary, although the afterglow lifetime of UOP nanoprobes is shorter than that of inorganic long afterglow nanoprobes, it is still much longer than the background fluorescence, and can also be observed with inexpensive devices such as mobile phone cameras, eliminating the use of expensive optical components and detection equipment, which is of positive significance for the development of portable POCT sensing devices. Compared with inorganic long afterglow nanoprobes, UOP nanoprobes do not contain heavy metals, are more environmentally friendly, have better biocompatibility, but their chemical stability and light bleaching resistance are weaker than those of inorganic long afterglow nanoprobe materials.

4.2.3 Chemiluminescent Nanoprobes

Chemiluminescence refers to the direct emission of light when the products generated after a chemical reaction are in an excited state and transition back to the ground state, or indirect emission of light by exciting fluorescent molecules through chemiluminescence resonance energy transfer (CRET). The reactions that induce chemiluminescence are usually oxidation-reduction reactions with reactive oxygen species (ROS) as the oxidant. If ROS is produced by photosensitizers excited by external light, this type of chemiluminescence is photo-induced chemiluminescence. The chemiluminescence used for detection either does not require external excitation light, or is photo-induced chemiluminescence with an excitation wavelength shorter than the emission wavelength, so it can also solve the problem of background fluorescence interference in the sensing detection process. Currently, many chemiluminescent nanoprobes have been applied in the field of biomedical sensing and detection.

4.2.3.1 Luminol Chemiluminescent Nanoprobes

Luminol is the earliest applied chemiluminescent agent. When it is oxidized by ROS under alkaline conditions, it generates an excited state of aminophthalate ion, which emits blue light when it transitions back to the ground state. Luminol is often used in the field of criminal investigation to detect bloodstains. An et al. prepared a nanoprobe based on luminol. This probe uses self-assembled PEG nanospheres as the substrate, loaded with luminol and dihydro porphyrin (Ce6). The overexpressed H_2O_2 at the tumor site in mice oxidizes luminol, and the excited state product generated excites the fluorescent agent and photosensitizer Ce6 through CRET, causing it to emit light imaging and simultaneously play a role in photodynamic therapy. The average hydrated particle size of this nanoprobe is 171 nm, and the chemiluminescence has two peaks at 450 nm and 675 nm. Luminol chemiluminescent nanoprobes have the advantages of clear principles, mature technology, cheap materials, and simple preparation. However, the luminous intensity of luminol is affected by the pH value of the solution medium. For example, luminol emits weak light in neutral solutions and strong light in alkaline solutions, but strongly alkaline solution media can cause protein denaturation, thus limiting its biomedical applications.

4.2.3.2 Oxalate Derivative Chemiluminescent Nanoprobes

Oxalate esters are also a commonly used chemiluminescent agent, often used in glow sticks, and have many applications in the field of biomedical detection. Singh and others prepared a nanoprobes based on oxalate esters, which loaded the chemiluminescent agent bis(2,4,5-trichlorophenyl) oxalate (Bis(2-carbopentyloxy-3,5,6-trichlorophenyl) oxalate, CPPO) and the anthracene derivative fluorescent agent BDSA. The overexpressed H_2O_2 at the inflammation site in mice oxidizes CPPO, and the excited state product generated excites the fluorescent agent BDSA through CRET, thereby achieving luminescent imaging of the inflammation site. The average particle size of this nanoprobe is about 15 nm, and the emission spectrum is a wide peak covering 500–800 nm, with a maximum emission wavelength of 584 nm. The chemiluminescent nanoprobe of oxalate ester derivatives has the advantages of high luminescence intensity and wide application scenarios. Moreover, as an indirect chemiluminescent agent, its emission spectrum depends on the CRET receptor, so it is easy to adjust. However, the water solubility of oxalate ester derivatives is poor, and when used for biomedical detection, it needs to be loaded on a highly hydrophilic nanomatrix.

4.2.3.3 1,2-Dioxetane-Based Chemiluminescent Nanoprobes

In the chemiluminescent process involving ROS, the chemiluminescent agent is usually first oxidized by ROS to generate a peroxide bond, and then the peroxide bond breaks to generate an excited state product. Substances containing peroxide bonds, such as 1,2-dioxetanes, can also be used directly as chemiluminescent agents. As shown in Fig. 4.12, Liu and others used mesoporous SiO_2 nanospheres to load the chemiluminescent agent 3-(2-spiroadamatane)-4-methoxy-4-(3-phosphoryloxy)-phenyl-1,2-dioxetane (AMPPD) and the fluorescent agent Rhodamine B. The peroxide bond of AMPPD is specifically cleaved by Alkaline Phosphatase (ALP), and the excited state product generated excites the fluorescent agent Rhodamine B to emit light, thereby detecting the content of ALP. The average particle size of this probe is about 60 nm, the pore size is about 3 nm, and the maximum chemiluminescence wavelength is 580 nm, which is the same as the maximum fluorescence emission wavelength of Rhodamine B. 1,2-Dioxetane-based chemiluminescent nanoprobes have the advantages of clear principles and strong specificity, but the types of substances that can be detected are limited, and their chemical stability is relatively poor.

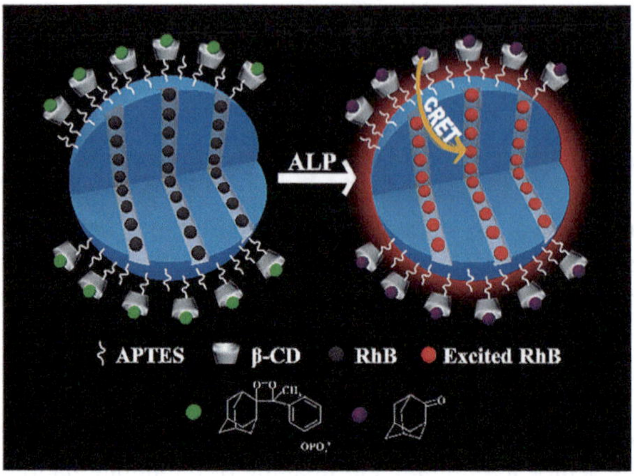

Fig. 4.12 The structure and luminescence mechanism diagram of the chemiluminescent nanoprobe of mesoporous SiO_2 loaded with Rhodamine B and AMPPD (RhB@mSiO_2-β-CD@AMPPD)

4.2.3.4 Sulfur Oxide Cyclohexene-Based Chemiluminescent Nanoprobes

Thioxanthene-based chemiluminescent agents are rarely reported, but they exhibit excellent luminescent properties. Su and others used paraffin oil as the base material, loaded with octabutoxyphthalocyanine palladium, 2-(4-(N,N-dimethylamino)phenyl)-3-phenyl-1,4-thioxanthene, and perylene. Among them, octabutoxyphthalocyanine palladium is a photosensitizer that produces singlet oxygen under light, 2-(4-(N,N-dimethylamino)phenyl)-3-phenyl-1,4-thioxanthene is a thioxanthene-based chemiluminescent agent, known as the Photoenergy Cache Unit (PCU), which can react with singlet oxygen to produce excited oxidation products. Perylene is a fluorescent agent that acts as a CRET receptor to accept the energy of the PCU oxidation product and emit light. The average particle size of this nanoprobe is about 150 nm, which can be excited by a 730 nm laser to produce singlet oxygen, and the chemiluminescence peak is the same as the fluorescence of perylene. The singlet oxygen produced under 730 nm excitation light can oxidize the PCU, and the excited product excites perylene to emit light, the lifetime of which depends on the temperature. For example, the temperature at the inflammation site is higher than at normal sites, so this probe material can be used to image the luminescence lifetime of inflamed tissue in mice. Thioxanthene-based chemiluminescent nanoprobe has the advantages of strong luminescence and high chemical stability. Moreover, as an indirect chemiluminescent agent, its emission spectrum is determined by the CRET receptor, so it is easy to adjust. However, the synthesis process of thioxanthene-based chemiluminescent agents is complex, the purification is difficult, the yield is low, and the price is high. Chemiluminescent nanoprobes can eliminate the interference of background fluorescence in biological samples, which is a clear advantage, but the current types of chemiluminescent materials are not many, and the luminescence intensity is generally weaker than fluorescence, phosphorescence and other luminescent materials, and most lack covalent bonding groups with the nano-matrix, so the load is mainly physical adsorption, and the stability is poor.

4.2.4 Upconversion Luminescent Nanoprobe

Upconversion luminescence refers to the emission of light with a wavelength shorter than the excitation wavelength. Because the background fluorescence wavelength is longer than the excitation light, upconversion luminescent nanoprobes can effectively avoid the interference of background fluorescence in the detection process of biological samples. $NaYF_4$ is the most commonly used substrate for upconversion luminescent nanoprobes, because the phonon energy of the $NaYF_4$ lattice is the lowest, and the non-radiative dissipation caused by vibration is the smallest. Two types of trivalent rare earth ions (RE^{3+}) need to be doped in the matrix, one as a sensitizer, absorbing photons and being excited, and the other as an activator, absorbing the energy transferred by multiple sensitizing ions to reach a higher excited state, thereby achieving upconversion luminescence. Common sensitizers include Yb^{3+} and Nd^{3+} ions, and common activators include Er^{3+}, Tm^{3+}, and Ho^{3+} ions. Jie and others prepared $NaYF_4$:Tm/Yb@$NaYF_4$. The upconversion nanoparticles are then coated with a layer of hollow mesoporous SiO_2 as a drug-loading cargo hold, and the surface is coated with a lipid bilayer doped with azobenzene derivatives as a light-controlled hatch. Under low-intensity 980 nm near-infrared light irradiation, the probe only emits visible light, causing the azobenzene structure to isomerize from cis to trans, thus closing the hatch; under high-intensity near-infrared light of the same wavelength, the probe emits both visible and ultraviolet light, causing the azobenzene to continuously undergo isomerization and violently flip, pushing open the lipid bilayer and causing the hatch to open for drug release. This probe achieves both in vivo imaging and controlled drug release. The average hydrated particle size of this nanoprobe is about 120 nm, the pore size is about 4 nm, it emits visible light with a wavelength of 450–475 nm under low-intensity 980 nm near-infrared light irradiation, and it emits ultraviolet light with a wavelength of 345–360 nm under high-intensity 980 nm near-infrared light irradiation.

Although upconversion nanoprobe can eliminate background interference by using emitted light with a shorter wavelength than the excitation light, its conversion efficiency is generally very low, which has become a major bottleneck limiting its widespread application. In addition, the synthesis of upconversion micronanoprobe materials involves the use of fluorides that pose significant threats to health and the environment, and their safety in use is also an important consideration for their biomedical applications.

4.2.5 *Luminescent Energy Transfer Nanoprobe*

Luminescent resonance energy transfer refers to the process where a molecule in an excited state (donor) transfers energy to another nearby molecule (acceptor) through a non-radiative method. The donor transitions back to the ground state in this process, its luminescence is quenched, and the acceptor is excited and emits light when it transitions back to the ground state, with a luminescence spectrum identical to the fluorescence emission spectrum of the acceptor. If the energy source for the donor's excitation is light, this energy transfer is called Fluorescence Resonance Energy Transfer (FRET). If the energy source for the donor's excitation is a chemical reaction, this energy transfer is called Chemiluminescence Resonance Energy Transfer (CRET). FRET and CRET require a certain degree of overlap between the emission spectrum of the donor and the absorption spectrum of the acceptor, and the physical distance between the two is a few nanometers or less, with the energy transfer efficiency inversely proportional to the sixth power of the separation distance.

Luminescent resonance energy transfer requires the distance between the donor and acceptor to be sufficiently close, and this characteristic can be used to develop immunological probes based on luminescent energy transfer. Rusanen and others have prepared a FRET immunological probe, as shown in Fig. 4.13, using two antibodies of the antigen to be tested to label Eu chelate and AF647 respectively. Eu chelate serves as the FRET donor and AF647 as the FRET acceptor. When the antigen to be tested is present, a sandwich is formed, bringing the donor and acceptor closer together, triggering FRET, and thus detecting the amount of antigen to be tested. This method requires the antigen to be tested to be small enough, with a size of a few nanometers, such as the spike protein of the new coronavirus successfully detected by Rusanen's FRET immunological probe. Various indirect luminescent chemiluminescent probes have applied the CRET process. However, luminescent

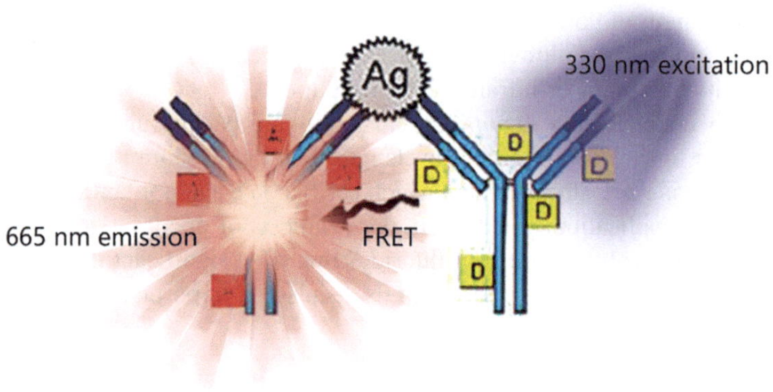

Fig. 4.13 Eu chelate serves as the FRET donor, AF647 as the FRET acceptor. When the antigen to be tested is present, a sandwich is formed, bringing the donor and acceptor closer together, triggering FRET to achieve detection

resonance energy transfer requires the distance between the donor and acceptor to be at the nanometer level, so it cannot be used to detect larger biomolecules.

4.2.6 *Luminescence Quenching Nanoprobes*

There are many causes of luminescence quenching, and the most commonly used in nanoprobes is luminescence quenching based on energy transfer, which is similar to the luminescence energy transfer nanoprobe mentioned above. Gold nanoparticles (AuNPs) can quench fluorescence. Rosebrock used AuNPs to cross-link with CdSe/CdS quantum dots to form a three-dimensional network, thereby quenching the fluorescence of the quantum dots. This combination can be used for luminescence quenching immunoassay nanoprobes. CuS is also a luminescence quencher. Chen and others used a peptide chain to connect CuS and near-infrared inorganic long afterglow material $Zn_{1.1}Ga_{1.8}Ge_{0.1}O_4:Cr^{3+}$ nanospheres to obtain a luminescence quenching nanoprobe. At this time, CuS will quench the luminescence of the long afterglow nanospheres. For example, MMP is an enzyme overexpressed at tumor sites, which can specifically hydrolyze the peptide chain connecting the two, causing CuS and the long afterglow luminescent sphere to separate, relieving quenching, and thus the probe can be successfully used for tumor imaging.

The key to the detection of luminescence energy transfer nanoprobes and luminescence quenching nanoprobes is to bring the donor and acceptor closer together through the linker, and the specificity of the linker modified on the surface of the probe material to the analyte should be strong, because the sensitivity and accuracy of the detection analysis results highly depend on the binding ability and efficiency of the linker.

4.3 Biomedical Detection Application of Optical Nanoprobes

Optical nanoprobe materials are widely used in the field of biomedicine, which can be used for biochemical diagnostic detection and biomedical imaging. The following will briefly introduce the biomedical applications of optical nanomaterials.

4.3.1 *Optical Nanoprobes for In Vitro Detection*

The use of optical nanoprobes for in vitro detection refers to labeling the optical nanoprobe on the biomolecule (antibody or aptamer, etc.) that specifically binds to the analyte, specifically binding with the analyte in the biological sample, and then detecting the presence and content of the analyte through the collection of the optical signal of the probe.

4.3.1.1 Detection of Large Biological Molecules

The most common method for in vitro detection of large biological molecules with optical nanoprobes is the sandwich method. The usual mechanism of the sandwich method is: the probe and the substrate are modified with two antibodies of the analyte, the binding sites of the two antibodies are different, when the analyte is present in the sample, the probe, the analyte, and the substrate form a sandwich, so that the probe is fixed on the substrate, after washing away the unbound probe, the content of the analyte can be measured through the optical properties of the probe.

Lei and others used PS nanospheres modified with carboxyl groups loaded with Eu chelates as time-resolved fluorescence nanoprobes for detecting the bladder cancer marker Cyfra21-1 on test strips in urine. As shown in Fig. 4.14, the mouse-derived anti-Cyfra21-1 antibody is linked to the probe through an amide bond, another mouse-derived anti-Cyfra21-1 antibody is loaded on the T line of the test strip, and sheep anti-mouse IgG is loaded on the C line. When Cyfra21-1 is present in the sample, Cyfra21-1 chromatographs on the test strip, first binding with the probe, and then being captured by the antibody on the T line, forming a sandwich on the T line. Excess unbound probes are captured by sheep anti-mouse IgG on the C line. The content of Cyfra21-1 is measured by the ratio of time-resolved fluorescence intensities of the T line and C line.

In addition to being used for in vitro detection on substrates such as test strips, the sandwich method can also be used for homogeneous wash-free in vitro detection. For example, as mentioned in Sect. 4.2 of this chapter, Rusanen and others

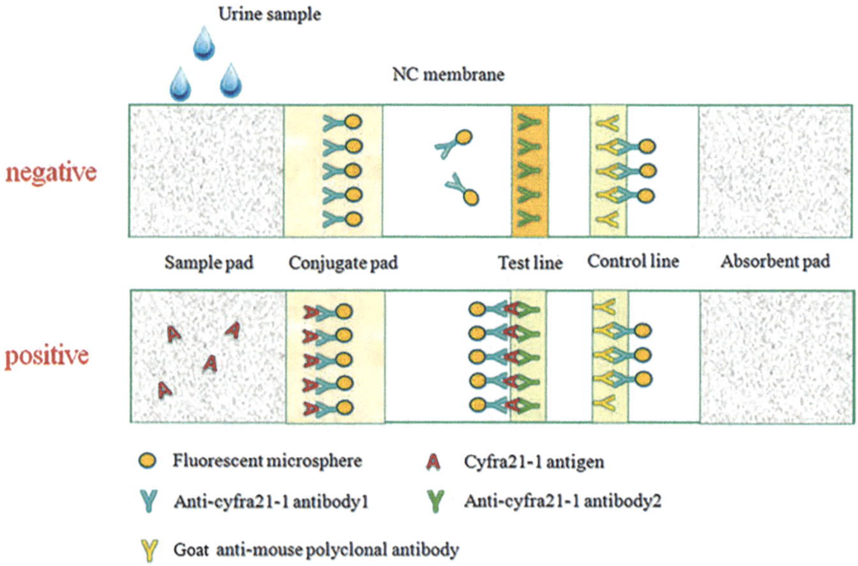

Fig. 4.14 Schematic diagram of the mechanism of time-resolved fluorescence nanoprobe and its test strip for detecting Cyfra21-1

used FRET immunoassay probes to detect the spike protein of the new coronavirus: the two antibodies of the antigen to be tested are labeled with Eu chelates and AF647 respectively, with Eu chelates as the FRET donor and AF647 as the FRET acceptor. When the antigen to be tested is present, the formation of a sandwich brings the donor and acceptor closer, triggering FRET, thereby detecting the content of the antigen to be tested.

4.3.1.2 Detection of Small Molecules

The most common method for detecting small molecules in vitro with optical nanoprobes is the competition method. The usual mechanism of the competition method is: the antibody that can bind to both the probe and the test substance is fixed on the substrate. This antibody can bind to both the probe and the test substance, and its affinity for the test substance is stronger than that for the probe. When the test substance is present in the sample, it competes with the probe. After washing away the unbound probe, the content of the test substance is measured by the decrease in the signal of the probe. Wu and others successfully used the AlphaLISA competition method to detect aflatoxin AFB1: as shown in Fig. 4.15, AFB1 is connected to BSA to obtain the complete antigen AFB1-BSA, which is biotinylated. The donor sphere is coated with streptavidin, and the acceptor sphere is labeled with AFB1 monoclonal antibody. The binding of streptavidin to biotin and the binding of antigen to antibody bring the donor and acceptor spheres closer to form aggregates, producing a luminescent signal; when a positive sample is added, AFB1 competes with the complete antigen AFB1-BSA, reducing the number of donor and acceptor sphere aggregates and weakening the luminescence, thereby achieving the detection of AFB1.

4.3.2 Optical Nanoprobes for In Vivo Imaging

Optical nanoprobes interact with specific molecules at the target site, thereby enriching the optical signal at the target site and achieving medical imaging within the organism. For example, the H_2O_2 content at tumor sites and inflammation sites

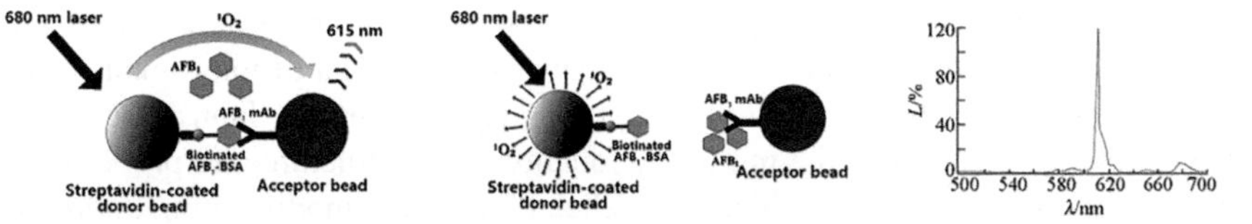

Fig. 4.15 Schematic diagram of the mechanism of Wu and others using the AlphaLISA competition method to detect aflatoxin AFB1

is much higher than that in normal tissues, and chemiluminescent probes with H_2O_2 as the reactant are often used for imaging at tumor sites and inflammation sites. For example, the oxalate ester derivative chemiluminescent nanoprobe prepared by Singh and others, the overexpressed H_2O_2 The derivative of oxalic acid ester CPPO generates excited state products that excite the fluorescent agent BDSA through CRET, thereby realizing luminescent imaging of the inflammation site.

Tumor sites are also rich in some enzymes that are rarely distributed in normal tissues, such as matrix metalloproteinases (Matrix Metalloproteinase, MMP). The specific hydrolysis reaction of these enzymes to substrates can be used to design in vivo imaging nanoprobes. For example, Chen and others connected near-infrared inorganic long afterglow material $Zn_{1.1}Ga_{1.8}Ge_{0.1}O_4:Cr^{3+}$ nanospheres with CuS nanoparticles using a peptide chain. The CuS quenches the luminescence of the inorganic long afterglow material. The MMP at the tumor site hydrolyzes the peptide chain used as a linker, relieving the quenching of the luminescence of the inorganic long afterglow material, thereby realizing medical imaging of the tumor location.

4.4 Chapter Summary

Optical nanoprobes utilize optical properties such as fluorescence, phosphorescence, upconversion luminescence, chemiluminescence, fluorescence resonance energy transfer, and fluorescence quenching to achieve accurate, sensitive, convenient, and simple in vitro detection and in vivo imaging. Their biomedical applications are extremely wide. However, current optical nanoprobes still face many challenges and there are some problems to be solved. For example, some optical nanoprobes, especially upconversion luminescence nanoprobes and chemiluminescence nanoprobes, have very low luminescence efficiency; optical nanoprobes, as nanoscale materials, their highly active surfaces easily cause them to aggregate and settle, and various non-specific bindings also exist widely, affecting their sensing detection efficiency; and many optical nanoprobe materials still have problems such as poor monodispersity of prepared particles, uneven size distribution and luminescent performance, high biological toxicity, and poor water solubility.

The application prospects of optical nanoprobes in the field of bio-intelligent sensing are broad, so how to deal with and solve the above problems is the future key research and development direction. Quantum dots, UOPs, rare earth compounds and other high-efficiency, long-life probe materials are currently hot topics in the field of optical nanoprobes. And the diversified design of micro-nano structures, such as core-shell structure, mesoporous layer wrapping, Janus asymmetric structure, etc., can realize the multifunctionality of the structure and properties of optical nanoprobe materials, which is beneficial for surface modification, improving monodispersity, water solubility and binding reaction specificity, etc. Combining optical nanoprobes with new self-propelled micro-nano motors and other controllable active carriers can accurately control the movement behavior of probe materials through means such as magnetic fields, solve the problems of nanoprobe material aggregation, sedimentation and various non-specific bindings, and can also improve the incubation efficiency of the probe and the test object, and also provide a very potential development direction for the development of intelligent micro-nano probe materials and sensing detection technology.

Bibliography

1. Ma Huimin, Su Meihong, Liang Shuquan. Triazine optical probes and label analysis [J]. Analytical Chemistry, 2003, (10): 1256-60.
2. Jie Xu. Construction and biological detection application of intelligent responsive optical probes [D]; Chongqing University, 2020.
3. Du Baoan, Liu Cheng, Cao Yuhong, et al. Quantitative detection of trace mercury using ZnS:Mn quantum dots as phosphorescent probes [J]. Spectroscopy and Spectral Analysis, 2014, 34(04): 1070-4.
4. Yan Jixiang. Study on the mechanism of laser-induced fluorescence [J]. Journal of Beijing University of Technology, 2000, (02): 229-31.
5. Ma Yazhi, Yang Xiaofeng, Chen Yuzhe, et al. Construction and application of pure organic room temperature phosphorescent systems [J]. Science Bulletin, 2016, 61(Z1): 453-66.
6. Wei Yinan, Li Yuanzong, Chang Wenbao, Ci Yunxiang. Application of fluorescence resonance energy transfer technology in bioanalysis [J]. Analytical Chemistry, 1998, (04): 477-84.
7. Zhang Zhiyi, Zhou Tao, Gong Weili, et al. Application and research progress of fluorescence resonance energy transfer technology in life sciences [J]. Journal of Electron Microscopy, 2007, (06): 620-4.
8. He Hanwei, Zhou Ke, Zhao Xiong, Huang Boyun. Research progress on upconversion materials of infrared-visible light [J]. Journal of Chinese Rare Earths, 2003, (02): 123-8.
9. Yang Jianhu. Upconversion luminescence of rare earth ions and research progress [J]. Progress in Physics, 2003, (03): 284-98.
10. Zhao Suling. Research on upconversion luminescence of rare earth ions [J]. Semiconductor Optoelectronics, 2000, (04): 241-4.
11. Xie Juan. Research on the design, synthesis, growth mechanism and optical properties of rare earth upconversion luminescent nanomaterials [D]; Nanjing Normal University, 2017.
12. Lei Hong, Huang Zhenli, Wang Hezhou. Physical properties and applications of two-photon absorption of organic materials [J]. Physics, 2003, (01): 19-26.
13. Chen Hongqi. Research on new methods of fluorescence energy transfer analysis based on rare earth luminescent nanoparticles [D]; Shanghai Jiaotong University, 2012.
14. Wu Changfeng. Photon avalanche upconversion of TiO2:Mo system [J]. Acta Physica Sinica, 2003, (06): 1540-4.
15. Chen Xueyuan, Luo Zundu. Four-level model of photon avalanche upconversion [J]. Acta Optica Sinica, 2000, (01): 19-23.
16. Yang Xiaoliang. Research on the efficiency and mechanism of energy upconversion of rare earth doping [D]; Xiangtan University, 2001.
17. Li Xiaozhou. Research and application of chemiluminescent systems [D]; Jilin University, 2006.
18. Yin Dongguang He, Liu Yibing, et al. Luminescent properties of several major chemiluminescent substances and their chemiluminescent immunoassay systems [J]. Labeled Immunoassays and Clinical Medicine, 2002, (04): 225-30.
19. ARMSTRONG C M, RUTH L E, CAPOBIANCO J A, et al. Detection of Shiga Toxin 2 Produced by Escherichia coli in Foods Using a Novel AlphaLISA [J]. TOXINS, 2018, 10(11).
20. EGLEN R M, REISINE T, ROBY P, et al. The use of AlphaScreen technology in HTS: current status [J]. Current chemical genomics, 2008, 1: 2-10.

21. MA Q, WANG J, LI Z, et al. Recent Progress in Time-Resolved Biosensing and Bioimaging Based on Lanthanide-Doped Nanoparticles [J]. Small, 2019, 15(32): 1804969.
22. ZHANG K Y, YU Q, WEI H, et al. Long-Lived Emissive Probes for Time-Resolved Photoluminescence Bioimaging and Biosensing [J]. Chemical Reviews, 2018, 118(4): 1770-839.
23. ZHENG W, ZHOU S, CHEN Z, et al. Sub-10 nm Lanthanide-Doped CaF2 Nanoprobes for Time-Resolved Luminescent Biodetection [J]. Angewandte Chemie International Edition, 2013, 52(26): 6671-6.
24. WANG M, YE H, YOU L, et al. A Supramolecular Sensor Array Using Lanthanide-Doped Nanoparticles for Sensitive Detection of Glyphosate and Proteins [J]. ACS Applied Materials & Interfaces, 2016, 8(1): 574-81.
25. XIAO M, SELVIN P R. Quantum Yields of Luminescent Lanthanide Chelates and Far-Red Dyes Measured by Resonance Energy Transfer [J]. Journal of the American Chemical Society, 2001, 123(29): 7067-73.
26. CHEN Y, CHI Y, WEN H, et al. Sensitized Luminescent Terbium Nanoparticles: Preparation and Time-Resolved Fluorescence Assay for DNA [J]. Analytical Chemistry, 2007, 79(3): 960-5.
27. YANG W, FU L M, WEN X, et al. Nanoprobes for two-photon excitation time-resolved imaging of living animals: In situ analysis of tumor-targeting dynamics of nanocarriers [J]. BIOMATERIALS, 2016, 100: 152-61.
28. ZHOU J, YANG Y, ZHANG C-Y. Toward Biocompatible Semiconductor Quantum Dots: From Biosynthesis and Bioconjugation to Biomedical Application [J]. Chemical Reviews, 2015, 115(21): 11669-717.
29. MA G. Background-Free In vivo Time Domain Optical Molecular Imaging Using Colloidal Quantum Dots [J]. ACS Applied Materials & Interfaces, 2013, 5(8): 2835-44.
30. CHEN C, ZHANG P F, GAO G H, et al. Near-Infrared-Emitting Two-Dimensional Codes Based on Lattice-Strained Core/(Doped) Shell Quantum Dots with Long Fluorescence Lifetime [J]. ADVANCED MATERIALS, 2014, 26(36): 6313-7.
31. CHEN Z H, LIANG R L, GUO X X, et al. Simultaneous quantitation of cytokeratin-19 fragment and carcinoembryonic antigen in human serum via quantum dot-doped nanoparticles [J]. BIOSENSORS & BIOELECTRONICS, 2017, 91: 60-5.
32. PAN Z, LU Y-Y, LIU F. Sunlight-activated long-persistent luminescence in the near-infrared from Cr^{3+} doped zinc gallogermanates [J]. Nature Materials, 2012, 11(1): 58-63.
33. CHEN L-J, SUN S-K, WANG Y, et al. Activatable Multifunctional Persistent Luminescence Nanoparticle/Copper Sulfide Nanoprobe for in Vivo Luminescence Imaging-Guided Photothermal Therapy [J]. ACS Applied Materials & Interfaces, 2016, 8(48): 32667-74.
34. SHI H, SONG L, MA H, et al. Highly Efficient Ultralong Organic Phosphorescence through Intramolecular-Space Heavy-Atom Effect [J]. The Journal of Physical Chemistry Letters, 2019, 10(3): 595-600.
35. XIE Z, ZHANG X, WANG H, et al. Wide-range lifetime-tunable and responsive ultralong organic phosphorescent multi-host/guest system [J]. Nature Communications, 2021, 12(1): 3522.
36. LI M, LING K, SHI H, et al. Prolonging Ultralong Organic Phosphorescence Lifetime to 2.5 s through Confining Rotation in Molecular Rotor [J]. Advanced Optical Materials, 2019, 7(10): 1800820.
37. YOU Y, HUANG K, LIU X, et al. Hydrophilic Ultralong Organic Nanophosphors [J]. Small, 2020, 16(8): 1906733.
38. NICOL A, KWOK R T K, CHEN C, et al. Ultrafast Delivery of Aggregation-Induced Emission Nanoparticles and Pure Organic Phosphorescent Nanocrystals by Saponin Encapsulation [J]. Journal of the American Chemical Society, 2017, 139(41): 14792-9.
39. AN H, GUO C, LI D, et al. Hydrogen Peroxide-Activatable Nanoparticles for Luminescence Imaging and In Situ Triggerable Photodynamic Therapy of Cancer [J]. ACS Applied Materials & Interfaces, 2020, 12(15): 17230-43.

40. XU X, AN H, ZHANG D, et al. A self-illuminating nanoparticle for inflammation imaging and cancer therapy [J]. Science Advances, 5(1): eaat2953.
41. SINGH A, SEO Y H, LIM C-K, et al. Biolighted Nanotorch Capable of Systemic Self-Delivery and Diagnostic Imaging [J]. ACS Nano, 2015, 9(10): 9906-11.
42. LIU X, FAN N, WU L, et al. Lighting up alkaline phosphatase in drug-induced liver injury using a new chemiluminescence resonance energy transfer nanoprobe [J]. Chemical Communications, 2018, 54(88): 12479-82.
43. SU X, WEN Y, YUAN W, et al. Lifetime-based nanothermometry in vivo with ultra-long-lived luminescence [J]. Chemical Communications, 2020, 56(73): 10694-7.
44. SUN L-D, DONG H, ZHANG P-Z, et al. Upconversion of Rare Earth Nanomaterials [J]. Annual Review of Physical Chemistry, 2015, 66(1): 619-42.
45. CHEN C, LI C, SHI Z. Current Advances in Lanthanide-Doped Upconversion Nanostructures for Detection and Bioapplication [J]. Advanced Science, 2016, 3(10): 1600029.
46. JIE X, YANG H, SU Y, et al. Time-Resolved Monitoring of Intracellular Processes with a Cyclical On–Off Photoswitchable Nanoprobe [J]. ACS Sensors, 2020, 5(1): 40-9.
47. LOU J, TANG X, ZHANG H, et al. Chemiluminescence Resonance Energy Transfer Efficiency and Donor–Acceptor Distance: from Qualitative to Quantitative [J]. Angewandte Chemie International Edition, 2021, 60(23): 13029-34.
48. RUSANEN J, KAREINEN L, SZIROVICZA L, et al. A Generic, Scalable, and Rapid Time-Resolved Förster Resonance Energy Transfer-Based Assay for Antigen Detection—SARS-CoV-2 as a Proof of Concept [J]. mBio, 12(3): e00902-21.
49. ROSEBROCK M, ZáMBó D, RUSCH P, et al. Spatial Extent of Fluorescence Quenching in Mixed Semiconductor–Metal Nanoparticle Gel Networks [J]. Advanced Functional Materials, 2021, 31(41): 2101628.
50. LEI Q F, ZHAO L L, YE S X, et al. Rapid and quantitative detection of urinary Cyfra21-1 using fluorescent nanosphere-based immunochromatographic test strip for diagnosis and prognostic monitoring of bladder cancer [J]. ARTIFICIAL CELLS NANOMEDICINE AND BIOTECHNOLOGY, 2019, 47(1): 4266-72.
51. WU Zhen, ZONG Jing, LI Chenxi, et al. Research on the detection method of aflatoxin based on photochemical chemiluminescence homogeneous immunoassay technology [J]. Optical Technology, 2019, 45(01): 117-23.

Chapter 5
Application of Magnetic Nanoparticles in Sensor Detection

5.1 Introduction to Magnetic Nanoparticles

5.1.1 Overview of Magnetic Nanoparticles

Nanoscience and technology, which involves the creation of materials from individual atoms and molecules, is one of the most important research fields in modern science. Researchers have used nanoscience and technology to achieve significant developments in life sciences and healthcare, profoundly impacting human society. Nanoparticles refer to particles with diameters between 1–100 nm. The concept of nanotechnology was first proposed by Nobel laureate Richard Feynman in 1959. Subsequently, researchers began studying nanoparticles. In 1984, German scientist Gleiter et al. first successfully produced iron nanoparticles using inert gas condensation, marking the official birth of nanoscience and technology. With the development of nanoscience and technology, nanomaterials, especially Magnetic Nanoparticles (MNPs), have attracted great research interest. MNPs are a new type of nanomagnetic material that not only possesses the four effects of ordinary nanoparticles (surface effect, quantum size effect, small size effect, and macro quantum tunneling effect) but also has special magnetic properties, such as superparamagnetism, high coercivity, low Curie temperature, and high magnetization rate. They have been successfully applied in areas such as magnetic resonance imaging, drug and gene delivery, cancer treatment, biomaterial separation, cell labeling, and imaging.

In recent years, MNPs synthesized under precise control conditions have controllable physicochemical properties and have been applied in biosensors. When combined with electrochemical, optical, piezoelectric, and magnetic devices, the sensitivity and stability of biosensors have been significantly improved. This has also shortened the biochemical reaction time and increased the detection throughput, opening up broad prospects in the field of biosensors.

X. Ma et al., *Functional Micro/Nanomaterials for Intelligent Sensing Devices*, https://doi.org/10.1007/978-981-96-4990-7_5

5.1.2 Sensing Characteristics of Magnetic Nanoparticles

The magnetic properties of MNPs have a direct impact on their sensing applications. Typically, MNPs exhibit optimal performance when they possess high saturation magnetization, superparamagnetism, stable magnetism, and surface properties. In MNP-based sensing detection applications, MNPs with high saturation magnetization can respond strongly to an external magnetic field, have a high magnetization rate, and can mitigate the magnetophoresis phenomenon caused by rapid magnetization. This characteristic can be used for magnetic enrichment and rapid collection of analytes, improving sensor sensitivity and shortening response time. The superparamagnetism of MNPs manifests as being magnetized by an external magnetic field, and the magnetism disappears simultaneously when the external magnetic field is removed. When MNPs are applied to sensors, their stability is often affected. This is because the iron oxide nanomaterials used to prepare MNPs have different magnetic and redox properties. Many redox substances exist in biological systems, which can react with MNPs to form different forms of iron oxide and produce toxic by-products. The magnetic performance will be poor and it is easy to be polydispersed in structure, thereby reducing the saturation magnetization strength and uniformity. Therefore, in the sensing detection applications of MNPs, how to obtain MNPs with high saturation magnetization, superparamagnetism, stable magnetism, and surface properties is crucial, and these properties are often affected by the structural properties of MNPs. To optimize performance, it is usually necessary to precisely adjust the size, shape, crystallinity, and functionalization of MNPs. The magnetism of superparamagnetic MNPs is very sensitive to changes in the size, shape, and crystallinity of MNPs. This is because the superparamagnetic relaxation time τ depends on the inherent relaxation ability τ_0, and is exponentially related to the magnetic crystal anisotropy (K_1) and nanoparticle volume (V), as shown by the Arrhenius kinetic formula (Eq. (5.1)):

$$\tau = \tau_0 e^{(-K_1 V / kT)} \tag{5.1}$$

Therefore, when preparing MNPs, the structure of MNPs should be considered to ensure the best magnetic performance in sensing detection applications. The following mainly introduces the structural features that affect the performance of MNPs in different sensing detection applications.

5.1.3 Structural Features Affecting Magnetic Performance

5.1.3.1 Uniformity

With the emergence of heterogeneous cell analysis and single-molecule sensing, the changes in signals generated in the detection applications of electrochemical, optical, or magnetic sensor devices based on MNPs should all come from changes within the analyzed sample, rather than from the changing characteristics of non-uniform MNPs. Therefore, it is necessary to ensure that MNPs have the same magnetic, electrical, optical, and surface properties to more accurately achieve quantitative detection, especially as we approach the limit of single-molecule detection.

The uniformity of MNPs is an important indicator to improve sensor performance. In the polydisperse MNP samples we use daily, there are both small particles with lower than expected saturation magnetization and large particles that aggregate due to irreversible magnetization. This non-uniformity can have a significant impact on the performance of the sensor. In a Magnetic Particle Imaging (MPI) study, In MPI applications, the detected signal is generated by the saturation magnetization value of each MNP. Experiments show that MNPs with pure single-crystal magnetite crystals and a 5% size difference have higher signal strength, spatial resolution, and longer blood half-life than multi-dispersed and polycrystalline particles. Therefore, to obtain MNPs with uniform magnetism, the key is to use single-crystal, size and shape uniformly dispersed MNPs.

5.1.3.2 Crystallinity

The crystallinity of MNPs is of great significance for their application in sensor detection. For example, in imaging techniques that depend on the size of the magnetic moment of each MNP, the presence of antiphase and grain boundaries in polycrystalline MNPs can prevent the vectors of individual magnetic moments from adding up in a relatively weak external magnetic field, thereby affecting the signal strength. Using single-crystal MNPs can solve this problem well. In single-crystal MNPs below the superparamagnetic critical size, all magnetic moments are arranged in one direction and add up linearly, forming a single domain. This allows the saturation magnetization strength of a single MNP to be maximized even under the action of a weak external magnetic field, enhancing the sensitivity of the sensor. Therefore, in practical applications, preparing suitable MNPs according to the different requirements of the crystallinity of MNPs in the application scenario can play a key role in improving the detection performance of the sensor.

5.1.3.3 Size

For applications that need to utilize the superparamagnetic properties of MNPs, such as Magnetic Resonance Imaging (MRI), MNPs need to produce a single superparamagnetic domain under a critical size. This is because if the size of the MNPs is above the critical size, permanent magnetization will occur when the external magnetic field is removed due to the domain being too large. However, if a larger magnetic signal or faster response speed is needed, it is necessary to ensure that the MNPs are below the critical size to maximize the size. Therefore, it is crucial to finely control the size of each MNP in the sample.

As can be seen in formula (5.1), in addition to the superparamagnetic relaxation time being exponentially related to the volume of the nanoparticles, slight changes in the size of the MNPs can also significantly change the saturation magnetization strength and coercivity in the sample. In a study by Thanh and his colleagues, increasing the size of Co ferrite from 7.5 nm to 10.6 nm led to a significant increase in saturation magnetization strength from 51 emu g^{-1} to 64 emu g^{-1}. Precise control synthesis of MNPs size can achieve a significant increase in saturation magnetization strength. For example, by gradually increasing the size of iron core oxide shell nanoparticles, Tilley and colleagues precisely increased the diameter from 10 nm to 11–13 nm. Since each size of MNPs has a high degree of uniformity, each MNP in the sample basically has the same strong magnetic response, allowing the saturation magnetization strength to increase from 104 emu g^{-1} to 117–176 emu g^{-1} while still retaining superparamagnetic characteristics. Since the goal of using MNPs in sensors is usually to maximize the saturation magnetization strength while maintaining superparamagnetism, the size of the MNP needs to precisely match the critical size. To achieve this goal, it is necessary to finely control the structure of the MNPs during the synthesis process.

5.1.4 Synthesis Methods of Magnetic Nanoparticles

The properties of MNPs largely depend on their structural characteristics, so their synthesis and preparation must be designed to obtain particles with appropriate physicochemical properties. Typically, MNPs can be divided into pure metal oxides and magnetic nanocomposites. The most commonly used MNPs in the field of biomedicine are Fe, Co, Ti, Ni alloys and iron oxides, ferrites ($BaFe_{12}O_{19}$, $CoFe_2O_4$) and so on. MNPs have a series of unique and superior physical and chemical properties. With the development of synthesis technology, many studies have been devoted to the synthesis of MNPs in the past few decades, and a series of shape-controllable, stable, and monodisperse MNPs have been successfully produced. The preparation methods of MNPs include physical methods, biological methods, and chemical methods. The physical method is represented by the mechanical ball milling method. The nanoparticles produced have a wide size distribution and are generally not suitable for the development of biosensors. Compared with the physical method, the nanoparticles prepared by the biological method show more obvious advantages, especially in terms of biocompatibility. However, the bacterial culture relied on by this method is difficult, and the particle extraction process is very cumbersome, so the preparation of MNPs mainly relies on chemical methods. Several commonly used chemical synthesis methods include co-precipitation, thermal decomposition, microemulsion, hydrothermal synthesis, etc., all of which can be used for the synthesis of high-quality MNPs. The following mainly introduces these four methods in detail (Fig. 5.1).

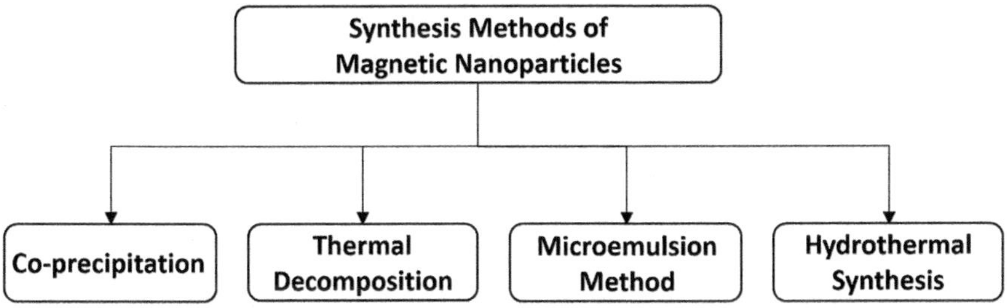

Fig. 5.1 Synthesis methods of MNPs

5.1.4.1 Co-precipitation

Co-precipitation is a method to prepare MNPs by simultaneously hydrolyzing different valence ions in aqueous solution under certain conditions. It is a common method for preparing various magnetic oxide nanoparticles. The size, shape, and composition of MNPs largely depend on the type of salt used (such as chlorides, sulfates, nitrates), the ratio of Fe^{2+}/Fe^{3+}, reaction temperature, pH value, and ion strength of the medium. Through this synthesis method, the performance of MNPs is fully controllable and reproducible under fixed synthesis conditions. Since magnetite is a ferromagnet, it is easily oxidized. Magnetite particles are not very stable under normal environmental conditions and are easily oxidized to hematite or dissolved in acidic media. Therefore, magnetite particles can be deliberately dispersed in acidic media, and then ferric nitrate is added to oxidize them into hematite particles. The obtained hematite particles have strong chemical stability in alkaline and acidic media. However, even if the magnetite particles are initially formed and then transformed into hematite with strong stability, the challenge of synthesizing MNPs by co-precipitation is to control the particle size to achieve a narrow particle size distribution. The particles prepared by co-precipitation are often quite dispersed. It is well known that the short burst of nucleation and subsequent slow controlled growth are key to producing monodisperse particles. Therefore, controlling these processes is crucial for preparing monodisperse MNPs.

In recent years, significant progress has been made in the research of preparing monodisperse MNPs of different sizes using organic additives as stabilizers or reducing agents by co-precipitation. For example, studies have shown that magnetite nanoparticles with sizes of 4–10 nm can stably exist in 1 wt% polyvinyl alcohol (PVA) aqueous solution. When PVA with 0.1 mol% carboxyl group is used as a stabilizer, magnetite nanoparticles are precipitated in the form of chain clusters. This result indicates that the selection of an appropriate surfactant is key to stabilizing such particles. In an alkaline medium, in the presence of trisodium citrate, magnetite can be prepared by oxidizing with ferric nitrate at 90 °C for 30 min to produce size-adjustable maghemite nanoparticles. By adjusting the molar ratio of citrate ions to metal ions (Fe^{2+} and Fe^{3+}), the particle size can be controlled between 2–8 nm.

The advantages of synthesizing MNPs by co-precipitation method are simple reaction principle, relatively cheap equipment and raw materials, suitable for mass production. Moreover, the nanoparticles synthesized by this method can be directly dissolved in water, without the need for subsequent surface functionalization. However, it is usually difficult to obtain well-crystallized monodisperse colloids, and there are many factors affecting the particle size and magnetic properties of MNPs that need to be considered during the preparation process. The process parameters need to be strictly controlled, which limits its further application in various fields.

5.1.4.2 Thermal Decomposition

Inspired by the synthesis of high-quality semiconductor nanocrystals and oxides through thermal decomposition in non-aqueous media, similar methods for controlling the size and shape of magnetic particles have also been developed. This is the thermal decomposition method: a method of preparing nanoparticles by heating and decomposing organic metal compounds in high boiling point organic solvents. The organic metal precursors used in this method include metal acetylacetonate salts [$M(acac)_n$] (where M = Fe, Mn, Co, Ni, Cr; n = 2 or 3; acac = acetylacetone), metal copper iron salts [M^xCup_x] (where M = metal ion; Cup = N-nitrosophenylhydroxyl amine, $C_6H_5N(NO)O^-$) or carbonyl. Fatty acids, oleic acid, and hexadecylamine are often used as surfactants. The ratio of organic metal compounds, surfactants, and solvents is a decisive parameter for controlling the size and shape of MNPs. Reaction temperature, reaction time, and aging period are also crucial for precise control of the size and shape of MNPs.

If the metal in the precursor is zero-valent, such as iron pentacarbonyl, it will initially form metal through thermal decomposition. By controlling the temperature at 100 °C, it is decomposed in a mixture of octyl ether and oleic acid, and then a mild oxidant trimethylamine oxide $(CH_3)_3$ NO is added at high temperature, monodisperse γ-Fe_2O_3 nanocrystals can be formed, with a size of about 13 nm. When $Fe(acac)_3$ is decomposed in phenyl ether by 1,2-hexadecanediol, and oleic acid, the precursor with a cationic metal center decomposes directly to form Fe_3O_4. Peng and his colleagues shared a general decomposition method for synthesizing size and shape controllable magnetic oxide nanocrystals based on the thermal decomposition of metal fatty acid salts in non-aqueous solutions. The reaction system generally consists of metal fatty acid salts, corresponding fatty acids (capric acid, lauric acid, myristic acid, oleic acid, palmitic acid, stearic acid), hydrocarbon solvents (ODE, eicosane, tatracontane or a mixture of ODE and dotriacontane) and activating agents. Almost monodisperse Fe_3O_4 nanocrystals can be synthesized, with sizes adjustable within the range of 3–50 nm and controllable shapes, as shown in Fig. 5.2. This method has also been successfully applied to the synthesis of other magnetic nanocrystals, such as Cr_2O_3, MnO, Co_3O_4 and NiO. By changing the reactivity and concentration of the precursor, the size and shape of the nanocrystals can be controlled. The reaction rate can be adjusted by changing the chain length and

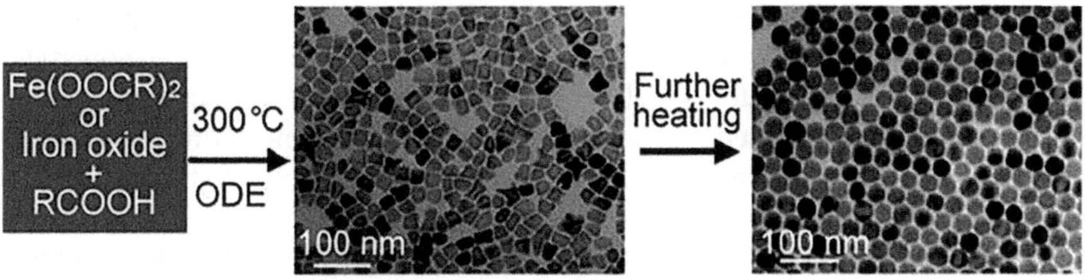

Fig. 5.2 Formation process of Fe_3O_4 nanocrystals (the middle and right panels are TEM images of synthesized nanocrystals at different reaction times)

concentration of the fatty acid. Generally speaking, the shorter the chain, the faster the reaction. Alcohols or primary amines can accelerate the reaction and lower the reaction temperature.

Hyeon and his colleagues also used a similar thermal decomposition method to prepare monodisperse iron oxide nanoparticles. They used non-toxic and cheap iron chloride and sodium oleate to in situ generate iron oleate complex, which then decomposed in different solvents (1-hexadecene, octyl ether, 1-octadecene, 1-eicosene or trioctylamine) at temperatures between 240–320 °C, producing particles with diameters ranging from 5–22 nm, the specific size depends on the decomposition temperature and aging period. In this synthesis, aging is a necessary step to form iron oxide nanoparticles. The resulting nanoparticles can be dispersed in various organic solvents, such as hexane and toluene. Sequential decomposition of iron pentacarbonyl and iron oleate complex at different temperatures produces monodisperse iron nanoparticles with sizes between 6–15 nm. The above nanoparticles can be dispersed in organic solvents. However, water-soluble MNPs are more popular in biotechnology. For this, LaMer et al. demonstrated a very simple synthesis method for water-soluble MNPs. This method uses $FeCl_3 \cdot 6H_2O$ as the iron source, and 2-pyrrolidone as the coordinating solvent, to prepare water-soluble Fe_3O_4 nanocrystals under reflux (245 °C) conditions. When the reflux time is 1, 10, and 24 h, the average particle size can be controlled at 4, 12, and 60 nm, respectively. As the reflux time increases, the particle morphology changes from the early spherical shape to the long-time cubic shape. Hu et al. developed a one-pot synthesis method for water-soluble magnetite nanoparticles, which prepares by adding a terminal polyethylene glycol as a surface coating agent under similar reaction conditions. These nanoparticles provide potential application prospects for magnetic resonance imaging contrast agents for cancer diagnosis.

The thermal decomposition method can also be used to prepare metal nanoparticles. Compared with metal oxides, the advantage of metal nanoparticles lies in their greater magnetization strength. Under nitrogen at 170 °C, $Fe(CO)_5$ and polyisobutylene were thermally decomposed in decalin to synthesize iron nanoparticles. By adjusting the ratio of $Fe(CO)_5$ and polyisobutylene, the particle size can be controlled within the range of 2–10 nm, with a polydispersity of about 10%. The thickness of the polymer layer around the iron nanoparticles is about 7.0 nm. However, these iron particles are still easily oxidized when exposed to air, which can cause a slight increase in particle size. In the process of synthesizing cobalt nanoparticles by the thermal decomposition method, the shape and size of the cobalt nanoparticles are controllable. Alivisatos and his colleagues reported a synthesis method for cobalt nanoparticles, which is achieved by the thermal decomposition of cobalt carbonyl precursors. Chaudret et al. described the synthesis of cobalt nanorods and nickel nanorods by high-temperature reduction of non-carbonyl organometallic complexes. For example, monodisperse ferromagnetic cobalt nanorods are synthesized by the decomposition of $Co(h_3\text{-}C_8H_{13})(h_4\text{-}C_8H_{12})$ in anisole at 150 °C, in the presence of a mixture of hexadecylamine and fatty acids (such as lauric acid, octanoic acid or stearic acid) under the action of H_2. The thermal decomposition method overcomes the shortcomings of the co-precipitation method. The synthesized MNPs have a narrow particle size distribution, good crystallinity, good monodispersity, controllable size and morphology. However, so far, the cost of this synthesis method is still relatively high, not suitable for large-scale production, which to some extent limits its further application.

5.1.4.3 Microemulsion Method

Microemulsion is a method of preparing nanoparticles using a ternary system of water (electrolyte aqueous solution), oil (hydrocarbon) and surfactant as the reaction environment. This synthesis method has good universality, can synthesize various types of nanomaterials by adjusting the amount and type of surfactant, co-surfactant, solvent, or changing the reaction conditions, and can also effectively control the size of nanoparticles by controlling the size of the microemulsion, which is also considered one of the important methods for synthesizing MNPs. Using microemulsion technology, with n-butanol as the co-surfactant and octane as the oil phase, metal cobalt, cobalt/platinum alloy and gold-coated cobalt/platinum nanoparticles can be synthesized in the reverse micelles of hexadecyltrimethylammonium bromide. MFe_2O_4 (where M = Mn, Co, Ni, Cu, Zn, Mg or Cd, etc.) is an important magnetic material, widely used in the electronics field. Spinel ferrites can be synthesized in microemulsions and reverse micelles. For example, a study synthesized controllable $MnFe_2O_4$ nanoparticles with a particle size of about 4–15 nm in a toluene-water reverse micelle with sodium dodecylbenzenesulfonate (NaDBS) as the surfactant. This synthesis method first uses $Mn(NO_3)_2$ and $Fe(NO_3)_3$ form a transparent aqueous solution. The NaDBS aqueous solution is added to the metal salt solution, followed by the addition of a large amount of toluene to form a reverse micelle. The volume ratio of water to toluene determines the size of the $MnFe_2O_4$ nanoparticles. Woo et al. reported the preparation of iron oxide nanorods through a sol-gel reaction in a reverse micelle formed by oleic acid and benzyl ether, using $FeCl_3{\cdot}6H_2O$ as the iron source and propylene oxide as the proton scavenger. During the reflux or heating process of tetrahydronaphthalene, the phase of the nanorods can be controlled by changing the reaction temperature, atmosphere, and hydration state of the gel. By mixing a sodium dodecyl sulfate solution with ferric chloride or cobalt acetate solution, in-situ generated dodecyl sulfate iron and cobalt are prepared into cobalt ferrite fluid through a methylamine reaction. The size of the cobalt ferrite particles decreases with the decrease in the total reactant concentration and the increase in the concentration of sodium dodecyl sulfate. The average size of the particles can range from 2 to 5 nm, but this synthesis method has a high degree of dispersion, about 30–35%.

Using microemulsion technology, nanoparticles can be prepared into spheres, or they can be prepared into elliptical cross-sections or tubular shapes. Although many types of MNPs have been synthesized in a controllable manner using microemulsion methods, the size and shape of the particles usually vary within a relatively large range. In addition, compared with other methods such as thermal decomposition and co-precipitation, the working window for synthesizing nanoparticles in microemulsions is usually quite narrow, and the yield is low. To synthesize a considerable amount of material, a large amount of solvent is needed. Therefore, this is not a very efficient process and it is difficult to scale up.

5.1.4.4 Hydrothermal Synthesis

Hydrothermal reaction is a method of synthesizing nanoparticles by reacting in a closed system with water as the solvent. Under a certain temperature and the self-pressure of water, the closed reaction system has a high-temperature and high-pressure reaction environment, and the chemical reaction occurs in such an environment, eventually obtaining the desired product. Li et al. reported a generalized hydrothermal method for synthesizing various different nanocrystals through a liquid-solid reaction. This system consists of metal linoleate (solid), ethanol linoleate liquid phase, and water-ethanol solution under hydrothermal conditions at different reaction temperatures. This method is based on the general phase transfer and separation mechanism that occurs on the liquid phase, solid phase, and solution interface during the synthesis process. Taking Fe_3O_4 and $CoFe_2O_4$ nanoparticles as examples, very uniform nanoparticles with diameters of about 9 and 12 nm can be prepared, as shown in Fig. 5.3. Li et al. also reported a method for synthesizing monodisperse, hydrophilic, single-crystal ferrite microspheres through a hydrothermal reduction method. A mixture consisting of $FeCl_3$, ethylene glycol, sodium acetate, and polyethylene glycol is vigorously stirred to form a clear solution, then sealed in a stainless steel autoclave lined with polytetrafluoroethylene, heated to 200 °C and maintained for 8–72 h. In this way, the obtained monodisperse ferrite spheres have sizes ranging from 200–800. The range can be adjusted within nm. Li et al. ingeniously used a mixture of multiple components such as ethylene glycol, sodium acetate, and polyethylene glycol to guide the synthesis. Ethylene glycol acts as a high-boiling-point reducing agent, used to produce monodispersed metal or metal oxide nanoparticles, sodium acetate acts as an electrostatic stabilizer to prevent particle agglomeration, and polyethylene glycol acts as a surfactant to prevent particle agglomeration.

Table 5.1 briefly summarizes the advantages and disadvantages of the above four synthesis methods. From the perspective of simplicity of synthesis, co-precipitation

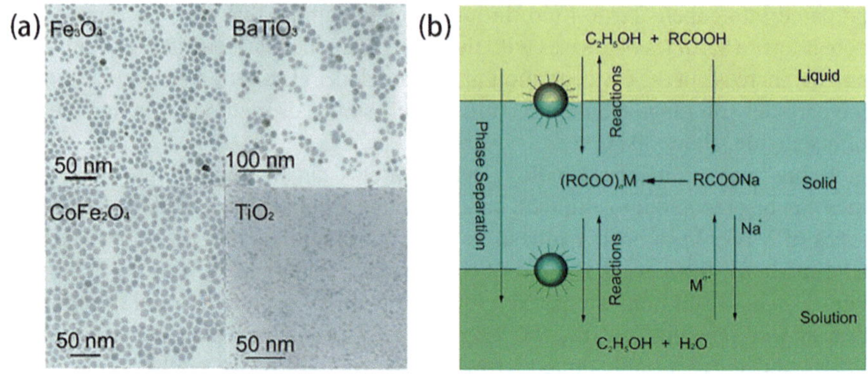

Fig. 5.3 (**a**) Shows the TEM images of magnetic and dielectric nanocrystals; (**b**) shows the liquid-solid-container (LSS) phase transfer synthesis strategy

is a better synthesis method. In terms of size and morphology control of nanoparticles, thermal decomposition is currently the best developed method. As an alternative, microemulsion can also be used to synthesize monodispersed nanoparticles of various shapes. However, this method requires a large amount of solvent. Hydrothermal synthesis is a relatively less explored method for synthesizing MNPs, although it can synthesize high-quality nanoparticles. So far, co-precipitation and thermal decomposition are the best methods for research because they can be used for large-scale preparation of MNPs.

The synthesis of MNPs has made substantial progress in the past few decades. However, how to synthesize high-quality MNPs in a controllable manner and a detailed understanding of their synthesis mechanism are still challenges to be faced in the next few years. The synthesis of oxide or metal MNPs usually requires the use of toxic or expensive precursors, the reaction usually takes place in the organic phase, the temperature is high, and the dilution is high. The obtained nanoparticles can usually only be dispersed in organic solvents and cannot be dispersed in the aqueous phase. Therefore, finding a simple synthesis route for water-soluble metal oxides or even metal nanoparticles with controllable size and shape will still be an active research field.

5.2 Application of Magnetic Nanoparticles in Sensor Detection

The detection of some bioactive substances such as proteins, viruses, DNA, and bacteria is an important direction in the development of modern biotechnology. The application of nanotechnology in biosensors has promoted the improvement of performance indicators such as sensitivity, signal-to-noise ratio, and detection time in biosensors. MNPs, due to their unique physicochemical properties and with the

Table 5.1 Comparison of various synthesis methods

Synthesis methods	Synthesis condition	Temperature (°C)	Reaction time	Solvent	Size distribution	Topographic control	Yield
Co-precipitation	Very simple, surrounding conditions	20–90	Minutes	Water	Narrow	Not good	High
Thermal decomposition	Complex, inert gas protection	100–380	Hours to days	Organic solvent	Very narrow	Very good	High
Microemulsion method	Complex, surrounding conditions	20–50	Hours	Organic solvent	Narrow	Good	Low
Hydrothermal synthesis	Simple, high pressure	100–300	Hours to days	Organic solvent/water	Very narrow	Very good	Average

advancement of preparation technology, are widely used in various fields. In sensor detection applications, MNPs can be directly applied to sensors, integrated into sensor materials, or dispersed in samples to achieve detection of various analytes. Based on different sensing principles, the sensing applications based on MNPs are mainly divided into electrochemical sensing, optical sensing, piezoelectric sensing, and magnetoresistive sensing. We will introduce and discuss these categories in the following sections, with a focus on the application of magnetoresistive sensors.

5.2.1 MNPs-Based Electrochemical Sensors

Electrochemical sensors are detection devices based on electrochemical principles. Their function is to convert the interaction between chemical or biological molecules into detectable electrical signals (current, voltage, and impedance). Electrochemical sensors have the advantages of fast speed, high sensitivity, low cost, easy miniaturization, and operation, and have broad application prospects in clinical, environmental, biological, and pharmaceutical fields. According to their working principles, they can be divided into amperometric, potentiometric, voltammetric, chemoresistive, and capacitive types. Generally, by fixing antibodies, enzymes, tissues, and DNA bio-recognition elements on the working electrode surface, electrochemical immunosensors and enzyme, tissue, DNA biosensors are designed. To improve the sensitivity of electrochemical sensors, people have tried to use MNPs for signal amplification. MNPs can be applied to electrochemical sensors by contacting the electrode surface, transferring redox-active substances to the electrode surface, and forming thin films on the electrode surface. There are several different detection modes for electrochemical biosensors based on MNPs, such as voltammetry, amperometry, potentiometry, electrochemiluminescence, and electrochemical impedance methods, which are mainly used for the detection and quantitative analysis of analytes. The most commonly used detection method is voltammetry. Figure 5.4 shows an example of an electrochemical sensor based on voltammetry.

Due to the superparamagnetism of Fe_3O_4, its biocompatibility with antibodies and enzymes, and the convenience of preparation, it is widely used in the development of biosensors. However, the magnetic dipole attraction of Fe_3O_4 and its large surface area to volume ratio can lead to the problem of clustering in biological solutions. This problem can be solved, and biocompatibility can be enhanced by functionalized MNPs. Various functionalized MNPs have been widely used, such as core-shell Au-Fe_3O_4, core-shell Au-Fe_3O_4@SiO_2, core-shell structure Fe_3O_4@SiO_2, Au-Fe_3O_4 composite nanoparticles, etc. Among them, the core-shell structure of Fe_3O_4@SiO_2 is the most widely used in biosensors because it helps the stability of MNPs in solution and enhances the binding of ligands on the surface of MNPs. Core-shell Fe_3O_4@SiO_2 is also widely used for electrode surface modification due to its good conductivity, large surface area, and more electroactive sites, which can provide enhanced mass transfer and easier access to active sites, thereby improving signal analysis capabilities and sensitivity.

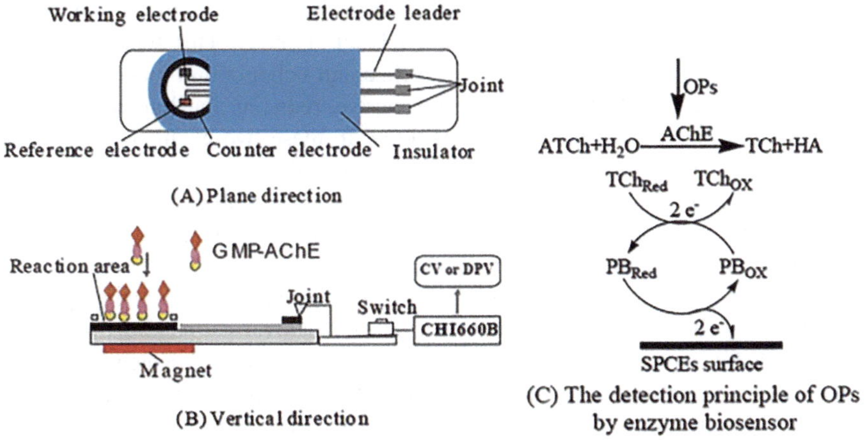

Fig. 5.4 Schematic diagram of an electrochemical (voltammetric type) biosensor. (**a**) Is a plan view of the sensor; (**b**) is a side view of the sensor; (**c**) is the principle of detecting organophosphorus pesticides using the sensor

Carbon materials such as Carbon Nano Tubes (CNTs) are also widely used for functionalizing MNPs due to their large specific surface area, chemical and thermal stability, controllable nanostructure, and optoelectronic properties. Arvand et al. synthesized a magnetic core-shell modified multi-walled carbon nanotube (MWCNTs) nanocomposite Fe_3O_4@SiO_2, and is used to prepare a modified carbon paste electrode for uric acid measurement (Fe_3O_4@SiO_2/MWCNTs-CPE). The electrochemical sensing characteristics of the MNPs modified CPE (Fe_3O_4@SiO_2/MWCNT-CPE), unmodified CPE and MWCNT-CPE were studied by cyclic voltammetry. The study shows that the anodic peak current of the MNPs modified CPE is 2.7 times higher than that of the MWCNT-CPE and 4.6 times higher than that of the unmodified CPE. Compared with the other two electrodes (MWCNT-CPE and unmodified CPE), the core-shell Fe_3O_4@SiO_2/MWCNT has fast electron transfer kinetics and a larger electroactive surface area, thereby improving the sensitivity of the sensor. Au-Fe_3O_4 composite nanoparticles are also widely used due to their simple preparation, large specific surface area, good biocompatibility, strong adsorption capacity, and good conductivity. For example, Gan et al. used MNPs composite material to modify the screen printed carbon electrode. In this device, acetylcholinesterase coated Au-Fe_3O_4 MNPs were synthesized and adsorbed on the surface of the carbon nanotube modified screen printed carbon electrode. This biosensor can be used to detect the pesticide residues in cabbage, and its performance is comparable to that of the gas chromatography-flame photometric detector (GC-FPD). In addition, due to the superparamagnetism of Au-Fe_3O_4 MNPs, under the action of an external magnetic field, the surface of this biosensor can be easily renewed and has the advantages of electrode integration, simple operation, low sample consumption, simple experimental design, and reduced manufacturing cost. Zamfir et al. developed an electrochemical impedance immunosensor for the detection of ochratoxin a based on the surface of the gold working electrode modified with iron oxide carboxyl MNPs with anti-ochratoxin a monoclonal antibody. This sensor uses iron oxide carboxyl modified MNPs to immobilize anti-ochratoxin a monoclonal antibodies, making the electrode easy to regenerate, reducing the impedance of the system, and thereby improving its sensitivity. In these two examples, MNPs are concentrated on the surface of the electrode material, which improves sensitivity and stability, and by releasing MNPs and replacing them with new MNPs, the electrode can be easily updated. Electrochemical immunosensors currently use MNPs as affinity labels or immobilization carriers.

Quantum dots, such as CdS, CdSe or core/shell type ZnS/CdSe, due to quantum confinement effects have optical and electronic characteristics, making them excellent labels for improving the sensitivity of the transducer surface coated with MNPs and magnetic capture probes, and have always been the most concerned in electrochemical applications. Zhou et al. used magnetic particles as capture probes and quantum dots as signal labels, and adopted a sandwich immunoassay strategy to develop a method for detecting α -An electrochemical immunosensor for Alpha-fetoprotein (AFP). Figure 5.5 shows the preparation process of magnetic capture probe Au-Fe_3O_4/primary AFP antibody (Ab1) and CdSAu/secondary AFP antibody (Ab2) signal label. Ab1 is first fixed on the surface of Au-Fe_3O_4 nanospheres through Au-S bonds. The product of Ab1 fixed on the surface of Au-Fe_3O_4 captures AFP (antigen) from the solution. Finally, protein-labeled CdS-Au nanoparticles are introduced into the immune response of the AFP exposed part. The electrochemical immunosensor is constructed with Au-Fe_3O_4/Ab1/AFP/Ab2/CdS-Au. In the solution, Fe_3O_4 MNPs modified electrode has almost no electrochemical signal, while Au-Fe_3O_4 MNPs modified electrode has a slight signal enhancement. Therefore, compared with the unmarked system (Au-Fe_3O_4/Ab1/AFP), by adding CdS-Au as a label, the signal of the immunosensor is further enhanced. When using CdS-Au

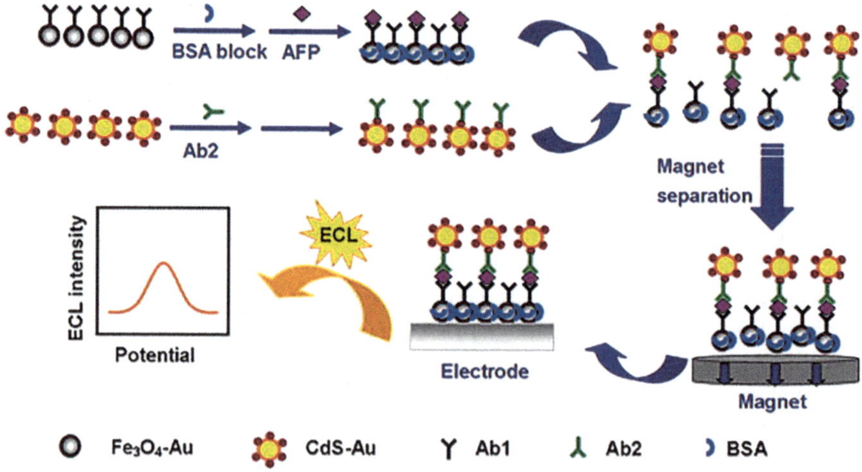

Fig. 5.5 Preparation process of electrochemical immunosensor based on MNPs labeling

composite film instead of CdS nanoparticles, the electrochemical signal increased by 2.5 times. This increase can be attributed to the catalytic activity of Au nanoparticles, which improves conductivity and sensitivity. The performance of this immunosensor in detecting human serum alpha-fetoprotein is comparable to ELISA, with potential for clinical application.

5.2.2 Optical Sensors Based on MNPs

Optical sensor devices have the characteristics of low signal-to-noise ratio, small interference, and low manufacturing cost, and have been applied to the detection of various analytes such as clinical samples, environmental samples, and food samples. Optical sensor devices are mainly divided into fluorescence spectroscopy, interference spectroscopy, reflection spectroscopy, chemical spectroscopy, and scattering spectroscopy according to their detection principles. Among them, sensor devices based on chemical spectroscopy usually need to enhance the emission intensity to improve the selectivity of the application. To overcome this disadvantage, MNPs can be used as catalysts, biomolecule carriers, and separation tools in chemical spectroscopy detection to enhance signal response. In optical devices based on MNPs, the detection mode generally used is Surface Plasma Resonance, Surface Plasmon Resonance (SPR) and fluorescence spectroscopy. For instance, Wang and colleagues developed an immunosensor that combines SPR technology with Magnetic Nanoparticles (MNPs) for the detection and manipulation of beta-human chorionic gonadotropin (β-hCG). As shown in Fig. 5.6, this method is based on a grating-coupled SPR sensor chip, which is functionalized with antibodies that recognize the target analyte (β-hCG). MNPs are coupled with antibodies to form MNP-antibody complexes, and a magnetic field is used to capture changes in the refractive index of the complexes on the sensor surface. At the same time, MNPs can also serve as carriers for rapidly transporting analytes on the sensor surface, thereby enhancing the response of the SPR sensor. Compared to conventional SPR that directly detects, the use of MNPs and the application of a magnetic field on the sensor surface greatly improves sensitivity. This enhancement is due to the larger mass and higher refractive index of MNPs. When analyzing analytes such as viruses or bacterial pathogens, this can overcome the problem of low sensitivity caused by diffusion hindering the transfer of mass on the sensor surface in SPR biosensor technology.

Analytical signals related to fluorescence intensity can also be enhanced using MNPs (such as Fe_3O_4). Agrawal and colleagues developed a microfluidic immunosensor chip with annular microchannels for detecting E. coli. This method first binds Fe_3O_4 MNPs with antibodies, and then captures antigens in the microchannels. Under the influence of reverse magnetic flow, the captured MNPs form stacked structures at the detection site, increasing the retention time of the antigen at the capture site and the efficiency of antigen capture, thereby enhancing the intensity of the fluorescence signal.

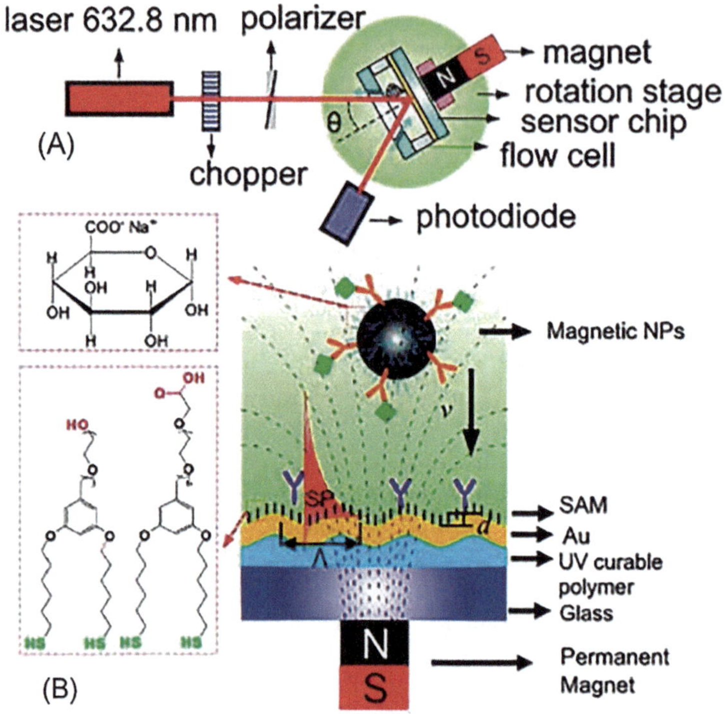

Fig. 5.6 (**a**) Optical sensor device; (**b**) MNP-enhanced grating-coupled SPR sensor chip

5.2.3 Piezoelectric Sensors Based on MNPs

Piezoelectric sensor devices mainly consist of Quartz Crystal Microbalance (QCM) and Surface Acoustic Wave (SAW). QCM is a quartz crystal disc with metal electrodes on both sides, which vibrates under the influence of an electric field. The vibration frequency depends on the cut and thickness of the disc. QCM utilizes the piezoelectric properties of quartz crystal resonators to convert changes in the mass on the surface of the quartz crystal electrode into frequency changes in the output electrical signal of the quartz crystal oscillation circuit. QCM piezoelectric sensor devices have the advantages of small size, large function, and low cost, and can respond quickly to extremely small mass changes. However, these devices have the disadvantage that as the surface area to volume ratio increases, instability leads to a reduction in size, resulting in increased noise. At the same time, QCM is easily affected by atmospheric humidity and is difficult to use for the determination of analytes in solutions. MNPs with piezoelectric properties provide a special conduction mechanism that can easily eliminate these problems. Zhou and colleagues developed a QCM immunosensor for detecting C-reactive Protein (CRP). The sandwich-type immune response is marked by a primary CRP antibody (MNs-CRPAbl) capture probe (silica-coated magnetic Fe_3O_4 nanoparticles, Fe_3O_4@SiO_2), CRP and signal tags (AuNPs-HRP/HRP-CRP Ab2) are composed. The capture probe containing MNPs (MNs-CRPAb1) enhances the analytical signal due to the magnetic separation and fixation on the electrode surface. In addition, the magnetic beads (Fe_3O_4@SiO_2) used for labeling CRP Abl have the advantages of a monodisperse size distribution and easy preparation of the labeled conjugate. Experimental results show that the performance of the QCM immunosensor in the detection of CRP in human serum is comparable to the ELISA method.

5.2.4 *Based on MNPs Magnetic Resistance Sensor*

Magnetic Resistance (MR) refers to the change in the resistance of a material with the change in the external magnetic field. Magnetic resistance sensors were first used to improve hard disk storage efficiency. Since the late 1990s, magnetoelectronics has become one of the hot research topics in the development of biosensors, and it has emerged and developed in the further application of biomarker detection. Most magnetoelectronic materials have limitations such as large volume, low sensitivity, and high power consumption. However, these limitations can be overcome by using MR-based sensors. By combining MNPs tags and magnetic resistance-based sensors, the existing magnetoelectronic detection methods have been well extended to better meet current needs.

5.2.4.1 Magnetic Resistance Detection Technology

The magnetoresistance effect is a phenomenon in which the resistance value of magnetoresistive materials changes under the action of an external magnetic field. Lord Kelvin completed the first measurement of magnetoresistance in 1857 and mathematically quantified it with the following equation:

$$MR = \frac{\Delta\rho}{\rho} = \frac{R_{\max} - R_{\min}}{R_{min}} \tag{5.2}$$

where ρ is the resistivity of the material, R_{max} and R_{min} are the resistances of the material under saturated magnetic field and zero magnetic field, respectively. Quantitative detection is achieved by changing the resistance of magnetoresistive materials by changing the external magnetic field. The basic principles of anisotropic magnetoresistance (AMR), giant magnetoresistance (GMR), and tunneling magnetoresistance (TMR) are introduced below, and Fig. 5.7 shows the working principles and design schemes of these three magnetoresistances.

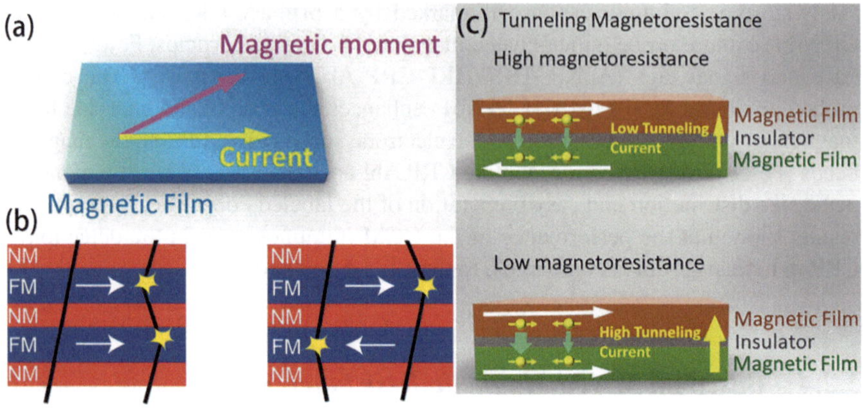

Fig. 5.7 Working principles and design schemes of three types of magnetoresistance (**a**) AMR; (**b**) GMR; (**c**) TMR

AMR

Ferromagnetic metals and their alloys have AMR effects, that is, the resistivity changes with the change in the angle between the direction of the external magnetic field and the direction of the external current. Its microscopic mechanism is based on the anisotropy of spin-orbit coupling and spin scattering. Spin-orbit coupling refers to the influence of the orbital motion of electrons on their spin orientation. When electrons orbit around the nucleus, a current loop is formed. Under the action of the magnetic field generated by the current loop, the spin orientation of the electron will be affected. At room temperature, the AMR values of conventional Fe, Co, and Ni metals are generally 0.2–2%, while the AMR values of NiFe and NiCo metals are 4–7%. Although the value of AMR is small, it played an indispensable role in the early development of magnetoelectronics.

GMR

The GMR effect, discovered by Albert Fert and Peter Grunberg in the late 1980s, is achieved through exchange-coupled multilayers composed of alternating ferromagnetic and non-magnetic layers. The physical origin of GMR is due to the opposite direction of the electron's spin and the material's magnetization, which affects the number of electrons passing through due to electron scattering. In ferromagnetic materials, there are only two directions of spin for conducting electrons: up-spin and down-spin. In the case of opposite magnetization directions of two ferromagnetic layers, electrons with different spin directions have difficulty passing through, resulting in a large resistance. GMR sensors appeared later, have excellent performance similar to AMR, but with higher magnetoresistance (typical GMR can be an order of magnitude higher than AMR), which expands the measurement range and application scope, making it more competitive than AMR.

Although the exchange-coupled multilayer structure can exhibit a large magnetoresistance effect, the higher saturation magnetic field can lead to lower magnetic field sensitivity, limiting its application in actual devices. Spin-valve sensors are a more advanced type of GMR sensor compared to exchange-coupled GMR stacks. The so-called spin-valve structure is a non-magnetic conductive layer existing between two ferromagnetic metals, thereby breaking the constraints of magnetic field conditions on GMR material applications. The free layer is a soft magnetic material that can significantly change the magnetization direction under a lower external magnetic field, thereby affecting the spin state of the electron. The role of the non-magnetic isolation layer is to eliminate the coupling effect between the two ferromagnetic layers. Currently, spin-valve structure read heads have become mainstream in computer hard disk drives. With the increasing requirements for biomarker detection, GMR, due to its low cost, high reliability, and high sensitivity, has been widely used and studied in POCT technology.

TMR

The mechanism of TMR is spin-dependent tunneling effect. TMR sensors are similar in structure to spin-valve sensors, but replace the non-magnetic conductive layer with insulating oxides (usually Al_2O_3 or MgO). The wave-particle duality of electrons in quantum mechanics provides the basis for electrons to pass through the insulating barrier. By changing the magnetization direction of the two ferromagnetic layers with an external magnetic field, the value of the tunneling current (resistance) perpendicular to the insulating barrier can be controlled. In Magnetic Tunneling Junctions (MTJs), there is basically no exchange coupling between the two ferromagnetic layers, so only a small external magnetic field can reach the maximum TMR, which makes its magnetic field sensitivity very high. In 1975, Julliere first observed the TMR effect in the Fe/Ge-O/Co system with a magnetoresistance of 14% at 4.2 K, and gave a phenomenological model to explain it. However, the low temperature conditions and lower TMR values limited its application. In 1995, Miyazaki et al. found that the tunneling resistance value of amorphous aluminum oxide barrier material at room temperature was as high as 18%, which quickly aroused researchers' interest in the TMR effect. Compared with the in-plane current mode of the GMR structure, the current in the MTJs structure is perpendicular to the film surface, which also makes the size of the junction have a great impact on the resistance. TMR sensors usually have higher magnetoresistance values and are particularly suitable for low magnetic field applications, so they have broad application prospects in magnetic random access memory, spin tunneling read heads, and biomedical fields.

5.2.4.2 Typical Applications Based on Magnetoresistance Sensing

The detection technology based on magnetoresistive sensing has high signal-to-noise ratio, high integration, and cheap components, which has broad application prospects in the fields of biomedical diagnosis, food safety, and environmental monitoring. The magnetic microarray, adapted from the traditional optical microarray format, is a pioneering advancement in the detection of biomarkers using magnetic sensing solutions. As shown in Fig. 5.8a, increasing the number of sensors while maintaining a relatively small overall array is an advantage brought by the sensor array. In addition, the Wheatstone bridge is an instrument that can accurately measure resistance. By fabricating two MR sensors on the Wheatstone bridge components, the output of the entire circuit can be detected, thereby improving sensitivity. The introduction of these concepts further strengthens the development of magnetoresistive sensor applications. The following is a summary of the application progress of magnetoresistive sensors in biomedical diagnosis, food safety detection, and environmental monitoring.

Biomedical Diagnosis Based on Magnetoresistive Sensors

Many diseases have a long incubation period, therefore, early accurate diagnosis helps in the treatment of diseases. The high sensitivity and low cost of magnetic resonance imaging make it an ideal solution for biomedical diagnosis. In the early days, many researchers were committed to the concept validation of MR sensors for biomedical diagnosis. However, there are still some problems with MR technology used for biomedical sensing. The actual field equipment cannot meet the requirements of complex sample additional processing steps (need to meet the direct analysis of complex samples such as blood, urine, sweat). Srinivasan et al. proposed a

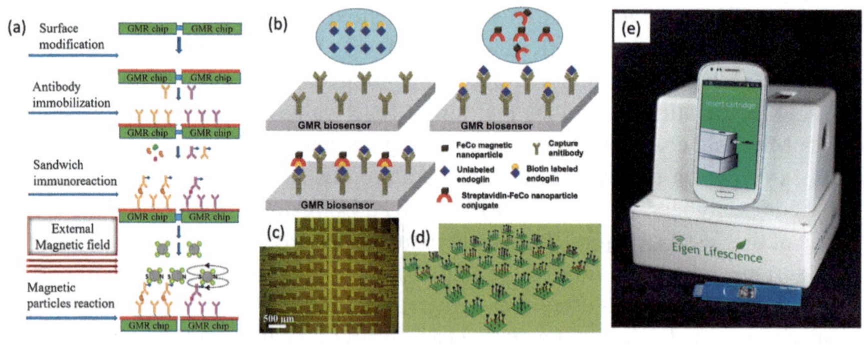

Fig. 5.8 (**a**) Immunoassay method based on GMR; (**b**) Three-layer competitive detection scheme; (**c**) 8 × 8 array chip sensor image; (**d**) Schematic diagram of multiple protein arrays on the chip; (**e**) Diagnostic platform composed of ink cartridge, reader, and smartphone

three-layer competitive analysis method (as shown in Fig. 5.8b), and used GMR sensors to quantify endorphins in unprocessed human urine samples. The three-layer competition scheme first uses capture antibodies to modify the surface of the GMR sensor, then mixes the labeled and unlabeled endorphins with the sensor surface, and captures the antibodies through competition. This system has high detection specificity and a three-order dynamic range. In addition, as shown in Fig. 5.8c, d, Wang et al. assembled a nanoscale GMR sensor array on a 16 × 16 mm chip, and developed a sensitive sensing system that can be used for real-time reading of serum and direct multiple measurements. They also confirmed that they can quantitatively detect pregnancy-associated plasma protein a (PAPP-A), proprotein convertase subtilisin/kexin type 9 (PCSK9) and oncogenic suppressor 2 (ST2). The detection limit of ST2 antigen reached 40 pg/mL, and the dynamic detection range of the three proteins increased by 4 orders of magnitude. Krishna et al. developed a method for detecting influenza A virus that combines anti-viral nucleoprotein (NP) monoclonal antibodies and MNPs, and used GMR biosensors for quantitative analysis. The detection limit of this system reached 1.5×10^2 $TCID_{50}$/mL. In addition, as shown in Fig. 5.8e, Choi et al. developed an immunodiagnostic feature diagnostic platform (EDP) system based on GMR spin valve. This system can detect multiple biomarkers in less than 15 min with a drop of blood for immunodiagnosis. It simultaneously detects human immunoglobulin G and M (IgG and IgM) antibodies, with sensitivities of 10 ng/mL and 50 ng/mL, respectively. Unlike standard clinical laboratories, the EDP is portable and can be used for POCT. By connecting with a smartphone, the electronic data processing system uploads patient information and data to the central network, greatly simplifying the medical workflow.

Microfluidics refers to the science and technology of using microchannels (ranging in size from tens to hundreds of micrometers) to process or manipulate tiny fluids. As shown in Fig. 5.9a, Gao et al. developed a method for detecting d-dimer as a thrombus biomarker using a GMR sensor integrated with microfluidics. The detection limit is 5 ng/mL. As shown in Fig. 5.9b, c, Gao et al. integrated a microfluidic device with a GMR sensor chip to develop an instrument that can simultaneously detect 12 types of tumor markers. This device controls the speed and direction of the microfluid under the pressure of a stepping motor, improving the efficiency of biochemical reactions. The detection limits for alpha-fetoprotein (AFP), carcinoembryonic antigen (CEA), cytokeratin 19 fragment (CYFRA21-1), neuron-specific enolase (NSE), free beta-human chorionic gonadotropin (free β-hCG), squamous cell carcinoma (SCC), pepsinogen I (PG I), pepsinogen II (PG II), total prostate-specific antigen (total PSA), free prostate-specific antigen (free PSA), thyroglobulin (Tg), and carbohydrate antigen 19–9 (CA19–9) are 0.52 ng/mL, 0.27 ng/mL, 0.25 ng/mL, 0.5 ng/mL, 0.3 ng/mL, 1 ng/mL, 0.5 ng/mL, 2 u/mL, 0.02 ng/mL, 0.07 ng/mL, 0.3 ng/mL, and 1 ng/mL, respectively. As a multi-analyte analysis system, this integrated device reduces the unit cost of each test and alleviates the burden on patients. In recent years, microfluidic technology built directly on integrated circuits has made significant progress. However, the combination of integrated circuits and microfluidic chips can lead to size mismatch problems. To address this

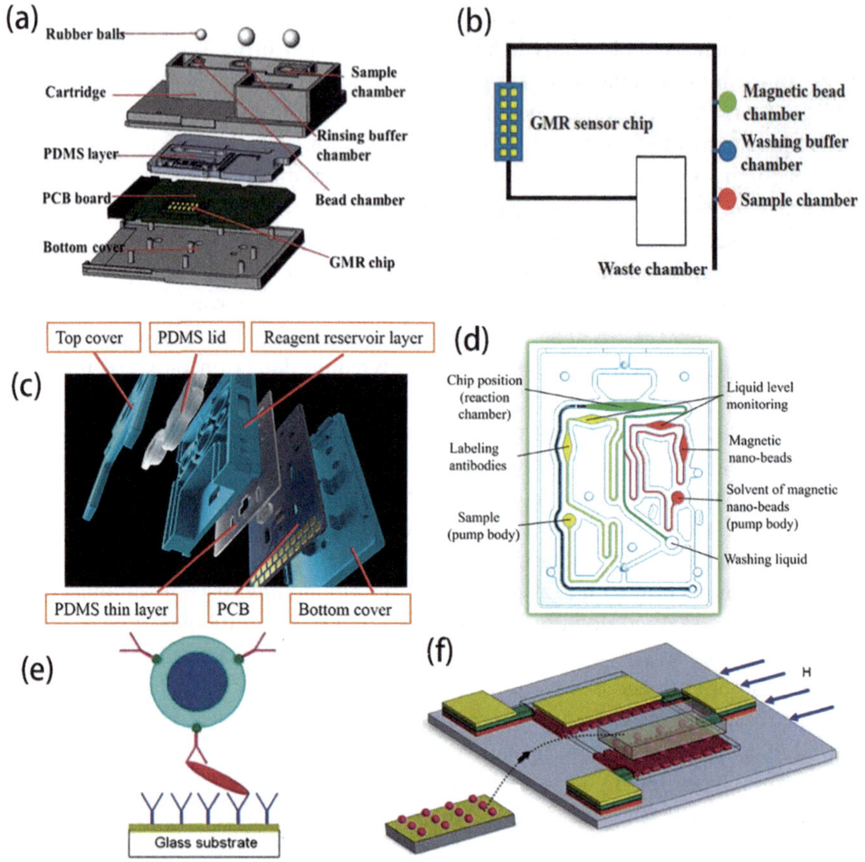

Fig. 5.9 (**a**) Schematic diagram of the test card for detecting d-dimer; (**b**) structure of the microchannel system; (**c**) schematic diagram of the test card for detecting 12 types of tumor markers; (**d**) GMR sensor; (**e**) image of capturing analytes on the sensor surface; (**f**) microfluidic GMR sensor

issue, Muluneh and Issadore reported a microfluidic chip system integrated with commercial integrated circuits and GMR sensing circuits. First, micrometer-scale fluidic channels are integrated on the surface of the integrated circuit using soft lithography to achieve high-precision control. Then, pre-designed 50 μm through-holes are used to connect these polydimethylsiloxane (PDMS) components to a larger PDMS chip to complete external connections. This structure greatly reduces the inherent challenge of size mismatch between the microfluidic system and sensors in integrated circuits, and promotes the integration of these two fields into commercial and practical platforms.

MEMS represents an industrial technology that combines microelectronics and mechanical engineering. Its working range is within the millimeter scale, greatly promoting the development of biosensors. As shown in Fig. 5.9d–f, Sun et al. used MEMS technology to make miniature GMR sensors. Each chip sensor consists of

200 spin-valve strips connected in series, with an effective electrical area of 300 × 5 μm for each spin-valve strip. The sandwich immunoassay was used to quantitatively detect the content of carcinoembryonic antigen on the surface of the self-assembled monolayer-modified gold film, with a detection limit as low as 10 pg/mL. In addition, they successfully completed the non-contact detection of prostate-specific antigen using the same system, with a detection limit as low as 0.1 ng/mL. The non-contact measurement ensures that the sensor does not immerse in the chemical solution, eliminating the cleaning step, and the sensor can be reused immediately. At the same time, MEMS technology miniaturizes the system, meeting the requirements for portability, and has broad application prospects in biomedical detection based on the specific combination of analytes and labels.

Immunochromatographic Test Strips (ICTS) are ideal tools for on-site detection, which can effectively quantify analytes in different samples. As shown in Fig. 5.10a, a typical ICTS consists of a sample pad, a binding pad, a nitrocellulose membrane, an absorption pad, and a paper card shell. By detecting the magnetic field strength generated by the functional magnetic signal probe on the ICTS, a new method for quantitative detection of target biomarkers based on magnetic resonance biosensors has been established. Magnetic ICTS based on MR provides greater functionality than other sensors. Most biological specimens, such as urine, serum, and blood, lack detectable magnetic components, so there is almost no background interference. For example, Ryu et al. combined MNPs with protein G, fixed anti-cardiac marker troponin I (cTnI) antibodies, and used GMR sensors for lateral blood flow

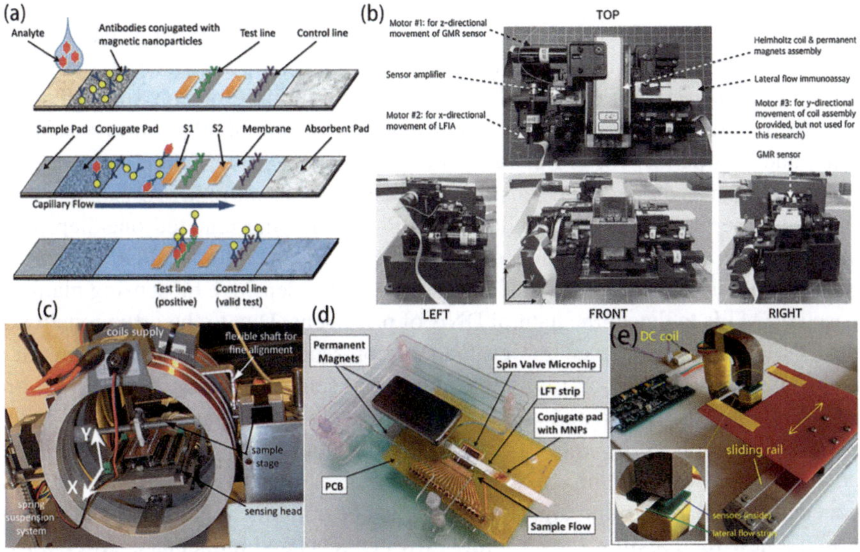

Fig. 5.10 (**a**) Structure and reaction principle of immunochromatographic test paper; (**b**) quantitative image detection GMR platform; (**c**) overall view of the contact scanning GMR system; (**d**) magnetic detector of Xintong technology composed of a microchip and two magnetic resonance sensors; (**e**) non-contact scanning TMR sensor

measurement. Compared with the conventional random fixation method, the sensitivity is significantly improved, and the detection limit can reach 0.01 ng/mL. Further, by using a GMR sensor to scan the surface of the ICTS and maintaining a constant speed through the motor, quantitative analysis of MNPs is performed, improving the detection limit, as shown in Fig. 5.10b, c. In addition, Chicharo et al. developed a low-power compact reading platform with a spin-valve GMR for POCT applications. However, the platform scanning will damage the surface of the paper, affecting the accuracy of the test results. To solve this problem, in another study. Lei et al. designed a non-contact scanning TMR sensor, as shown in Fig. 5.10e, in which two active TMR elements are arranged in parallel, and a differential sensing configuration is formed in a vertical magnetic field. The prototype has high sensitivity and strong anti-interference ability, and also improves the detection speed. By detecting human chorionic gonadotropin (HCG), the detection limit is 25 mIU/mL. The detection based on MR and ICTS is simple, fast, and user-friendly, promoting its application in medical diagnosis.

Food Safety Detection Based on Magnetoresistive Sensors

Food safety has increasingly become a major public health concern. Therefore, finding a highly sensitive biological detection solution suitable for POCT has become an important technical challenge in recent years. Wu et al. successfully applied the TMR-based biosensor to the detection of Escherichia coli O157:H7. The detection concentration range is 10^5–10^{10} CFU/mL, the detection limit is 100 CFU/mL, and the detection time is less than 5 h. Mak et al. designed an 8×8 array, each sensor has an effective area of 90×90 mm spin valve sensor. Subsequent experiments showed that the system can simultaneously detect multiple mycotoxins (aflatoxin B1 (AFB1), zearalenone, and HT-2), with a detection limit of 50 pg/mL for zearalenone and 333 pg/mL for HT-2. In addition, Koets et al. used MNPs as detection labels, which can quickly and sensitively detect multiple antibiotic resistance genes of foodborne pathogen Salmonella on GMR sensors, and realized one-step measurement on a small integrated detection platform suitable for POCT devices. Sharma and his team demonstrated the use of a highly sensitive biosensing platform based on MTJs to detect the natural DNA of pathogens. Due to the orthogonality of the DNA probe, this device can evaluate the presence of multiple pathogens in a single analysis. MR sensors, as simple and highly sensitive detection platforms, are becoming increasingly mature in the field of food safety and have become an indispensable solution in the field of food safety detection. In addition to using more advanced hardware to improve the capabilities of detection instruments, some excellent machine learning algorithms can also be combined. Nakamoto et al. proposed a new food texture sensor with GMR elements and inductance. After waveform feature analysis of food reflecting food texture using principal components, the sensor is divided into different clusters. In addition, the sensor also uses Support Vector Machine (SVM) to improve the accuracy of the system.

Environmental Monitoring Based on Magnetoresistive Sensors

Magnetoresistive sensors are also used for environmental monitoring. Trace pollutants in the environment can accumulate in the body through the food chain, therefore, it is necessary to control any possible pollution and harm in the environment. Mercury, as a heavy metal pollutant in the natural environment, is constantly threatening people's health. Thymine (TT) has a high selectivity for mercury ions. As shown in Fig. 5.11a, Wang et al. used T-Hg^{2+}-T coordination chemistry combined with GMR biosensing scheme to achieve the detection of Hg^{2+}. Experiments show that the detection limit of this method in buffer and natural water bodies is both 10 nM, which has great potential for the detection of pollutants in water. In addition, ricin can contaminate water sources and cause poisoning in humans, which is also a

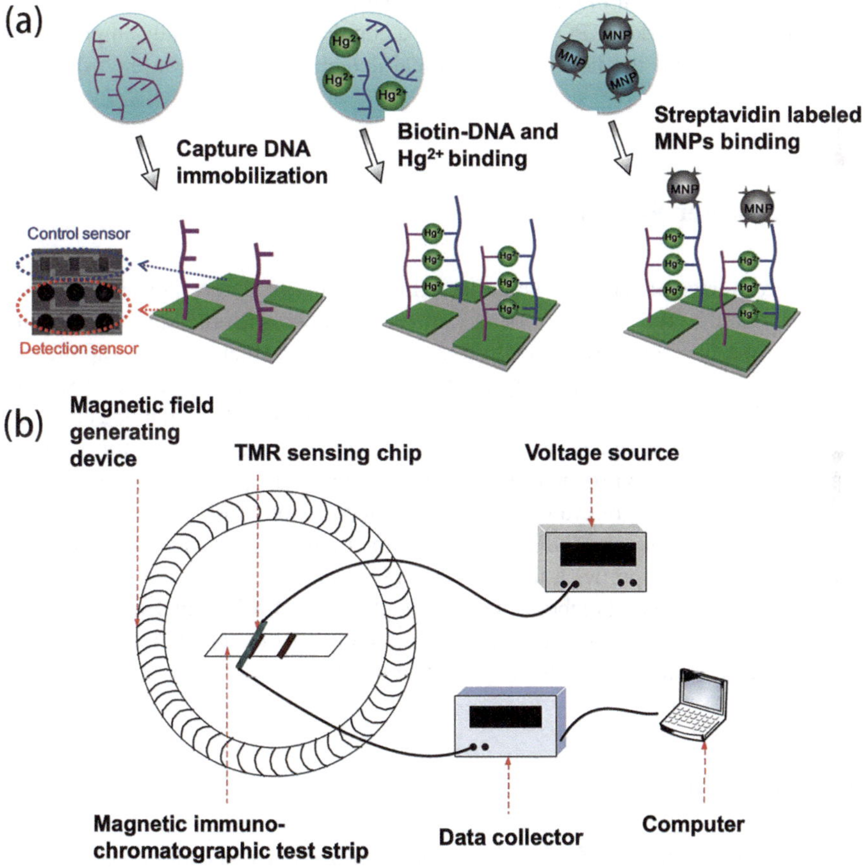

Fig. 5.11 (**a**) GMR sensor for Hg2þ detection; (**b**) Schematic diagram of the TMR sensor detection platform based on immunochromatography

major issue in environmental monitoring. As shown in Fig. 5.11b, Mu et al. effectively coupled the TMR sensitive sensor with ICTS to establish a quantitative method for detecting ricin, with a detection limit of 1 ng/mL. This system overcomes the interference of complex environmental factors during the detection process and meets the analysis requirements of simulated samples such as water, soil, food, and blood.

Magnetoresistive sensors combined with MNPs have solved the shortcomings of traditional optical signal sensors. To complete the detection of biomarkers, the most commonly used detection method is currently the GMR effect, but the maximum resistance change rate of GMR at room temperature is less than 20%, which limits further research on its detection limit. The maximum change rate of TMR can exceed 500%, but this will increase the difficulty and cost of manufacturing. On the other hand, compared with other POCT detection devices, the price of MR sensors is expensive, which hinders its commercial development. However, it can be foreseen that with the continuous development of magnetic resonance sensors in the detection of biomarkers, POCT detection technology will gradually improve, and will greatly benefit the fields of biomedical detection, food safety, environmental monitoring, etc.

5.3 Chapter Summary

The performance of sensors, such as stability, sensitivity, and response characteristics, is related to the properties of the substance to be measured, and also depends on the properties of the assembled and modified interface materials. The application of nanomaterials has significantly improved the detection performance of sensors, and has good application prospects in medicine, food, environmental detection, etc. MNPs, as a new type of nanomaterial, have created a broader prospect for the development of sensors with their unique physicochemical properties. This chapter mainly introduces the relevant content of the application of MNPs in sensor detection. First, it introduces MNPs in depth from the three aspects of the concept of MNPs, the main properties affecting the performance of MNPs, and the main synthesis methods. Then, it elaborates on the application of MNPs in sensor detection, aiming to let readers understand the current status of the application of MNPs in sensor detection and the future development trend.

Bibliography

1. LU L T, DUNG N T, TUNG L D, et al. Synthesis of magnetic cobalt ferrite nanoparticles with controlled morphology, monodispersity and composition: the influence of solvent, surfactant, reductant and synthetic conditions [J]. 2015, 7(46): 19596-610.
2. MCGRATH A, CHEONG S, HENNING A, et al. Size and shape evolution of highly magnetic iron nanoparticles from successive growth reactions [J]. 2017, 53(84): 11548-51.

3. JANA N R, CHEN Y, PENG X J C O M. Size-and shape-controlled magnetic (Cr, Mn, Fe, Co, Ni) oxide nanocrystals via a simple and general approach [J]. 2004, 16(20): 3931-5.
4. PARK J, AN K, HWANG Y, et al. Ultra-large-scale syntheses of monodisperse nanocrystals [J]. 2004, 3(12): 891-5.
5. HU F, WEI L, ZHOU Z, et al. Preparation of biocompatible magnetite nanocrystals for in vivo magnetic resonance detection of cancer [J]. 2006, 18(19): 2553-6.
6. PUNTES V F, ZANCHET D, ERDONMEZ C K, et al. Synthesis of hcp-Co nanodisks [J]. 2002, 124(43): 12874-80.
7. CORDENTE N, RESPAUD M, SENOCQ F, et al. Synthesis and magnetic properties of nickel nanorods [J]. 2001, 1(10): 565-8.
8. BNNEMANN H, BRIJOUX W, BRINKMANN R, et al. A size-selective synthesis of air stable colloidal magnetic cobalt nanoparticles [J]. 2003, 350(14): 617-24.
9. WOO K, LEE H J, AHN J P, et al. Sol–gel mediated synthesis of Fe2O3 nanorods [J]. 2003, 15(20): 1761-4.
10. WANG X, ZHUANG J, PENG Q, et al. A general strategy for nanocrystal synthesis [J]. 2005, 437(7055): 121-4.
11. ARVAND M, HASSANNEZHAD M J M S E C M B A. Magnetic core–shell Fe3O4@SiO2/MWCNT nanocomposite modified carbon paste electrode for amplified electrochemical sensing of uric acid [J]. 2014, 36(1): 160-7.
12. GAN N, YANG X, XIE D, et al. A Disposable Organophosphorus Pesticides Enzyme Biosensor Based on Magnetic Composite Nano-Particles Modified Screen Printed Carbon Electrode [J]. 2010, 10(1): 625-38.
13. ZAMFIR L G, GEANA I, BOURIGUA S, et al. Highly sensitive label-free immunosensor for ochratoxin A based on functionalized magnetic nanoparticles and EIS/SPR detection [J]. 2011, 159(1): 178-84.
14. ZHOU H, GAN N, LI T, et al. The sandwich-type electrochemiluminescence immunosensor for α-fetoprotein based on enrichment by Fe3O4-Au magnetic nano probes and signal amplification by CdS-Au composite nanoparticles labeled anti-AFP [J]. 2012, 746(none).
15. WANG Y, DOSTALEK J, KNOLL W J A C. Magnetic nanoparticle-enhanced biosensor based on grating-coupled surface plasmon resonance [J]. 2011, 83(16): 6202-7.
16. AGRAWAL S, PAKNIKAR K, BODAS D J M E. Development of immunosensor using magnetic nanoparticles and circular microchannels in PDMS [J]. 2014, 115: 66-9.
17. ZHOU J, GAN N, LI T, et al. Ultratrace detection of C-reactive protein by a piezoelectric immunosensor based on Fe3O4@ SiO2 magnetic capture nanoprobes and HRP-antibody co-immobilized nano gold as signal tags [J]. 2013, 178: 494-500.
18. BALASUBRAMANIAN SRINIVASAN Y L, YING JING, CHENGGUO XING, JOEL SLATON,, CHEMISTRY J-P W J A. A three-layer competition-based giant magnetoresistive assay for direct quantification of endoglin from human urine [J]. 2011, 83(8): 2996-3002.
19. WANG Y, WANG W, YU L, et al. Giant magnetoresistive-based biosensing probe station system for multiplex protein assays [J]. 2015.
20. KRISHNA V D, KAI W, PEREZ A M, et al. Giant Magnetoresistance-based Biosensor for Detection of Influenza A Virus [J]. 2016, 7(2472).
21. CHOI J, GANI A W, BECHSTEIN D, et al. Portable, one-step, and rapid GMR biosensor platform with smartphone interface [J]. 2016.
22. D Y Z G A, D L Z A, D W S H A, et al. An Integrated Giant Magnetoresistance Microfluidic Immuno-sensor for Rapid Detection and Quantification of D -dimer [J]. 2015, 43(6): 802-7.
23. YANG G, WHA B, LEI Z C, et al. Multiplex measurement of twelve tumor markers using a GMR multi-biomarker immunoassay biosensor - ScienceDirect [J]. 2019, 123: 204-10.
24. MULUNEH M, ISSADORE D J L O A C. A multi-scale PDMS fabrication strategy to bridge the size mismatch between integrated circuits and microfluidics [J]. 2014, 14(23): 4552-8.
25. SUN X C, LEI C, GUO L, et al. Giant magneto-resistance based immunoassay for the tumor marker carcinoembryonic antigen [J]. 2016.
26. RYU Y, JIN Z, KANG M S, et al. Increase in the detection sensitivity of a lateral flow assay for a cardiac marker by oriented immobilization of antibody [J]. 2011, 5(3): 193.

27. CHICHARO A, CARDOSO F, CARDOSO S, et al. Dynamical Detection of Magnetic Nanoparticles in Paper Microfluidics With Spin Valve Sensors for Point-of-Care Applications [J]. 2014, 50(11): 1-4.
28. HUAMING, LEI, KAN, et al. ContactlessMeasurement of Magnetic Nanoparticles on Lateral Flow Strips Using TunnelingMagnetoresistance (TMR) Sensors in Differential Configuration [J]. 2016.
29. WU Y, LIU Y, ZHAN Q, et al. Rapid detection of Escherichia coli O157: H7 using tunneling magnetoresistance biosensor [J]. 2017, 7(5): 056658.
30. MAK A C, OSTERFELD S J, YU H, et al. Sensitive giant magnetoresistive-based immunoassay for multiplex mycotoxin detection [J]. 2010, 25(7): 1635-9.
31. KOETS M, WIJK T, EEMEREN J, et al. Rapid DNA multi-analyte immunoassay on a magneto-resistance biosensor [J]. 2009, 24(7): 1893-8.
32. SHARMA P P, ALBISETTI E, MASSETTI M, et al. Integrated platform for detecting pathogenic DNA via magnetic tunneling junction-based biosensors [J]. 2017, 242: 280-7.
33. NAKAMOTO H, NISHIKUBO D, OKADA S, et al. Food Texture Classification Using Magnetic Sensor and Principal Component Analysis; proceedings of the Third International Conference on Computing Measurement Control & Sensor Network, F, 2017 [C].
34. WANG W, WANG Y, TU L, et al. Magnetic detection of mercuric ion using a giant magnetoresistance-based biosensing system [J]. 2014, 86(8): 3712.
35. MU X H, LIU H F, TONG Z Y, et al. A new rapid detection method for ricin based on a tunneling magnetoresistance biosensor [J]. 2019, B284(APR.): 638-49

Chapter 6
Sensors Based on One-Dimensional Nanomaterials

In 1991, Sumio Iijima successfully prepared carbon nanotubes by arc evaporation. The diameter of the carbon nanotubes is about tens of nanometers, and the length is at the micron level. This research announced the birth of one-dimensional nanomaterials. Subsequent research followed up, and more and more one-dimensional nanomaterials were gradually discovered. In addition to nanotubes, researchers also prepared a series of one-dimensional nanomaterials such as nanowires, nanorods, and nanobelts. Due to their ultra-high specific surface area and unique optical and electrical properties, one-dimensional nanomaterials can be widely used in biosensors, environmental protection, safety monitoring, and other sensor devices.

6.1 Synthesis of One-Dimensional Nanomaterials

The synthesis of one-dimensional nanomaterials is mainly realized by the generation of solid structures in one dimension. Its synthesis strategies include: (1) One-dimensional directional growth determined by the anisotropic crystallographic structure of solid materials; (2) Introducing a liquid-solid interface to reduce the symmetry of the seed crystal and achieve single-direction crystal growth; (3) Using a one-dimensional spatial template to form a one-dimensional nanostructure within a confined space; (4) Using a coating agent to control the growth rate of seed crystals in different directions; (5) Self-assembly of zero-dimensional nanomaterials; (6) High-dimensional material structure crushing. Among them, the first five methods are more common, and their preparation methods mainly include: vapor phase method, liquid phase method, template method, and self-assembly (as shown in Fig. 6.1).

X. Ma et al., *Functional Micro/Nanomaterials for Intelligent Sensing Devices*, https://doi.org/10.1007/978-981-96-4990-7_6

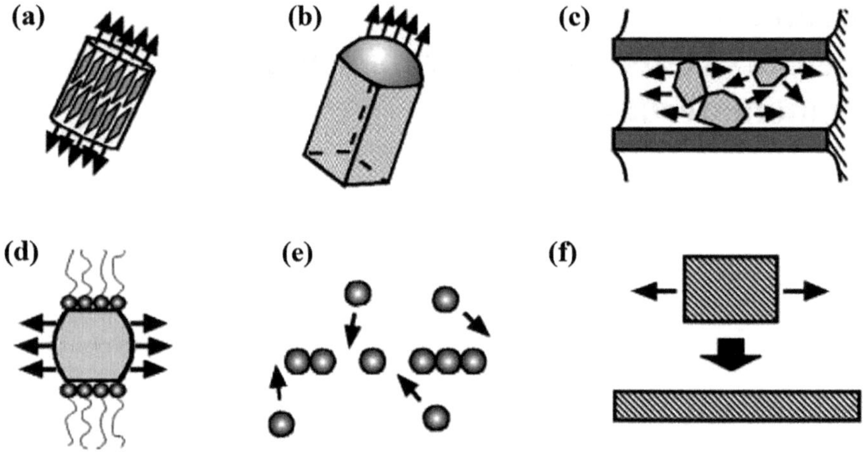

Fig. 6.1 Six strategies widely used for preparing one-dimensional nanomaterials. (**a**) Building one-dimensional nanomaterials by directional growth at both ends; (**b**) Introducing a liquid-solid interface to achieve single-direction crystal growth; (**c**) Preparing one-dimensional nanomaterials by the template method; (**d**) Controlling the formation of one-dimensional nanomaterials with a coating agent; (**e**) Building one-dimensional nanomaterials by self-assembly strategy; (**f**) Preparing one-dimensional nanomaterials by crushing high-dimensional material structures

6.1.1 Vapor Growth Method

The vapor growth method refers to the method in which the sample is sublimated into reaction vapor under high temperature or high pressure conditions, and then the vapor product is deposited on the substrate through physical or chemical reactions, and grows in a directional manner to form one-dimensional nanomaterials. The main synthesis mechanisms of the vapor growth method include Vapor-Liquid-Solid (VLS) and Vapor-Solid (VS) growth mechanisms.

6.1.1.1 VLS Growth Mechanism

The VLS growth mechanism, first reported in 1964 for the growth of Si single crystals, has been widely used in the preparation of various one-dimensional metal oxide nanomaterials. In the process of synthesizing one-dimensional nanomaterials by the VLS mechanism, the catalyst plays a key role. As shown in Fig. 6.2, in the VLS growth mechanism, metal nanoparticles (such as Au nanoparticles) are usually used as catalytic seeds to synthesize crystalline one-dimensional nanomaterials. Under certain temperature conditions, metal particles can form a liquid eutectic alloy with the reaction vapor. The reaction vapor dissolves into the metal particles in the form of solute. With the continuous supply of reaction vapor, the solute in the liquid eutectic alloy gradually saturates. After the solute precipitates, it forms a

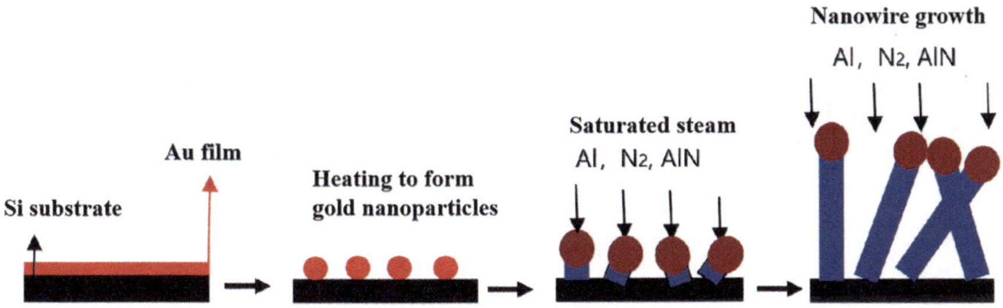

Fig. 6.2 Schematic diagram of preparing AlN nanowires based on the VLS growth mechanism

solid phase crystal on the surface of the liquid eutectic alloy, eventually growing into a one-dimensional nanomaterial. The structure of the one-dimensional nanomaterial formed during the VLS growth process is often related to the properties of the catalytic metal nanoparticles. Choosing the right catalytic metal particles is particularly important for the preparation of one-dimensional nanomaterials. A notable feature of the VLS mechanism is that the starting end of the grown nanowire is attached with catalyst particles. Therefore, the particle size of the catalyst largely determines the final diameter of the synthesized one-dimensional nanomaterial, while the reaction time affects the aspect ratio of the material. At present, researchers have successfully prepared various materials with one-dimensional nanostructures, such as pure substances, metal oxides, and metal carbides, based on the VLS growth mechanism. This synthesis method can be used to prepare quasi-one-dimensional nanomaterials with good structural controllability.

6.1.1.2 VS Growth Mechanism

Studies have shown that many one-dimensional materials can also be grown without the use of reaction catalysts, that is, one-dimensional nanomaterials can be prepared directly through the VS growth mechanism. In the VS growth process, gaseous reactants can be prepared by chemical reduction, gas phase reaction, or thermal evaporation, and then transported to the low-temperature zone to deposit on the substrate. Its growth mode is to use microscopic defects such as liquid-solid interface dislocations and twins as nucleation centers, and further grow into one-dimensional nanomaterials. In the VS growth mechanism, the saturation of the gas phase has a certain impact on the morphology and structure of the nanomaterials. For example, under low saturation conditions, whisker growth occurs, forming one-dimensional materials; block crystals are formed under higher saturation conditions; and powders are formed through uniform nucleation under supersaturated conditions. Therefore, when preparing one-dimensional nanomaterials based on the VS growth mechanism, the reactant vapor should be kept under low saturation conditions. Taking the synthesis of one-dimensional tantalum pentoxide (Ta_2O_5) material as an example (as shown in Fig. 6.3), the metal tantalum filament first forms tantalum vapor at

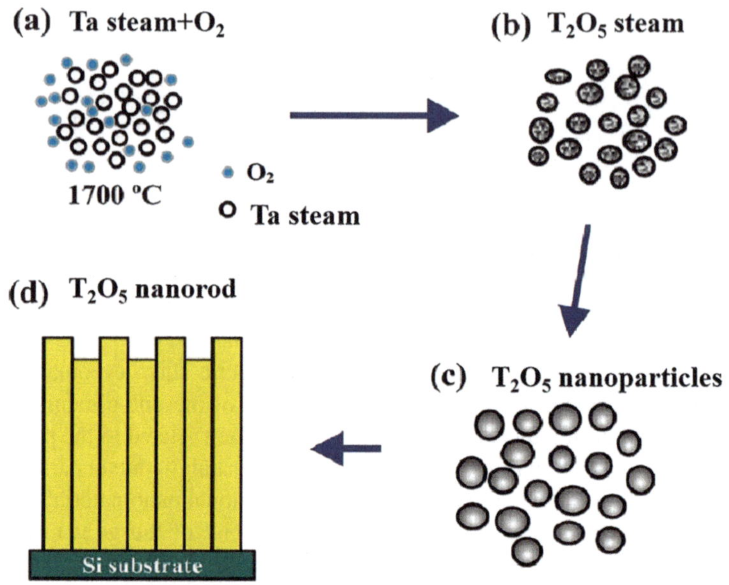

Fig. 6.3 Preparation of Ta_2O_5 one-dimensional nanorods using the VS growth mechanism

1700 °C, and the tantalum vapor forms Ta_2O_5 vapor after oxidation by oxygen. When the Ta_2O_5 vapor comes into contact with the low-temperature substrate, it condenses into small droplets, and the droplets eventually form Ta_2O_5 one-dimensional nanorods.

6.1.2 Liquid Phase Growth Method

The liquid phase growth method includes Solution-Liquid-Solid (SLS), coating liquid method, and solvent thermal chemical synthesis method.

6.1.2.1 SLS Method

The SLS growth mechanism was initially proposed by Trentler and others in 1995. It has now become a method with industrial application prospects for large-scale production of one-dimensional nanomaterials of semiconductors and metal oxides. This method is easy to control and has good repeatability of morphological structure. The SLS growth mechanism is similar to the VLS growth mechanism, and the SLS growth system is completely immersed in the solution to achieve one-dimensional material synthesis (as shown in Fig. 6.4). The catalytic effect still plays a crucial role in the material synthesis process, and metal particles are also used as

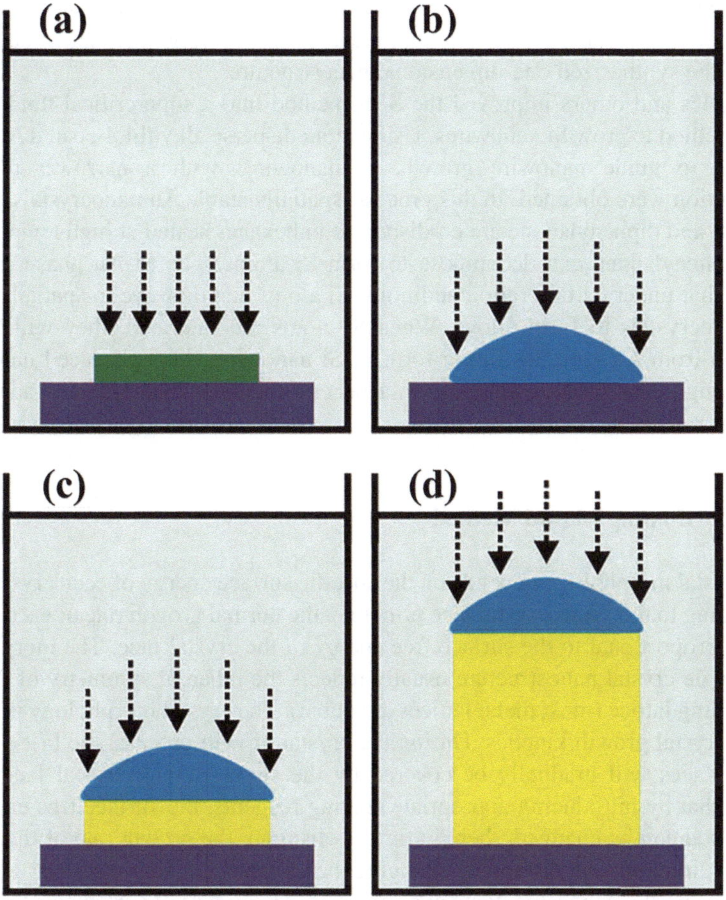

Fig. 6.4 Schematic diagram of the SLS growth mechanism. The SLS growth system is immersed in an active solution (indigo), and metal particles (green) as catalytic seeds can form a liquid eutectic alloy (blue) with the material precursor in the active solution. When the solute in the eutectic alloy is saturated, it will precipitate and grow on the substrate to form a one-dimensional nanostructure (yellow)

catalytic seeds to synthesize one-dimensional nanocrystals. Organic metal precursors in the solution are used to provide the elements required to form one-dimensional nanomaterials. At a certain temperature, metal particles become liquid eutectic alloys in the active solution, because the organic metal precursor as a solute dissolves into the metal particles. As the eutectic alloy saturates, the solute gradually deposits outside the liquid eutectic alloy to form a solid crystal phase, and as the precursor in the solution is gradually consumed, the crystal phase gradually grows into a one-dimensional nanostructure. Similar to the VLS growth mechanism, the diameter of the one-dimensional nanomaterial depends on the size of the liquid eutectic alloy, and the homogeneous liquid eutectic alloy can be used to grow homogeneous one-dimensional nanostructures. The catalyst activity, solute solubility, concentration, and system temperature affect the morphology and crystal structure of the synthesized one-dimensional nanostructure.

Holmes and others improved the SLS method into a supercritical fluid-liquid-solid method to grow Si nanowires. Using monodisperse alkylthiol-coated Au nanocrystals to guide nanowire growth, Si nanowires with a narrower diameter distribution were obtained. In this process, spatially stable Au nanocrystals, Si precursors, and diphenylsilane are co-dispersed in hexane, heated at high temperature, and diphenylsilane can decompose to form Si atoms. The Si-Au phase diagram shows that under certain ratio conditions, Si atoms can dissolve in spatially stable Au nanocrystals to form alloys. When Si atoms are saturated, they will be discharged from Au particles in the form of Si nanowires. The produced nanowires have a high degree of crystallinity, an aspect ratio greater than 100, and adjustable crystal orientation.

6.1.2.2 Coating Liquid Method

The crystal morphology depends on the specific surface energy of each crystal face. According to the Wulff crystal face principle, the normal growth rate of each crystal face is proportional to the surface free energy of the crystal face. The morphology of a single crystal nanostructure usually reflects the inherent symmetry of the corresponding lattice (most metal lattices are cubic). The crystal morphology is related to the crystal growth kinetics. During the crystal growth process, the fast-growing crystal faces will gradually be covered by the slow-growing crystal faces. This means that by introducing appropriate coating reagents, the surface free energy of the crystal can be changed, thereby further changing the growth rate of the crystal in each direction, and ultimately changing the shape of the grown crystal.

For example, Alivisatos, Peng and others used a mixture of surfactants to control the synthesis of CdSe semiconductor nanorods. By adding hexyl phosphonic acid to trioctyl phosphine oxide (commercially used as a stabilizer for the synthesis of CdSe quantum dots), it was found that the crystal mainly grows along the *c*-axis direction, eventually forming nanorods with a larger aspect ratio. However, as the CdSe precursor in the solution is consumed, the short-axis direction begins to grow rapidly, and the aspect ratio gradually decreases. At the same time, the authors also optimized the process conditions to obtain monodisperse CdSe nanorods, and also proved that strong Cd ligands are necessary, which can ensure a higher precursor concentration, thereby facilitating the formation of monodisperse CdSe nanorods with a larger aspect ratio.

6.1.2.3 Solvent Thermal Synthesis Method

The solvent thermal synthesis method is a method of preparing one-dimensional nanomaterials by controlling process parameters under the conditions of high temperature and high pressure supercritical conditions of the solvent, which can significantly improve the solubility and chemical reactivity of solids. For example, the precursor and a reagent that can regulate crystal growth are added to the solvent, and then the mixture is placed in a high-pressure reaction kettle to realize nanowire growth under supercritical conditions. The advantage of the solvent thermal synthesis method is that most solid materials become soluble under supercritical conditions. The solvent thermal synthesis method mainly includes hydrothermal synthesis and organic solvent thermal synthesis. For example, Vayssieres first prepared zinc oxide nanowires by the hydrothermal synthesis method using a nano ZnO film, with a nanowire diameter of about 10–20 nm and a length of up to 10 μm. Gao and others used $KMnO_4$ and NH_4Cl as raw materials to obtain α-MnO_2 nanowires with a diameter of 10–30 nm and a length of about 6–10 μm through simple hydrothermal synthesis. Wang and others reported a simple, repeatable, non-toxic solvent thermal synthesis method for preparing CdS nanowires. The preparation process does not require solid templates and complex surfactants. $CdCl_2$ and elemental S are added to the organic solvent, and then placed in a high-pressure reaction kettle to react at 200 °C for 12 h. Then cool, filter to obtain one-dimensional CdS nanomaterials. Transmission electron microscopy characterization found that CdS nanowires have a hexagonal structure, with a diameter of about 40 nm and a length of more than 10 microns.

6.1.3 Porous Template Method

At present, template-based synthesis technology has become the most commonly used technology for preparing one-dimensional nanostructures. The template method uses a confined template to control the nucleation morphology and growth direction of nanomaterials. The template uses materials with the ability to control pore size, and can use electrochemical deposition to fill the template. After the reaction is completed, the template needs to be removed by chemical reaction to prepare nanomaterials of the required shape and size. In the preparation process, it is necessary to first determine the structure of the one-dimensional nanomaterial to be synthesized (such as nanowires or nanorods), and then select the template material with the appropriate pore diameter and depth for the synthesis of one-dimensional nanomaterials. Based on the porous template method, a variety of one-dimensional nanomaterials such as metals, metal oxides, and conductive polymers can be synthesized. Common templates used for the synthesis of one-dimensional nanomaterials include: anodic aluminum oxide (AAO), nanochannel glass (NCG), porous polymer films, etc.

The properties of the template material, including pore size, pore uniformity, and pore density, have a significant impact on the growth and formation of one-dimensional nanomaterials. As early as 1970, O'Sullivan and Woods proposed a model describing the anodization of aluminum films in acid. Under certain anodization conditions, the resulting oxide film has densely arranged columnar pores, the size and spacing of which can be controlled by changing the formation conditions. The main driving force for pore formation in anodized aluminum is the electric field, which allows for the continuous growth of ordered channels, unaffected by crystal orientation. The surface and cross-sectional morphology of the AAO template and its application in the formation process of one-dimensional nanomaterials are shown in Fig. 6.5. In addition, two other materials of interest are NCG and porous polymer films. NCG is a small, high-density uniform pore matrix, also widely used as a template for nanowire growth. Tonucci and others have confirmed that using NCG films with diameters between 500 nm and 8 mm, 50 Å GaAs/AlGaAs quantum wells can be obtained by molecular beam epitaxy. Thurn-Albrecht and others created array templates from polystyrene (PS) and polymethyl methacrylate diblock copolymer P(S-b-MMA), annealing the diblock copolymer film to above the glass transition temperature between two electrodes to form a hexagonal channel matrix perpendicular to the film surface. After removing PMMA, the resulting polystyrene film has a pore diameter of 14 nm, with more than 1.9×10^{11} pores/cm^2.

The porous template method has the advantages of low cost and high repeatability, and can prepare high-density, neatly arranged nanowire matrices, which have potential applications in micro-electro-mechanical system (MEMS) devices. Nanowire matrices have been applied in magnetic, optical, electrical, lithium-ion batteries, and chemical and biological sensing.

6.1.4 Electrochemical Deposition

Electrochemical deposition is a relatively early technology, widely used for metal layer deposition on solid material surfaces. In recent years, electrochemical deposition technology has been widely used in the preparation of one-dimensional nanomaterials, including: Bi, Ag and ZnO nanowires, carbon and Cu nanotubes, Ga_2O_3-Al_2O_3 nanorods, etc. In the electrochemical deposition process, the template is attached to the cathode for electroplating. When an electric field is applied between the cathode and anode, the cations at the cathode are gradually attracted and reduced, leading to the growth of nanowires in the template pores. The length

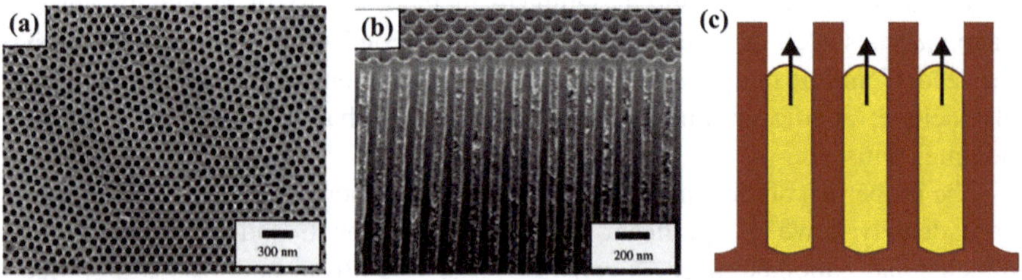

Fig. 6.5 (**a**, **b**) Are SEM top view and cross-sectional images of the anodized aluminum template, respectively; (**c**) one-dimensional nanomaterials are formed by material growth along the pore direction

of the nanowires depends on the length of the porous channels, which can be controlled during the electroplating process.

6.2 Properties of One-Dimensional Nanomaterials

Metal oxides have a wide range of electrical, chemical, and physical properties, which are usually highly sensitive to changes in their chemical environment. Therefore, research on sensors based on metal oxides is very extensive. At present, most commercial sensors are also based on metal oxides or metal oxide-doped materials to some extent. Traditional sensor manufacturing methods are mainly based on semiconductors, ion conduction, photoconductivity, piezoelectricity, thermoelectricity, and the luminescent properties of metal oxides. Currently, with the development of nanoscience and technology, chemical and biological sensors based on metal oxide nanowires are gradually gaining widespread attention due to their excellent electrochemical properties. One-dimensional oxide nanomaterials potentially have special sensing properties due to their unique material structure and physicochemical properties. This is mainly reflected in the following aspects:

1. Large surface-to-volume ratio. This property means that most of the atoms (or molecules) in such materials are distributed on the surface of the material and can fully participate in surface reactions;
2. The Debye length λ_D (a parameter characterizing the range of electric field action) of most semiconductor oxide nanowires is close to their radius over a wide range of temperatures and doping levels. Therefore, the electrical properties of nanowire materials are closely related to their surface properties. It can be imagined that nanowires can switch between insulating and highly conductive states as chemical reactions occur on their surfaces. This property may give sensors based on nanowires ultra-high sensitivity and selectivity. For example, it has been reported that the sensitivity of sensors based on a single In_2O_3 nanowire is 10^5 times higher than that of similar solid thin film devices. Through signal-to-noise ratio analysis, it was found that 10^3 molecules can be reliably detected on a device 3 μm long, and in principle, as few as 10 molecules can be detected when the conductive channel length is reduced to about 30 nm.
3. The average time (~10^{-12}–10^{-10} s) required for photoexcited carriers to diffuse from the inside of the oxide nanowire to its surface is significantly less than the recombination time of electrons to holes (~10^{-9}–10^{-8} s). This means that surface photoinduced redox reactions with near-unity quantum yields can usually be carried out on nanowires (as shown in Fig. 6.6). The response and recovery times of conductive sensors usually depend on the adsorption-desorption kinetics (determined by the working temperature), and the rapid diffusion of electrons and holes to the surface of nanomaterials can allow the analyte to be quickly photodesorbed in a few seconds at room temperature.

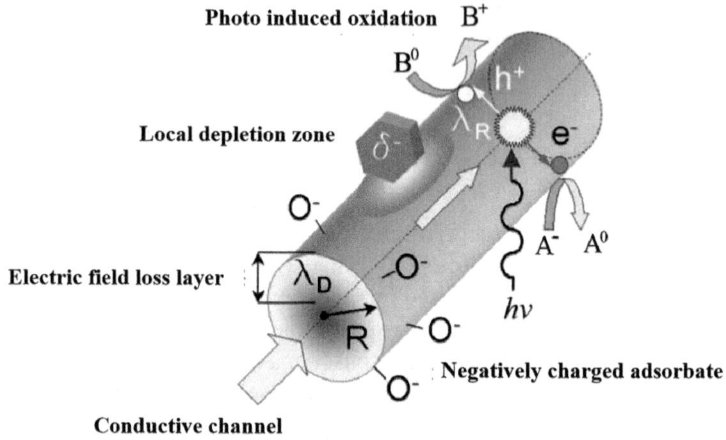

Fig. 6.6 Some electrical, chemical, and optical reaction processes occurring in metal oxides. When the material size is reduced to the nanometer level, it will be beneficial to the reaction process

4. Compared with multi-particle oxides used in sensors, semiconductor oxide nanowires usually have better definition in terms of stoichiometry and have higher crystallinity, which may reduce instability related to permeation or hopping conduction.
5. Nanowires can be easily configured as Field Effect Transistors (FETs), and can potentially be integrated with traditional devices and device manufacturing technologies.

6.3 One-Dimensional Nanomaterial Sensors

6.3.1 Chemical Gas Molecule Sensing

6.3.1.1 Gas Molecule Sensing Mechanism

Understanding semiconductor properties is a prerequisite for understanding the sensing mechanism of metal oxide semiconductors. Semiconductors can be divided into electron-type (n-type) semiconductors and hole-type (p-type) semiconductors, and different types of semiconductors have different chemical sensing mechanisms. As is well known, in silicon-based semiconductors, most carriers (i.e., electrons or holes) can be manipulated by doping donor or acceptor impurities. However, in wide-bandgap metal oxides, carriers are mainly caused by the deviation of metal and oxygen atoms from the chemical ratio in the absence of doping. For example, the p-type semiconductor characteristics of NiO can be explained by the lack of metal ions, while the n-type characteristics of SnO_2 are due to the formation of oxygen vacancies resulting in free electrons. In chemical or gas sensors based on metal

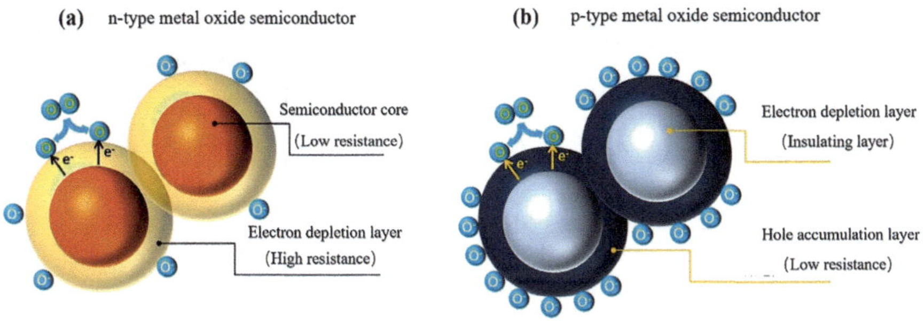

Fig. 6.7 (**a**) n-type and (**b**) p-type metal oxide semiconductor electron core-shell structure model diagrams

oxide semiconductors, high responses to specific molecules can be achieved by operating the sensor at a specific optimal temperature, which can change the concentration of most carriers and their chemical interactions with ambient gases. In addition, under high temperature (100–500 °C) conditions, oxygen molecules can be chemically adsorbed and ionized by capturing electrons on the semiconductor surface. The reactions of oxygen ion adsorption on the surface of metal oxides under different temperature conditions are as follows:

$$O_2\left(\text{gas}\right) \leftrightarrow O_2\left(\text{ads}\right) \tag{6.1}$$

$$O_2\left(\text{ads}\right) + e^- \leftrightarrow O_2^-\left(\text{ads}\right)\left(< 100°C\right) \tag{6.2}$$

$$O_2^-\left(\text{ads}\right) + e^- \leftrightarrow 2O^-\left(\text{ads}\right)\left(100 - 300°C\right) \tag{6.3}$$

$$O^-\left(\text{ads}\right) + e^- \leftrightarrow O^{2-}\left(\text{ads}\right)\left(> 300°C\right) \tag{6.4}$$

In the formula, gas represents the gaseous state, and ads represents the adsorbed state. When the temperature is below 100 °C, the adsorbed oxygen molecules capture electrons from the surface of the metal oxide, forming O_2^-, and at temperatures of 100–300 °C, O_2^- captures electrons from the surface of the metal oxide to form O^-. As the temperature further increases to above 300 °C, O^- can further capture electrons to form O^{2-}. This adsorption characteristic of oxygen ions gives metal oxides an electron core-shell structure. In n-type metal oxides, such as SnO_2, due to the adsorption of oxygen ions, an electron depletion layer (EDL) and shell are formed around the semiconductor core, while in p-type metal oxides, a hole accumulation layer (HAL) and shell are formed near the surface and around the insulating core. The core-shell structures of n-type and p-type semiconductors are shown in Fig. 6.7.

To further elaborate on the formation process of the EDL layer in n-type metal oxides, Fig. 6.8 provides the band diagrams before and after the chemical adsorption of oxygen. In a vacuum, there is no chemical adsorption on the surface, and the band is flat; when the metal oxide is exposed to air, oxygen adsorption occurs on the surface.

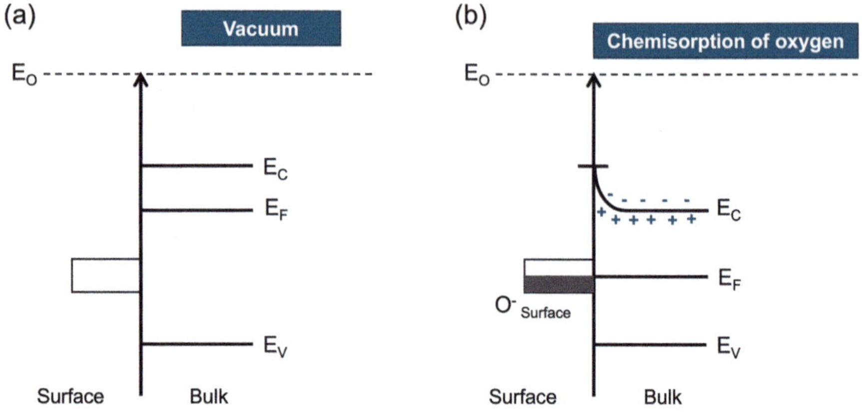

Fig. 6.8 The phenomenon of band bending is caused by the chemical adsorption of oxygen on the surface of n-type metal oxide semiconductors. (**a**) Vacuum environment; (**b**) Contact with oxygen conditions. E_0, E_C, E_V, and E_F represent the vacuum energy level, conduction band, valence band, and semiconductor Fermi energy level energy, respectively

These chemically adsorbed oxygen ions capture electrons from the conduction band of the metal oxide and induce band bending phenomena (upward band bending), forming a negative ion surface and EDL on the n-type metal oxide. On the other hand, if oxygen is chemically adsorbed on the surface of a p-type metal oxide, a HAL will be formed.

Interestingly, studies have shown that the number of oxygen ions adsorbed on the surface of n-type and p-type semiconductors is significantly different. Through temperature-controlled desorption experiments, the total oxygen desorption amount below 560 °C (V_{560}) of metal oxides was measured, and it was found that compared with n-type metal oxides (such as SnO_2, TiO_2, ZnO and Fe_2O_3), p-type metal oxides (MnO_2, NiO, Co_3O_4 and Cr_2O_3) have a large amount of adsorbed oxygen (V_{560} value is larger). For n-type transition metal oxides, due to their variable oxidation states, they often have a larger amount of oxygen ion adsorption. Considering the electronic shell structure of metal oxides, the resistance of chemical/gas sensors based on n-type metal oxides is determined by the contact between the shells. If R_{core} and R_{shell} are defined as the resistance of a semiconductor core and the inter-shell resistance, respectively, then the total resistance of the n-type metal oxide gas sensor will be the series of these two, while in the p-type metal oxide sensor, conduction can be explained by the parallel paths between the wide resistance core (R_{core}) and the narrow conductive shell (R_{shell}). Therefore, when n-type metal oxides are exposed to reducing gases (such as CO), electrons are transferred to the semiconductor, CO is oxidized by reacting with the surface-adsorbed oxygen ions, resulting in a decrease in the concentration of oxygen ions on the surface of the metal oxide, a decrease in sensor resistance, an increase in conductivity, which is proportional to the concentration of the gas analyte. The reaction that occurs on the surface of n-type metal oxides with CO can be represented as:

$$CO(gas) \rightarrow CO(ads) \tag{6.5}$$

$$CO(ads) + O^-(ads) \rightarrow CO_2(gas) + e^- \tag{6.6}$$

On the other hand, the interaction of n-type metal oxides with highly electronegative gases (i.e., oxidizing gases, such as NO_2) will cause an increase in sensor resistance, and the reaction equation is as follows:

$$NO_2(gas) \leftrightarrow NO_2(ads) \tag{6.7}$$

$$e^- + NO_2(ads) \leftrightarrow NO_2(ads) \tag{6.8}$$

When p-type metal oxides interact with reducing gases (such as CO), electrons flow into the p-type oxide material, reducing the hole concentration in the HAL layer, which in turn causes an increase in the sensor's resistance. According to literature reports, n-type metal oxide sensors usually have a higher response than p-type metal oxide sensors. Therefore, most of the current research on metal oxide sensors is focused on n-type, and the manufacture of highly sensitive p-type metal oxide chemical/gas sensors has always been a challenge. The conduction of p-type semiconductors mainly occurs in the narrow HAL, and the hole-electron combination (electrons provided by the reducing gas) does not significantly affect the overall resistance of the sensor. Therefore, the response of p-type metal oxide sensors is usually lower than that of n-type metal oxide sensors. Using S_p and S_n to represent the response of p-type and n-type metal oxide sensors to specific gas analytes, according to the research results of Hubner and others, the two have the following relationship:

$$S_p = \sqrt{S_n} \tag{6.9}$$

Although the sensitivity of p-type metal oxide sensors is lower, p-type metal oxides are still considered potential materials for new chemical gas sensors due to their unique physical/chemical properties. In addition, with the advancement of nanotechnology, the development of p-type gas sensors is also advancing. For example, NiO nanowires grown by the VLS mechanism, when integrated into conductive chemical sensors, exhibit excellent detection performance for H_2.

To determine the final sensing performance, in addition to the above-mentioned chemical reactions and conduction mechanisms, other important parameters such as surface morphology, nanocrystal properties, and specific surface area need to be considered. Therefore, predicting the sensing behavior of active materials is very challenging without experimental testing.

6.3.1.2 Gas Sensing Applications

In 1962, Seiyama and others found that the conductivity of ZnO would change significantly in the presence of active gases. Since then, the application of semiconductor metal oxides in gas sensing research has gradually emerged. In recent years, one-dimensional semiconductor nanomaterials (including nanowires, nanotubes, nanobelts, and nanorods) have been widely used in chemical/gas sensing research due to their large surface area and high crystallinity.

At present, one-dimensional nanomaterials used in gas sensors include SnO_2, ZnO, In_2O_3, TiO_2, WO_3, Fe_2O_3, Ga_2O_3, CuO, NiO and V_2O_5, etc. Among them, research based on SnO_2 and ZnO one-dimensional nanomaterials is more common, and related research involves metal oxide nanowires, nanotubes, nanobelts, and nanorods. The types of target gases detected include NO_2, NH_3, H_2, CO, H_2S, ethanol, oxygen, etc. (see Table 6.1). Law and others found that single-crystal SnO_2 nanobelts have a strong photoelectric response. Therefore, by using ultraviolet light with energy close to the SnO_2 band gap ($E_g = 3.6$ eV, 300 K), SnO_2 nanobelts can detect NO_2 gas at ppm levels. Li and Zhou reported a highly sensitive single In_2O_3 nanowire sensor for detecting NO_2 and NH_3 at room temperature. This device has a sensitivity (i.e., the ratio of resistance after exposure to resistance before exposure) of 10^6 for NO_2 and 10^5 for NH_3, which is four or five orders of magnitude higher than that of thin-film-based sensors. The response times (i.e., the time required for a change in resistance by an order of magnitude) for 100 ppm NO_2 and 1% NH_3 are 5 s and 10 s, respectively. The detection limit for NO_2 is 0.5 ppm, and for NH_3 it is 0.02%. Moreover, ultraviolet light irradiation can significantly promote the desorption of surface gas molecules, thereby shortening the

Table 6.1 Gas sensors and performance based on one-dimensional nanomaterials

Gas molecules	Materials	Detection limit (temperature)	Sensitivity (concentration)	Response time
NO_2	SnO_2 nanobelts	2 ppm (25 °C)	7 (100 ppm)	~1 min
NO_2	SnO_2 nanowires	<0.1 ppm (25 °C)	1 (10 ppm)	~1 min
NO_2	In_2O_3 nanowires	0.5 ppm (25 °C)	10^6 (100 ppm)	5 s
NO_2	In_2O_3 nanowires	0.02 ppm (25 °C)	0.8 (1 ppm)	15 min
H_2	SnO_2 nanobelts	0.2% (25 °C)	0.17 (0.2%)	N/A
H_2	SnO_2 nanowires	<0.1 ppm (200 °C)	4 (1 ppm)	~50 s
H_2	ZnO nanorods	200 ppm (25 °C)	0.04 (200 ppm)	30-40 s
H_2	VO_2 nanowires	N/A (50 °C)	1000 (100%)	~10 min
H_2	$WO_{2.72}$ nanowires	<100 ppm (25 °C)	22 (1000 ppm)	40 s
CO	SnO_2 Nanobelts	5 ppm (400 °C)	7 (250 ppm)	30 s
CO	SnO_2 nanowires	100 ppm (25 °C)	15 (500 ppm)	~10 min
CO	ZnO nanowires	<50 ppm (275 °C)	3200 (400 ppm)	~50 min
CO	NiO nanowires	N/A (150 °C)	0.25 (800 ppm)	~2 h
CO	CeO_2 nanowires	<10 ppm (25 °C)	2 (200 ppm)	~10 s
H_2S	SnO_2 nanowires	<1 ppm (150 °C)	6×10^6 (50 ppm)	N/A
H_2S	ZnO nanowires	N/A (25 °C)	8 (300 ppm)	~50 s
H_2S	In_2O_3 nanowires	1 ppm (25 °C)	1 (20 ppm)	48 s
Ethanol	ZnO Nanorods	10 ppm (300 °C)	45 (100 ppm)	~80 s
O_2	β-Ga_2O_3 nanowires	<50 ppm (25 °C)	20 (50 ppm)	1s

recovery time of the device. Based on this, the research team further optimized the performance of the single In_2O_3 nanowire sensor, reducing the detection limit for NO_2 to 20 ppb.

Cheng and Wang demonstrated a field-effect transistor (FET) style gas sensor constructed from SnO_2 nanobelts. The FET characteristics significantly changed when exposed to 0.2% H_2, with the channel conductivity linearly increasing by 17% under all gate voltages, due to the reaction of hydrogen with chemically adsorbed oxygen on the metal oxide surface, thereby increasing the electron concentration and conductivity of the nanobelt. Hernandez-Ramirez and Morante constructed a single SnO_2 nanowire sensor for CO detection. AC impedance detection showed that the CO detection threshold of the single SnO_2 nanowire sensor is less than 5 ppm, and the instability measured at 295 °C is less than 4%. Moreover, the self-heating single SnO_2 nanowire gas sensor they constructed exhibits ultra-low power consumption, such as the sensor's response to 0.5 ppm NO_2 without a heater (I_m = 10 nA) is equivalent to the response with a heater (T =175 °C). In addition, the authors also demonstrated that a single SnO_2 The equivalence between the thermal response of nanowire gas sensors and the room temperature ultraviolet light-assisted response. For example, the sensor's response to 0.5 ppm NO_2 under ultraviolet light irradiation at room temperature is equivalent to the response when operating at 175 °C under dark conditions.

In addition to the above-mentioned gas sensors based on SnO_2, In_{23}, and ZnO nanowires or nanobelts, single TiO_2, $WO_{2.72}$, NiO, and β-Ga_2 O_3 nanowires have also been used in gas sensing research. These one-dimensional nanoscale metal oxide materials have been proven to have good gas sensing performance. Introducing some metal nanoparticles (such as Pd, Pt, etc.) into one-dimensional metal oxide nanomaterials can effectively improve the gas detection sensitivity of the sensor. For example, Kolmakov and Moskovits demonstrated that the gas sensor performance of a single SnO_2 nanowire can be enhanced by decorating the surface with metal nanoparticles (such as Pd). For Pd nanoparticle-modified SnO_2 nanowires, the sensitivity to O_2 and H_2 increased by 500–1000%, attributed to the enhanced catalytic dissociation of surface adsorbates by Pd nanoparticles, allowing product atoms/ions to diffuse to the metal oxide surface. Liao et al. found that introducing Pt nanoparticles into a single CeO_2 nanowire sensor can significantly improve the sensor's sensitivity to CO, H_2, ethanol, gasoline, and H_2S gases. Compared with traditional metal oxide sensors, Pt nanoparticle-doped single CeO_2 nanowire sensors have a clear advantage in selectively detecting CO gas. Moskovits et al. designed an exceptionally sensitive H_2 sensor using a single VO_2 nanowire decorated with Pd nanoparticles. During the operation of the sensor, the metal oxide can recombine with atomic hydrogen dissociated from the Pd surface, resulting in an approximately 1000-fold increase in current.

6.3.2 *Biomolecule Sensors*

6.3.2.1 Biomolecule Sensing Mechanism

Electrochemical Sensing

Electrochemical and electrical sensing are the main types of biosensors based on one-dimensional nanomaterials. Electrochemical sensors mainly detect analytes based on electrochemical reactions. These sensors oxidize or reduce the analytes on the surface of the working electrode and convert the information of the analyte concentration into a proportional electrical signal. Typically, an electrochemical sensor includes three electrodes: a working electrode, a reference electrode, and a counter/auxiliary electrode. One-dimensional nanomaterials, due to their good electrical properties and high specific surface area, are generally used as reaction sites (i.e., working electrodes). Figure 6.9 show the basic construction of an electrochemical sensor based on one-dimensional nanomaterials. The analyte undergoes an oxidation-reduction reaction on the surface of the one-dimensional nanomaterial and transfers electrons to the working electrode. In biosensor applications, the surface of the one-dimensional nanomaterial is often functionalized to enhance the affinity of the material surface with biomolecules. The combination of one-dimensional nanomaterials with bioenzymes with redox activity provides a convenient platform for fundamentally understanding biological redox reactions and promotes the development of reagent-free biosensors.

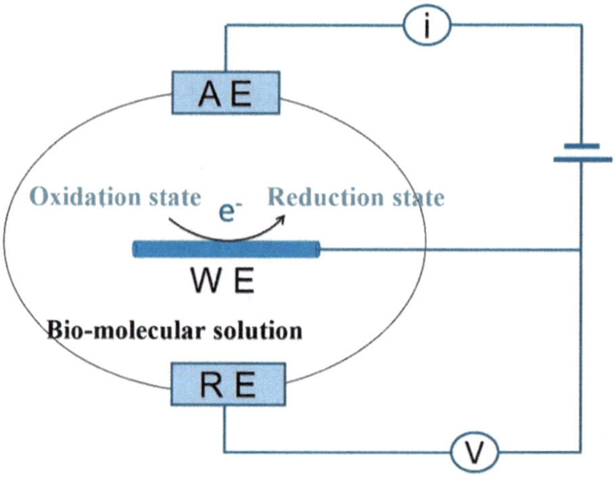

Fig. 6.9 Basic construction of a bio-electrochemical sensor based on one-dimensional nanomaterials. The one-dimensional nanomaterial is used as the working electrode (WE), AE represents the counter/auxiliary electrode, and RE represents the reference electrode

Electrical Sensing

Electrical sensing is achieved by monitoring the conductivity or resistivity changes when the one-dimensional nanomaterial comes into contact with the analyte. Electrical sensing mainly includes two pathways: firstly, when the analyte is introduced into the system, the electrical characteristics of the one-dimensional nanomaterial change, and the detection of the analyte can be achieved by real-time monitoring of the conductivity of the one-dimensional nanomaterial. This type of sensor is also known as a chemiresistor. In addition, one-dimensional nanomaterials can also be constructed into field-effect transistors to achieve electrical sensing. In both configurations, the one-dimensional nanomaterial is used as a channel connecting the source electrode and the drain electrode. This channel can be composed of a single one-dimensional nanomaterial or a network of one-dimensional nanomaterials.

The mechanism of biomolecular electrical sensing is similar to that of gas molecule sensing. When charged biomolecules are adsorbed on the surface of one-dimensional nanomaterials, it will cause disturbances in the local charge distribution of the material, rearrangement of the electronic energy bands, and changes in the interface dielectric constant and electrostatic gate control. These changes will cause changes in metal function, carrier concentration, and mobility, further affecting the conductivity of the sensor. For example, the combination of negatively charged biomolecules with p-type semiconductors increases the density of carriers (holes), thereby increasing the conductivity of the channel. Conversely, the combination of positively charged biomolecules with p-type semiconductors will consume the carriers of the material, thereby reducing the channel conductivity. Therefore, the conductivity of one-dimensional nanomaterials is related to the positive and negative charges and charge distribution of the adsorbed molecules.

The Debye length (λ_D) is a physical quantity related to the sensing mechanism of one-dimensional nanomaterials, expressing the distance that the external electric field can act after being shielded by the electrostatic effect of charge carriers. In biosensors, the Debye length is related to the charge accumulation and loss caused by the adsorption of the target analyte-bio-receptor. The Debye length penetrating the main structure of the one-dimensional nanomaterial is expressed as:

$$\lambda_D = \left(\frac{\varepsilon_r \varepsilon_0 k_B T}{pe^2} \right)^{1/2} \tag{6.10}$$

where, ε_r is the dielectric constant of the one-dimensional nanomaterial (F/m); ε_0 is the dielectric constant of free space; k_B is the Boltzmann constant (eV/K); T stands for temperature (K); p stands for carrier concentration (m^{-3}); e stands for elementary charge (approximately 1.6×10^{-19} C).

According to formula (6.10), when constructing high-sensitivity chemiresistive/chemFET sensors based on one-dimensional nanomaterials, the size of the one-dimensional nanomaterials should be as close as possible to λ_D. In principle, if λ_D is equal to the size of the nanomaterial, the Schottky barrier height (SBH) can completely block the conductive path. Moreover, since λ_D is proportional to $(1/p)^{1/2}$, a lower carrier concentration can ensure further accumulation or loss of carriers when biomolecules interact with the material surface. In addition, the conductivity of one-dimensional nanomaterials must be significantly higher than the conductivity of the solution to avoid leakage.

6.3.2.2 Biomolecular Sensing Applications

Over the past 20 years, a large number of studies on biomolecular sensing based on one-dimensional nanomaterials have emerged. The one-dimensional nanomaterials used involve non-metallic inorganics and their oxides (such as silicon nanowires, single-walled carbon nanotubes, graphene oxide nanomaterials), metal oxides (such as In_2O_3, SnO_2, ZnO, etc.), and organic polymers (such as polyethylene dioxythiophene, polypyrrole, polyaniline). The biomolecules detected include: antibiotics, adenosine triphosphate (ATP), DNA, glucose, proteins, uric acid, dopamine, bacteria, etc.

In recent years, some representative studies of one-dimensional nanomaterials used for biomolecular detection sensors can be seen in Table 6.2. For example, Wang et al. found that one-dimensional carbon nanotube electrochemical sensors show ultra-high sensitivity, capable of detecting 160 zmol protein molecules and 1.3 zmol DNA molecules. In 2001, Cui et al. designed a boron-doped silicon nanowire FET biosensor, which can detect changes in the picomolar level of streptavidin. Subsequently, researchers used functionalized carbon nanotubes to construct sensors that could push the detection limit of glucose and proteins to the molecular level. Gao et al. used silicon nanowires to construct FET biosensors, fixed DNA probes on the surface of the nanowires, and used rolling circle amplification technology to achieve rapid unmarked detection of specific DNA molecules. This technology greatly reduces non-specific DNA connections, significantly improves the signal-to-noise ratio, and can ultimately achieve detection of femtomolar or even attomolar level DNA samples.

6.4 Chapter Summary

This chapter introduces the application of one-dimensional nanomaterials in sensing detection, including the synthesis and preparation of one-dimensional nanomaterials, the properties of one-dimensional materials, and the application of one-dimensional materials in gas sensing and biomolecular sensing fields. Compared with the currently mature commercial sensing technology, the bio-sensing technology based on one-dimensional nanomaterials is still in the development stage.

Table 6.2 Biosensors and performance based on one-dimensional nanomaterials

Material	Analyte	Recipient	Detection range	Detection limit
Si	m-antibiotic	Biotin	4–12 nM	4 nM
	ATP	AbI	–	100 pM
	ssDNA	Amine	10 pM to 500 nM	10 pM
	DNA	DNA probe	0.1 fM to 10 nM	0.1 fM
	DNA	DNA probe	1 fM to 100 pM	1 fM
	Ddopamine	APTMS	1 fM to 1 nM	1 fM
	MMP-2	DNA	100 fM to 10 nM	100 fM
SWCNT	Glucose	GOx	–	0.1 mM
	ssDNA	ssDNA	–	100 nM
	Escherichia coli	Adapter body	–	3.1E3 CFU/mL
	Glucose	ConcanavalinA	1 pM to 1 nM	1 pM
	PA toxin	Adapter body	1–800 nM	1 nM
	Thrombin	Adapter body	2.6 fM to 2.6 nM	2.6 fM
	Antibiotin protein	Biotin	100 pM to 10 μM	100 pM
	Ab-cTnI	cTnI	10 pg/mL to 10 ng/mL	10 pg/mL
In_2O_3	PSA antigen	PSA antibody	–	0.14 nM
	Nprotein	Antibody mimetic protein	2–10 nM	2 nM
SnO_2	Glucose	GOx	50–300 mg/dL	50 mg/dl
ZnO	Biotin	Streptavidin	2.5–250 nM	2.5 nM
	VitaminB$_2$	Adapter body	1 pM to 0.1 mM	1 pM
	IgG	Anti-IgG	–	0.1 μg/mL
	Uric acid	Cy5-uricase	1 pM to 1 μM	1 pM
PPy	CA 125	Antibody	1–100 U/mL	1 U/mL
	MUC1	Adapter body	2.66–66.5 nM	2.66 nM
	CMV	Antibody	10 ng/mL to 100 μg/mL	10 ng/mL
PEDOT	Antibody	M13	20–100 nM	20 nM
	PSMA	M13	20–120 nM	56 nM
PANi	IgG	Antibody	5 ng/mL to 2.8 μg/mL	5 ng/mL
	Glucose	GOx	0.01–8 mM	0.7 μM
GO	DNA	ssDNA	1 pM to 100 nM	1 pM
	DNA	Peptide nucleic acid	10 fM to 1 nM	10 fM

SWCNT single-walled carbon nanotubes, *PEDOT* poly(3,4-ethylenedioxythiophene), *PPy* polypyrrole, *PANI* polyaniline, *CMV* cucumber mosaic virus, *GO* graphene oxide, *MMP-2* matrix metalloproteinase-2, *APTMS* 3-aminopropyltrimethoxysilane, *PSMA* prostate-specific membrane antigen, *IgG* immunoglobulin G, *PSA* prostate-specific antigen, *GOx* glucose oxidase, *ATP* adenosine triphosphate, *cTnI* cardiac troponin I, *Ab-cTnI* cardiac troponin I antibody

Although this technology has shown great potential in experimental research, it still needs to compete with existing mature sensing technologies in terms of sensor construction cost, detection sensitivity, and portability.

Bibliography

1. XIA Y, YANG P, SUN Y, et al. One-dimensional nanostructures: Synthesis, characterization, and applications [J]. Adv Mater, 2003, 15(5): 353-89.
2. YU L, LV Y, ZHANG X, et al. Vapor-liquid-solid growth route to AlN nanowires on Au-coated Si substrate by direct nitridation of Al powder [J]. J Cryst Growth, 2011, 334(1): 57-61.
3. DEVAN R S, LIN J H, HO W D, et al. Investigation of high-temperature phase transformation in one-dimensional Ta_2O_5 nanorods [J]. J Appl Crystallogr, 2010, 43: 1062-7.
4. TRENTLER T J, HICKMAN K M, GOEL S C, et al. Solution-Liquid-Solid Growth of Crystalline Iii-V Semiconductors - an Analogy to Vapor-Liquid-Solid Growth [J]. Science, 1995, 270(5243): 1791-4.
5. HOLMES J D, JOHNSTON K P, DOTY R C, et al. Control of thickness and orientation of solution-grown silicon nanowires [J]. Science, 2000, 287(5457): 1471-3.
6. DEVAN R S, PATIL R A, LIN J H, et al. One-Dimensional Metal-Oxide Nanostructures: Recent Developments in Synthesis, Characterization, and Applications [J]. Adv Funct Mater, 2012, 22(16): 3326-70.
7. WANG Z L. Transmission electron microscopy of shape-controlled nanocrystals and their assemblies [J]. J Phys Chem B, 2000, 104(6): 1153-75.
8. PENG X G, MANNA L, YANG W D, et al. Shape control of CdSe nanocrystals [J]. Nature, 2000, 404(6773): 59-61.
9. VAYSSIERES L. Growth of arrayed nanorods and nanowires of ZnO from aqueous solutions [J]. Adv Mater, 2003, 15(5): 464-6.
10. GAO Y Q, WANG Z G, WAN J X, et al. A facile route to synthesize uniform single-crystalline alpha-MnO2 nanowires [J]. J Cryst Growth, 2005, 279(3-4): 415-9.
11. WANG Q Q, XU G, HAN G R. Solvothermal synthesis and characterization of uniform CdS nanowires in high yield [J]. J Solid State Chem, 2005, 178(9): 2680-5.
12. P. O S J, C. W G. The morphology and mechanism of formation of porous anodic films on aluminium [J]. Proceedings of the Royal Society of London A Mathematical and Physical Sciences, 1970, 317(1531): 511-43.
13. TONUCCI R J, PEARSON D H, KATZER D S, et al. Nanoscale patterning using nanochannel glass replica films [J]. Superlattice Microst, 1996, 20(4): 627-32.
14. THURN-ALBRECHT T, SCHOTTER J, KASTLE C A, et al. Ultrahigh-density nanowire arrays grown in self-assembled diblock copolymer templates [J]. Science, 2000, 290(5499): 2126-9.
15. CHOI Y C, KIM J, BU S D. Template-directed formation of functional complex metal-oxide nanostructures by combination of sol-gel processing and spin coating [J]. Mat Sci Eng B-Solid, 2006, 133(1-3): 245-9.
16. LI C, ZHANG D H, LIU X L, et al. In2O3 nanowires as chemical sensors [J]. Appl Phys Lett, 2003, 82(10): 1613-5.
17. KOLMAKOV A, MOSKOVITS M. Chemical sensing and catalysis by one-dimensional metal-oxide nanostructures [J]. Annu Rev Mater Res, 2004, 34: 151-80.
18. KIM H J, LEE J H. Highly sensitive and selective gas sensors using p-type oxide semiconductors: Overview [J]. Sensor Actuat B-Chem, 2014, 192: 607-27.
19. KAUR N, SINGH M, COMINI E. One-Dimensional Nanostructured Oxide Chemoresistive Sensors [J]. Langmuir, 2020, 36(23): 6326-44.
20. BARSAN N, SIMION C, HEINE T, et al. Modeling of sensing and transduction for p-type semiconducting metal oxide based gas sensors [J]. J Electroceram, 2010, 25(1): 11-9.

21. HUBNER M, SIMION C E, TOMESCU-STANOIU A, et al. Influence of humidity on CO sensing with p-type CuO thick film gas sensors [J]. Sensor Actuat B-Chem, 2011, 153(2): 347-53.
22. CHOI K J, JANG H W. One-Dimensional Oxide Nanostructures as Gas-Sensing Materials: Review and Issues [J]. Sensors-Basel, 2010, 10(4): 4083-99.
23. LAW M, KIND H, MESSER B, et al. Photochemical sensing of NO_2 with SnO_2 nanoribbon nanosensors at room temperature [J]. Angew Chem Int Edit, 2002, 41(13): 2405-8.
24. ZHANG D H, LIU Z Q, LI C, et al. Detection of NO_2 down to ppb levels using individual and multiple In2O3 nanowire devices [J]. Nano Lett, 2004, 4(10): 1919-24.
25. CHENG Y, XIONG P, FIELDS L, et al. Intrinsic characteristics of semiconducting oxide nanobelt field-effect transistors [J]. Appl Phys Lett, 2006, 89(9).
26. HERNANDEZ-RAMIREZ F, TARANCON A, CASALS O, et al. High response and stability in CO and humidity measures using a single SnO_2 nanowire [J]. Sensor Actuat B-Chem, 2007, 121(1): 3-17.
27. PRADES J D, JIMENEZ-DIAZ R, HERNANDEZ-RAMIREZ F, et al. Ultralow power consumption gas sensors based on self-heated individual nanowires [J]. Appl Phys Lett, 2008, 93(12).
28. ROUT C S, KULKARNI G U, RAO C N R. Room temperature hydrogen and hydrocarbon sensors based on single nanowires of metal oxides [J]. J Phys D Appl Phys, 2007, 40(9): 2777-82.
29. FENG P, XUE X Y, LIU Y G, et al. Achieving fast oxygen response in individual beta-Ga_2O_3 nanowires by ultraviolet illumination [J]. Appl Phys Lett, 2006, 89(11).
30. TRESBACK J S, PADTURE N P. Low-temperature gas sensing in individual metal-oxide-metal heterojunction nanowires [J]. J Mater Res, 2008, 23(8): 2047-52.
31. PRAKASH J, SAMRITI, KUMAR A, et al. Novel rare earth metal–doped one-dimensional TiO_2 nanostructures: Fundamentals and multifunctional applications [J]. Materials Today Sustainability, 2021, 13.
32. KOLMAKOV A, KLENOV D O, LILACH Y, et al. Enhanced gas sensing by individual SnO_2 nanowires and nanobelts functionalized with Pd catalyst particles [J]. Nano Lett, 2005, 5(4): 667-73.
33. LIAO L, LU H B, LI J C, et al. The sensitivity of gas sensor based on single ZnO nanowire modulated by helium ion radiation [J]. Appl Phys Lett, 2007, 91(17).
34. BAIK J M, KIM M H, LARSON C, et al. Pd-Sensitized Single Vanadium Oxide Nanowires: Highly Responsive Hydrogen Sensing Based on the Metal-insulator Transition [J]. Nano Lett, 2009, 9(12): 3980-4.
35. PRADES J D, JIMENEZ-DIAZ R, HERNANDEZ-RAMIREZ F, et al. Equivalence between thermal and room temperature UV light-modulated responses of gas sensors based on individual SnO_2 nanowires [J]. Sensor Actuat B-Chem, 2009, 140(2): 337-41.
36. LUPAN O, CHAI G, CHOW L. Novel hydrogen gas sensor based on single ZnO nanorod [J]. Microelectron Eng, 2008, 85(11): 2220-5.
37. QIAN L H, WANG K, LI Y, et al. CO sensor based on Au-decorated SnO2 nanobelt [J]. Mater Chem Phys, 2006, 100(1): 82-4.
38. KUANG Q, LAO C S, LI Z, et al. Enhancing the photon- and gas-sensing properties of a single SnO_2 nanowire based nanodevice by nanoparticle surface functionalization [J]. J Phys Chem C, 2008, 112(30): 11539-44.
39. WEI T Y, YEH P H, LU S Y, et al. Gigantic Enhancement in Sensitivity Using Schottky Contacted Nanowire Nanosensor [J]. J Am Chem Soc, 2009, 131(48): 17690-5.
40. LIAO L, MAI H X, YUAN Q, et al. Single CeO_2 nanowire gas sensor supported with Pt nanocrystals: Gas sensitivity, surface bond states, and chemical mechanism [J]. J Phys Chem C, 2008, 112(24): 9061-5.
41. KUMAR V, SEN S, MUTHE K P, et al. Copper doped SnO_2 nanowires as highly sensitive H_2S gas sensor [J]. Sensor Actuat B-Chem, 2009, 138(2): 587-90.
42. ZENG Z M, WANG K, ZHANG Z X, et al. The detection of H_2S at room temperature by using individual indium oxide nanowire transistors [J]. Nanotechnology, 2009, 20(4).

43. ZHAO S K, SHEN Y B, YAN X X, et al. Complex-surfactant-assisted hydrothermal synthesis of one-dimensional ZnO nanorods for high-performance ethanol gas sensor [J]. Sensor Actuat B-Chem, 2019, 286: 501-11.
44. BUITRAGO E, FERNANDEZ-BOLANOS M, RIGANTE S, et al. The top-down fabrication of a 3D-integrated, fully CMOS-compatible FET biosensor based on vertically stacked SiNWs and FinFETs [J]. Sensor Actuat B-Chem, 2014, 193: 400-12.
45. KULKARNI G S, ZHONG Z H. Detection beyond the Debye Screening Length in a High-Frequency Nanoelectronic Biosensor [J]. Nano Lett, 2012, 12(2): 719-23.
46. NAIR P R, ALAM M A. Design considerations of silicon nanowire biosensors [J]. Ieee T Electron Dev, 2007, 54(12): 3400-8.
47. WANG J, LIU G D, JAN M R. Ultrasensitive electrical biosensing of proteins and DNA: Carbon-nanotube derived amplification of the recognition and transduction events [J]. J Am Chem Soc, 2004, 126(10): 3010-1.
48. CUI Y, WEI Q Q, PARK H K, et al. Nanowire nanosensors for highly sensitive and selective detection of biological and chemical species [J]. Science, 2001, 293(5533): 1289-92.
49. GAO A R, ZOU N L, DAI P F, et al. Signal-to-Noise Ratio Enhancement of Silicon Nanowires Biosensor with Rolling Circle Amplification [J]. Nano Lett, 2013, 13(9): 4123-30.
50. GAO A R, LU N, WANG Y C, et al. Enhanced Sensing of Nucleic Acids with Silicon Nanowire Field Effect Transistor Biosensors [J]. Nano Lett, 2012, 12(10): 5262-8.
51. CUI Y, LAUHON L J, GUDIKSEN M S, et al. Diameter-controlled synthesis of single-crystal silicon nanowires [J]. Appl Phys Lett, 2001, 78(15): 2214-6.
52. WANG W U, CHEN C, LIN K H, et al. Label-free detection of small-molecule-protein interactions by using nanowire nanosensors [J]. P Natl Acad Sci USA, 2005, 102(9): 3208-12.
53. BUNIMOVICH Y L, SHIN Y S, YEO W S, et al. Quantitative real-time measurements of DNA hybridization with alkylated nonoxidized silicon nanowires in electrolyte solution [J]. J Am Chem Soc, 2006, 128(50): 16323-31.
54. LI B R, CHEN C W, YANG W L, et al. Biomolecular recognition with a sensitivity-enhanced nanowire transistor biosensor [J]. Biosens Bioelectron, 2013, 45: 252-9.
55. LI B R, HSIEH Y J, CHEN Y X, et al. An Ultrasensitive Nanowire-Transistor Biosensor for Detecting Dopamine Release from Living PC12 Cells under Hypoxic Stimulation [J]. J Am Chem Soc, 2013, 135(43): 16034-7.
56. CHOI J H, KIM H, CHOI J H, et al. Signal Enhancement of Silicon Nanowire-Based Biosensor for Detection of Matrix Metalloproteinase-2 Using DNA-Au Nanoparticle Complexes [J]. Acs Appl Mater Inter, 2013, 5(22): 12023-8.
57. BESTEMAN K, LEE J O, WIERTZ F G M, et al. Enzyme-coated carbon nanotubes as single-molecule biosensors [J]. Nano Lett, 2003, 3(6): 727-30.
58. TANG X W, BANSARUNTIP S, NAKAYAMA N, et al. Carbon nanotube DNA sensor and sensing mechanism [J]. Nano Lett, 2006, 6(8): 1632-6.
59. SO H M, PARK D W, JEON E K, et al. Detection and titer estimation of Escherichia coli using aptamer-functionalized single-walled carbon-nanotube field-effect transistors [J]. Small, 2008, 4(2): 197-201.
60. CELLA L N, CHEN W, MYUNG N V, et al. Single-Walled Carbon Nanotube-Based Chemiresistive Affinity Biosensors for Small Molecules: Ultrasensitive Glucose Detection [J]. J Am Chem Soc, 2010, 132(14): 5024-+.
61. CELLA L N, SANCHEZ P, ZHONG W W, et al. Nano Aptasensor for Protective Antigen Toxin of Anthrax [J]. Anal Chem, 2010, 82(5): 2042-7.
62. LIU S, ZHANG X Y, LUO W X, et al. Single-Molecule Detection of Proteins Using Aptamer-Functionalized Molecular Electronic Devices [J]. Angew Chem Int Edit, 2011, 50(11): 2496-502.
63. CHANG J B, MAO S, ZHANG Y, et al. Single-walled carbon nanotube field-effect transistors with graphene oxide passivation for fast, sensitive, and selective protein detection [J]. Biosens Bioelectron, 2013, 42: 186-92.

64. RAJESH, SHARMA V, PURI N K, et al. Label-free detection of cardiac troponin-I using gold nanoparticles functionalized single-walled carbon nanotubes based chemiresistive biosensor [J]. Appl Phys Lett, 2013, 103(20).
65. LI C, CURRELI M, LIN H, et al. Complementary detection of prostate-specific antigen using ln(2)O(3) nanowires and carbon nanotubes [J]. J Am Chem Soc, 2005, 127(36): 12484-5.
66. ISHIKAWA F N, CHANG H K, CURRELI M, et al. Label-Free, Electrical Detection of the SARS Virus N-Protein with Nanowire Biosensors Utilizing Antibody Mimics as Capture Probes [J]. Acs Nano, 2009, 3(5): 1219-24.
67. DHOBALE S, JOSHEE P, DEORE G, et al. Nanostructured glucose-oxidase immobilized SnO_2 thin films for glucose sensing [J]. Appl Phys Lett, 2011, 98(7): 073704.
68. CHOI A, KIM K, JUNG H I, et al. ZnO nanowire biosensors for detection of biomolecular interactions in enhancement mode [J]. Sensor Actuat B-Chem, 2010, 148(2): 577-82.
69. HAGEN J A, KIM S N, BAYRAKTAROGLU B, et al. Biofunctionalized Zinc Oxide Field Effect Transistors for Selective Sensing of Riboflavin with Current Modulation [J]. Sensors-Basel, 2011, 11(7): 6645-55.
70. YU R M, PAN C F, WANG Z L. High performance of ZnO nanowire protein sensors enhanced by the piezotronic effect [J]. Energ Environ Sci, 2013, 6(2): 494-9.
71. LIU X, LIN P, YAN X Q, et al. Enzyme-coated single ZnO nanowire FET biosensor for detection of uric acid [J]. Sensor Actuat B-Chem, 2013, 176: 22-7.
72. BANGAR M A, HANGARTER C M, YOO B, et al. Magnetically Assembled Multisegmented Nanowires and Their Applications [J]. Electroanal, 2009, 21(1): 61-7.
73. HUANG J Y, LUO X L, LEE I, et al. Rapid real-time electrical detection of proteins using single conducting polymer nanowire-based microfluidic aptasensor [J]. Biosens Bioelectron, 2011, 30(1): 306-9.
74. CHARTUPRAYOON N, RHEEM Y, NG J C K, et al. Polypyrrole nanoribbon based chemiresistive immunosensors for viral plant pathogen detection [J]. Anal Methods-Uk, 2013, 5(14): 3497-502.
75. ARTER J A, TAGGART D K, MCINTIRE T M, et al. Virus-PEDOT Nanowires for Biosensing [J]. Nano Lett, 2010, 10(12): 4858-62.
76. ARTER J A, DIAZ J E, DONAVAN K C, et al. Virus-Polymer Hybrid Nanowires Tailored to Detect Prostate-Specific Membrane Antigen [J]. Anal Chem, 2012, 84(6): 2776-83.
77. LEE H W, YOON Y, PARK S, et al. Selective dispersion of high purity semiconducting single-walled carbon nanotubes with regioregular poly(3-alkylthiophene)s [J]. Nat Commun, 2011, 2.
78. ZHAI D Y, LIU B R, SHI Y, et al. Highly Sensitive Glucose Sensor Based on Pt Nanoparticle/Polyaniline Hydrogel Heterostructures [J]. Acs Nano, 2013, 7(4): 3540-6.
79. CHEN T Y, PHAN T K L, HSU C L, et al. Label-free detection of DNA hybridization using transistors based on CVD grown graphene [J]. Biosens Bioelectron, 2013, 41: 103-9.
80. CAI B J, WANG S T, HUANG L, et al. Ultrasensitive Label-Free Detection of PNA-DNA Hybridization by Reduced Graphene Oxide Field-Effect Transistor Biosensor [J]. Acs Nano, 2014, 8(3): 2632-8.

Chapter 7
Two-dimensional Nanomaterials and Smart Sensor Devices

In 2004, two-dimensional (2D) graphene nanosheets were successfully prepared from highly oriented pyrolytic graphite by mechanical exfoliation, sparking a surge in 2D material research. In recent years, with the continuous development of 2D material synthesis technology, a variety of emerging 2D materials have emerged, such as Metal-Organic Frameworks (MOFs), Transition Metal Dichalcogenides (TMDs), Hexagonal Boron Nitride (h-BN), Black Phosphorus (BP), layered metal oxides, and Covalent Organic Frameworks (COFs), etc. Due to their unique microstructures and physical and chemical properties, they have shown important application potential in the fields of electronic devices, photonic devices, energy conversion, information storage, and smart sensing.

Compared with other materials, 2D materials have a variety of unique properties: (1) They have a large specific surface area. The high specific surface area allows 2D materials to have a higher interaction area with external substances. For example, in the field of sensing, this will increase the contact area between the two-dimensional material and the analyte, which can improve the high sensitivity detection of low concentration analytes; (2) The conductivity of 2D materials has a wide range, from insulating to nearly metallic. By introducing defects, doping or functionalization in the 2D material structure, and changing the number of atomic layers, the band gap of 2D materials can be controlled, thereby controlling their conductivity; (3) Some 2D materials exhibit excellent mechanical strength and significant optical transparency. For example, graphene oxide has excellent mechanical strength and flexibility, and its optical transparency can reach 90%; (4) The rich surface chemistry and conductive characteristics of 2D materials in response to the external environment make them very suitable for sensing detection in health and environmental monitoring, biomedical diagnosis and treatment, and other applications.

X. Ma et al., *Functional Micro/Nanomaterials for Intelligent Sensing Devices*, https://doi.org/10.1007/978-981-96-4990-7_7

7.1 Types of Two-dimensional Sensing Materials

In this section, we will introduce the classification, material structure, and physical and chemical properties of various 2D materials used in the field of sensor devices.

7.1.1 Graphene Family

In 2004, Andre Geim and Konstatin Novoselov prepared single-layer graphene from highly oriented graphite by mechanical exfoliation. Graphene nanosheets have a hexagonal lattice structure and, together with graphite (3D), carbon nanotubes (1D), and fullerenes (0D), are all allotropes of carbon. Due to its excellent physical and chemical properties, graphene has important applications in many fields. Graphene is a sp^2 hybrid structure of carbon atoms, where the 2s, 2px, and 2py electrons form covalent bonds through hybrid orbitals, and the remaining π electron on the 2pz orbital forms a large π bond perpendicular to the plane of graphene. The π electron can move freely within the crystal plane, giving graphene excellent electrical properties. Studies have shown that the carrier mobility in graphene layers at room temperature can be as high as 1×10^5 $cm^2/(V\ s)$, its hole and electron mobility are almost equal. The superconductivity of graphene stems from its special quantum tunneling effect. Studies have shown that during the process of graphene adsorbing different gas molecules, the conductivity can be regulated by changing the doping state. Due to its atomic thickness, the charge transfer through graphene is highly responsive to its surrounding environment. In addition, graphene has excellent light transmittance, with a visible light absorption rate of only 2.3% for single-layer graphene. As the number of layers increases, the visible light transmittance of graphene will dramatically decrease. Studies have shown that the interband light conversion and photo-transition of single-layer and bilayer graphene are closely related to the number of atomic layers. Besides having high transmittance in the visible light range, graphene also has high transmittance in the near-infrared and mid-infrared light regions.

Graphene is the hardest and strongest crystal structure among known materials. The strength of graphene is 100 times that of steel film of the same thickness, and its stress resistance is stronger than other 2D nanomaterials. The tensile strength and Young's modulus of single-layer graphene are 42 N/m^2 and 1.0 TPa, respectively. Due to its excellent mechanical strength and flexibility, graphene can be used in the manufacture and construction of wearable flexible electronic devices. The specific surface area of single-layer graphene can reach 2630 $m^2\ g^{-1}$, which is twice that of carbon nanotubes. The specific surface area of few-layer graphene measured by the BET method can reach 270–1550 $m^2\ g^{-1}$. By utilizing the high specific surface area of graphene, chemical sensors that can accurately detect individual molecules can be prepared, thereby achieving rapid detection of trace substances.

7.1.2 Transition Metal Dichalcogenides (TMDs)

Transition metal dichalcogenides belong to the MX_2 type of inorganic layered compounds, where the middle transition metal layer (M) forms a sandwich structure with two atomic layers of chalcogen element (X) atomic layers, connected by van der Waals forces between layers and strong covalent bonds within layers. Here, M represents IV, V, or VI group transition metals, such as Ti, Zr, Hf, V, Nb, Ta, Mo, W, Tc, Re, etc., and X represents chalcogen elements, including S, Se, Te, etc. Depending on the atomic arrangement, TMD materials can be divided into different crystal structure types, including trigonal prismatic (hexagonal, H), octahedral (tetragonal, T), and their distorted structures (T′). Taking MoS_2 as an example, as shown in Fig. 7.1, according to the coordination relationship between Mo and S atoms and the stacking order between layers, its crystal types mainly include 1H, 1T, 2H, and 3R, among which 2H-MoS_2 has a thermodynamically stable structure and has been proven to have the most application prospects in current research.

2D TMD materials have unique electrical and optical properties, which are mainly due to the quantum confinement effects that occur when 2D TMDs are transformed from bulk materials to single-layer or multi-layer two-dimensional materials, and their band gap changes from indirect to direct, as well as the surface effects brought about by the reduction in scale. Most semiconductor TMD materials have an indirect band gap in the bulk state, while they have a direct band gap in the single layer. For example, MoS_2 consists of several S-Mo-S layers, which are interconnected by weak van der Waals forces (vdWs). Bulk MoS_2 is an indirect band gap semiconductor (1.2 eV), while the single layer is a direct band gap semiconductor (1.8 eV). In addition, most TMDs have both metallic and semiconductor phases. At room temperature, the stable phase of MX_2 is the 2H phase, and the 1 T phase can be obtained by Li intercalation or electron beam irradiation. The tunable band gap in TMDs, along with strong photoluminescence (PL) and large exciton binding energy, make them highly promising for the construction of optoelectronic devices.

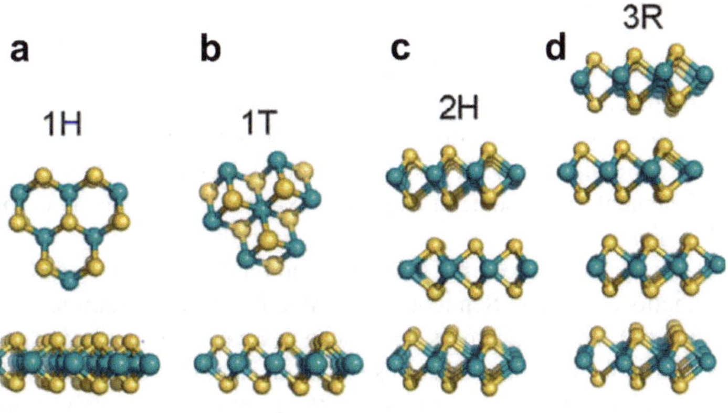

Fig. 7.1 Different crystal structures of MoS_2: (**a**) 1H; (**b**) 1T; (**c**) 2H; (**d**) 3R

Due to the layered structure of 2D TMD materials, the van der Waals gap between each adjacent layer and the larger specific surface area give them great advantages in the field of sensing applications. The high surface-to-volume ratio makes sensors based on 2D TMDs have higher sensitivity, selectivity, and lower power consumption. Existing studies have shown that two-dimensional materials such as MoS_2, WSe, ReS_2 have important application potential in gas sensing. In addition, the weakly bonded 2D TMD atomic layers are easy to isolate and stack with other TMDs. For example, heterostructures can be constructed by stacking atomic-thickness different 2D TMD layers, which have unique functional properties and excellent physicochemical performance.

7.1.3 Transition Metal Carbonitrides (MXene)

MXene is also a kind of emerging two-dimensional material, mainly transition metal carbides, nitrides, carbonitrides, etc., with the formula $M_{n+1}X_n$ (n = 1–3), where M is a transition metal (such as Sc, Ti, Zr, Hf, V, Ta, Nb, Cr, Mo, etc.), and X corresponds to carbon and/or nitrogen. More than 70 types of MXenes have been reported so far. The preparation of MXene usually involves the selective etching and delamination of the lamellar structure of the 3D MAX phase of layered bulk metal carbides, nitrides, or carbonitrides. As shown in Fig. 7.2, MAX phase compounds are $M_{n+1}AX_n$, where A is mainly located in groups IIIA and IVA of the periodic table. By selectively etching the A layer with corrosive liquid, 2D MXenes composed of M and X elements can be obtained. The formula of MXenes changes to $M_{n+1}X_nT_x$, where Tx corresponds to surface functional groups. The surface functional groups are mainly due to the acidic or alkaline solution (mainly hydrofluoric acid HF) used in the treatment process, and the high activity of the exposed M atoms themselves, which cause the surface to attach –OH, –F, –O or –Cl groups. The surface groups introduced during the preparation process have a great impact on the physical and surface chemical properties of MXenes. The hydrophilic functional groups on the surface make MXenes have good hydrophilicity. In addition, the elastic modulus of a single layer of $Ti_3C_2(OH)_2$ is about 300 GPa, which is smaller than the elastic modulus of graphene, but higher than most 2D oxides and layered clays. MXenes prepared by wet etching technology when mixed with water, the flexible film prepared has good flexibility, can be folded, and can support several times its own weight. The M in the MXenes material family can also be two or more transition metal elements at the same time, greatly expanding the diversity of the MXenes material family.

The electronic properties of MXenes depend on the properties of M, X and their surface functional groups. Semiconductor MXenes under certain stimuli (such as photoexcitation), active electrons and vacancies can be generated, and oxidative stress can be applied to these electrons or vacancies. They will react with the environment, which is a necessary process for the production of catalytic and reactive oxygen. The band gap of MXenes can be adjusted by multiple factors, so that the

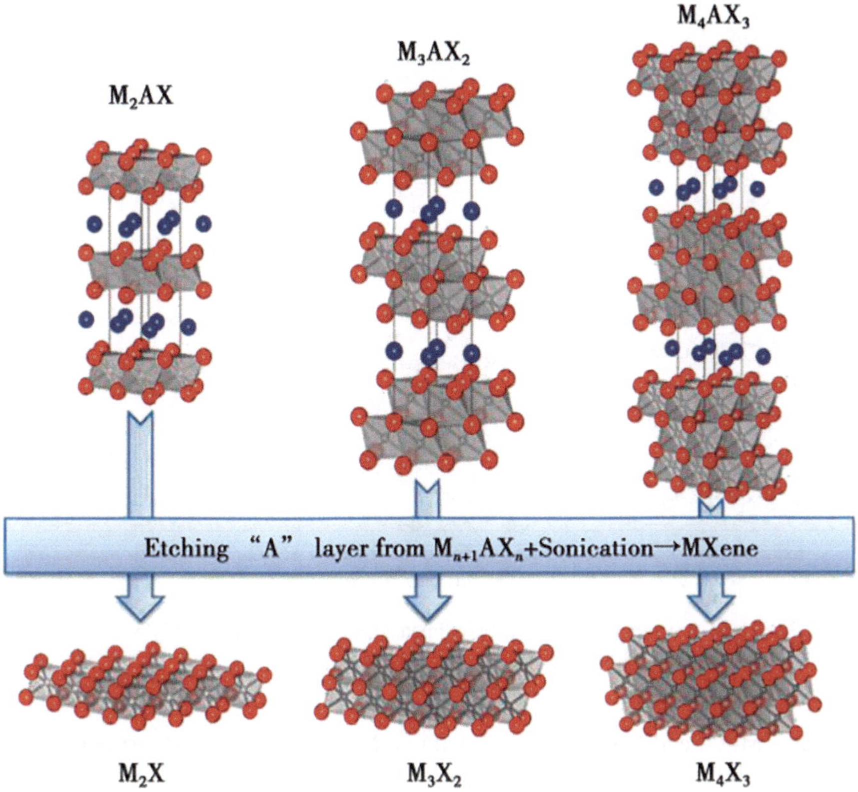

Fig. 7.2 Structure diagram of MAX phase and corresponding MXenes

material can respond intelligently to various types of stimuli. In terms of sensing applications, including electrochemical sensing, piezoelectric sensing, biosensing, etc., MXenes materials have been widely used. Biocompatible devices prepared based on MXenes can achieve fast, convenient, and label-free detection, such as MXenes-based bioenzyme sensors can be used to detect biomolecules, water pollutants, gas molecules, etc. These devices have high sensitivity and extremely low detection concentration limits.

7.1.4 Flaky Metal Oxides (MOXs)

Generally speaking, two-dimensional metal oxides (MOXs) can be divided into layered and non-layered structures. Layered MOXs include traditional layered MOXs, emerging hexagonal MOXs, and two-dimensional layered double metal hydroxides (LDHs). Figure 7.3 shows a schematic diagram of 2DMOXs with different structures. Among them, Fig. 7.3a, b show the traditional layered

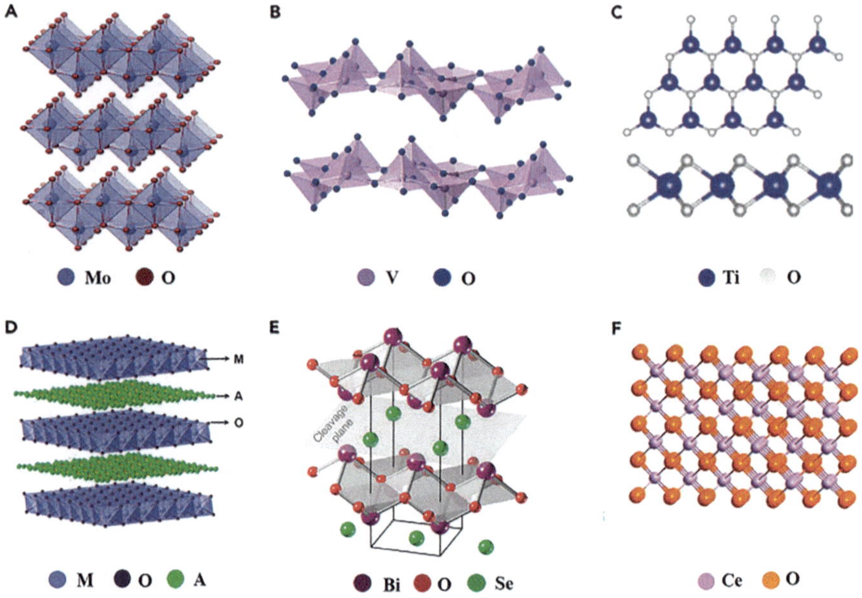

Fig. 7.3 Crystal structures of different two-dimensional metal oxides (**a**) MoO_3; (**b**) V_2O_5; (**c**) h-TiO_2; (**d**) LDHs; (**e**) BiO_2Se; (**f**) CeO_2

MOXs—MoO_3 and V_2O_5 crystal structures, MoO_3 has a layered structure, each layer is mainly composed of twisted MoO_6 octahedra in the orthorhombic crystal. While 2D V_2O_5 usually has an orthorhombic crystal structure, including sawtooth double chains composed of square VO_5 pyramids, bonded together with corner-sharing bridge oxygen atoms, the layered MOXs have strong chemical bonds within the layer and van der Waals forces between the layers. In recent years, a layered crystal structure with unique planar hexagonal coordination has been found in various metal oxides, including (Ti, Mn, Fe, Co, Ni, Cu, Al, Gd, Ge, etc.), which is significantly different from the traditional room temperature non-polar crystal structure. Figure 7.3c shows single-layer and few-layer hexagonal TiO_2 derived from the metal-gas interface, similar to the layered hexagonal TMC structure, each atomic layer is composed of hexagonal rings, which contain oxygen and metal atoms. Figure 7.3d is the typical crystal structure of LDH, where metal cations occupy the center of the octahedron, and the vertices containing hydroxyl ions share ions to form a two-dimensional layer. While in the layered MOX, atoms or ions are bound to the oxide layer by weak electrostatic forces. For example, Bi_2O_2 Se is composed of planar covalently bonded oxide layers (Bi_2O_2) is composed, with a Se square array with relatively weak electrostatic interaction sandwiched in the middle, as shown in Fig. 7.3e. Non-layered MOX is reorganized through chemical bonding in three dimensions, its atoms are connected by strong chemical bonds, thus forming a large number of unsaturated dangling bonds on the surface and edges, which have high energy and high activity. As shown in Fig. 7.3f, CeO_2 is a typical example of non-layered structure MOX, whose structure is formed by face-centered cubic cells of cations and anions occupying octahedral spaces.

Two-dimensional metal oxides have attracted great attention in many fields due to their unique physical and chemical properties, such as high photo-thermal response, temperature superconductivity, photo-luminescence, flexibility, unique catalytic ability, and lower plasmonic regulation ability.

The conductivity of ultra-thin 2D MOX has a very wide range, and the properties from insulation to conduction can be achieved by adjusting its band structure and doping level. Similar to graphene, 2D MOX with atomic thickness has higher optical transparency and mechanical flexibility, which make it widely used in flexible devices and optoelectronic devices.

7.1.5 Metal-Organic Framework Materials (MOF)

Metal-Organic Frameworks (MOF) are a series of spatial network structures composed of metal ions and organic ligands. So far, carbon materials, metals, transition metal oxides, transition metal carbides, transition metal dichalcogenides, transition metal phosphides, and their composites have been successfully prepared from MOF precursors. 2D MOFs, as a new type of two-dimensional material, have the advantages of both two-dimensional materials and porous materials. Compared with their 3D forms, 2D MOFs have excellent physical and chemical properties, such as ultra-thin layer thickness, high specific surface area, high aspect ratio, a large number of exposed unsaturated metal sites, adjustable chemical composition, and recognizable surface atomic structure. These characteristics provide necessary conditions and unique advantages for 2D MOFs in applications such as chemical gas sensing and electrochemical biosensing. 2D MOF nanosheets have a rich surface of unsaturated metal sites, which is conducive to promoting the contact of reactants with active sites, the rapid diffusion of reactants and products, the effective transfer and separation of photo-generated charge carriers, or the transfer of electrolytes and electrons. Non-enzyme sensors constructed from 2D MOFs, mainly based on the high electro-catalytic activity of two-dimensional MOF, can catalyze glucose, hydrogen peroxide, dopamine, ascorbic acid, and uric acid, etc., through oxidation-reduction reactions, to achieve the detection and quantitative analysis of substances related to human health status, which has high sensitivity, excellent selectivity, etc. Because of the poor conductivity of MOF materials, most MOF -based gas sensors are fluorescence, capacitance, impedance type sensing mechanisms.

7.1.6 Others

In addition to the aforementioned two-dimensional materials for sensor applications, other novel two-dimensional materials, such as covalent organic frameworks (COFs), boron nitride (BN), black phosphorus (BP), etc., can also be used in sensing. For instance, combining black phosphorus with surface plasmon resonance (SPR) technology can enhance the sensitivity and stability of sensing. Based on boron nitride nanosheets, researchers have developed various sensors for gas sensing, biosensing, mechanical sensing, etc., which have demonstrated excellent performance. However, for some new two-dimensional materials, exploration in the field of sensing is still limited and further research and development is needed.

7.2 Preparation of Two-Dimensional Sensing Materials

The synthesis route of two-dimensional nanomaterials is an important aspect that determines the final chemical, structural, and morphological characteristics of the active materials. For the field of sensor device applications, the controllability of material properties is an important factor determining device performance and stability. At present, many effective methods have been developed for the preparation of two-dimensional sensing nanomaterials that meet different needs. Generally, synthesis methods can be divided into top-down and bottom-up approaches. The top-down method mainly involves peeling layered bulk materials into single-layer or multi-layer two-dimensional crystal sheets, such as liquid phase exfoliation, mechanical exfoliation, etc. The bottom-up methods, such as chemical vapor deposition, atomic layer deposition, selective extraction, and wet chemical synthesis, are more commonly used for two-dimensional materials that do not have a typical layered structure. In this section, some typical methods for preparing two-dimensional sensing materials are briefly introduced.

7.2.1 Exfoliation Method

The exfoliation method mainly relies on external stimuli to disrupt the weak van der Waals forces between the layers of layered bulk materials, and peel off to obtain single-layer or multi-layer two-dimensional nanomaterials. The exfoliation process can be triggered by different driving forces, such as mechanical force, ultrasound, electrochemistry, or microwaves, etc.

7.2.1.1 Mechanical Exfoliation Method

The mechanical exfoliation method uses mechanical force to peel off layered bulk materials to obtain single-layer or multi-layer nanosheets, while retaining the crystal structure and characteristics. This method was first used by Novoselov and Geim for the exfoliation of graphene, using transparent tape to adhere and peel off from the graphite surface to obtain graphene. In addition, the versatility and low cost of this method make it very popular in the synthesis of two-dimensional materials, and it is very convenient for basic research. Researchers can transfer crystal discs to the substrate surface by rubbing the surface of the layered bulk material with the substrate interface, obtaining single-layer nanodiscs with good crystallinity. Although the nanosheets prepared by the mechanical exfoliation method have advantages such as high conductivity and low defects, the size and thickness uniformity are limited, and the sticky polymer can easily contaminate the surface of the single-layer two-dimensional material, so the feasibility of large-scale production is low. In addition, this method is only applicable to layered van der Waals solids and cannot obtain nanosheets of layered ionic solids and non-layered materials.

7.2.1.2 Ultrasonic Exfoliation Method

Ultrasonic exfoliation is an effective strategy for separating van der Waals solids into single or several layers of nanosheets, as shown in Fig. 7.4a. Compared with mechanical exfoliation, this method is more universally applicable and can achieve high yield and mass production. The ultrasonic exfoliation process generates a sonocavitation effect, which is the growth and rupture of bubbles or voids in the liquid phase due to pressure fluctuations, thereby causing the crystal layer to peel off and obtain nanosheets. The ultrasonic treatment time and appropriate solvents play a key role in the exfoliation process. In a good solvent, the exfoliated nanosheets can be stably dispersed to avoid re-aggregation. In recent reports, different types of organic solvents have been used as dispersion media for two-dimensional nanosheet materials. Using ultrasonic exfoliation, mass production

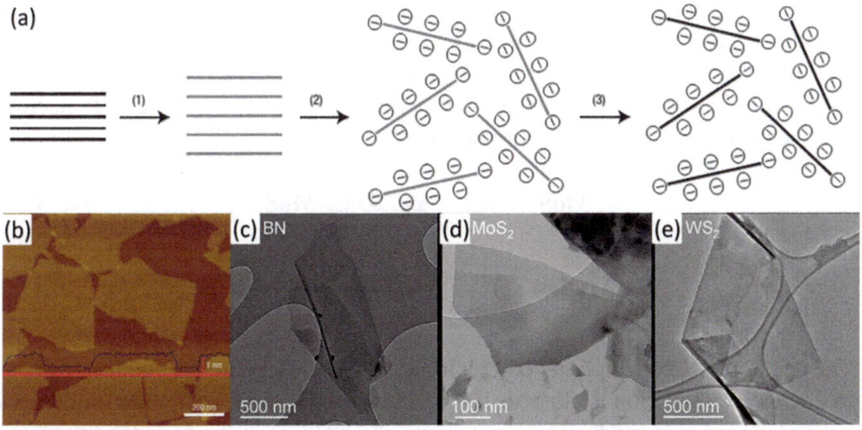

Fig. 7.4 (**a**) Schematic diagram of preparing graphene by ultrasonic exfoliation of vermiculite graphite: (1) Increase the interlayer spacing of oxidized graphite; (2) Obtain a stable GO colloid by ultrasonically exfoliating oxidized graphite in water; (3) The GO colloid is turned into a graphene colloid by hydrazine reduction; (**b**) AFM image of the obtained graphene nanosheets, the blue curve is the height profile taken along the red line (scale bar 1 nm); (**c–e**) TEM images of different nanosheets. (**c**) BN; (**d**) MoS_2; (**e**) WS_2

of various nanosheets can be achieved, such as graphene, BN, MoS_2, and WS_2. For example, by ultrasonically treating expanded graphite oxide in water or organic solvents, oxidized graphene nanosheets can be easily mass-produced. Coleman and others successfully obtained nanosheets of layered compounds by ultrasonic exfoliation in various solvents, such as BN, MoS_2, and WS_2 (Fig. 7.4c, d). The two-dimensional nanosheet suspension after ultrasonic exfoliation can be drop-coated or coated on the sensor substrate, further used to prepare the sensing layer of the device, can be mass-produced and used in many fields such as electronic devices, gas sensing, energy, and catalysis. However, this method has the disadvantage of being difficult to control the lateral diameter, thickness, and atomic layer number of the nanosheets, and may cause defects and fractures in the nanosheets during the exfoliation process.

7.2.1.3 Electrochemical Exfoliation

For 2D materials with charged main body layers, such as most layered metal oxides (excluding V_2O_5) and metal hydroxides, it is difficult to exfoliate by mechanical or ultrasonic means. Therefore, electrochemical exfoliation is more effective, that is, by infiltrating and expanding to carry out chemical reactions to weaken or destroy the strong interlayer interaction force. Electrochemical exfoliation is also widely used to exfoliate layered structure materials to obtain nanosheets, such as graphene, MoS_2, and WS_2, TiS_2, TaS_2, and ZrS_2. Equations (7.1) and (7.2) are the reaction process of lithium (Li) intercalation process to exfoliate MoS_2. This method uses Li^+ ions intercalation to expand MoS_2 interlayer distance, to disrupt its interlayer forces and peel off the nanosheet layers. In the test battery, the bulk material with a layered structure serves as the cathode, lithium as the anode, and electricity is maintained during the treatment process. After the lithium insertion is completed, the treated compound is washed with acetone and then ultrasonicated to obtain nanosheets.

$$\mathbf{MoS^2 + xLi^+ + xe^- \rightarrow Li_xMoS_2} \tag{7.1}$$

$$\mathbf{Li_xMoS_2 + xH_2 \rightarrow MoS_2 + \frac{x}{2}H_2 + xLiOH} \tag{7.2}$$

The reaction yield of single-layer transition metal dichalcogenides obtained by electrochemical exfoliation is close to 100%. However, there are still many challenges, such as the reaction needs to be carried out at high temperatures for a long time, and the lithium insertion needs to be finely controlled to obtain single-layer nanosheets, while also preventing the formation of metal nanoparticles and the precipitation of Li_2S. In addition, nanosheets of BN and metal selenides or tellurides have also been obtained by this method.

7.2.1.4 Ion Exchange Exfoliation

The above methods can peel off van der Waals solids to obtain ultra-thin nanomaterials, but there are certain difficulties when used to peel off layered ionic solids, such as $LiCoO_2$ or LDHs, etc. Because the interlayer of ionic solids has strong ionic bonds, ion exchange exfoliation is a common method to obtain their two-dimensional materials. For example, for the layered metal oxide $LiCoO_2$, a proton exchange reaction is carried out in HCl solution, then a tetrabutylammonium intercalation reaction is performed, and then ultrasonication is used to further improve the exfoliation effect.

7.2.2 Chemical Vapor Deposition (CVD)

CVD is a high-temperature chemical synthesis process that deposits the required materials on a substrate. It has been used to grow various two-dimensional nanomaterials, such as graphene, metal dichalcogenide co-compounds (MoS_2, etc.), boron nitride (BN), etc. Compared with mechanical or ultrasonic exfoliation, CVD technology can synthesize nanosheets with scalable size, controllable thickness, and excellent electronic properties. The main advantage of the CVD synthesis process is that it can obtain high-quality, high-purity two-dimensional nanomaterials with adjustable performance. This allows people to control the morphology, number of layers, crystallinity, and defects of two-dimensional nanostructures by adjusting the synthesis process parameters. Polycrystalline and single-crystal graphene nanosheets can be synthesized by CVD. Moreover, the number of layers of two-dimensional materials can also be controlled by adjusting CVD parameters. As shown in Fig. 7.5, the surface of the SiO_2 substrate is treated with an oxygen plasma process, achieving precise control of the number of layers of large-area MoS_2 thin films. Two-dimensional nanomaterials grown by CVD, such as MoS_2 and graphene, have shown great potential for use in gas sensors. In addition, CVD can directly grow high-quality 2D nanostructures on the sensor substrate to simplify the device manufacturing process.

7.2.3 Wet Chemical Synthesis

Wet chemical methods have the advantages of high yield, low cost, and large-scale production, and can be used to prepare various types of two-dimensional materials. Wet chemical strategies include various methods, such as water/solvent thermal synthesis, template synthesis, solvent evaporation, and self-assembly methods. Among them, water/solvent thermal method is a common method for synthesizing inorganic materials. Compared with other crystal growth methods, water/solvent

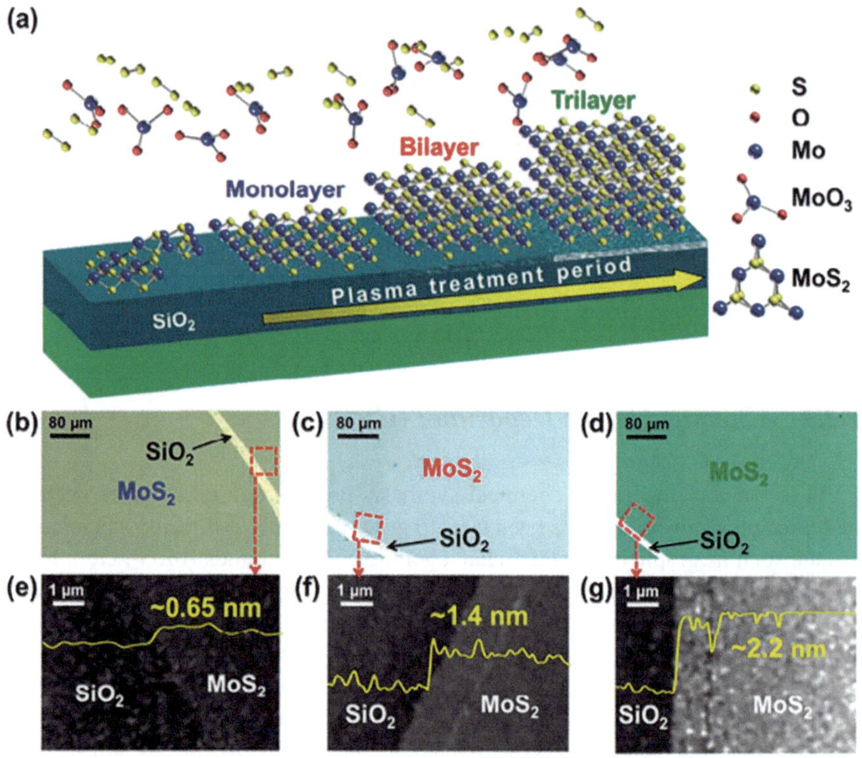

Fig. 7.5 (**a**) Schematic diagram of layer-controlled CVD growth of large-area MoS_2 thin films; (**b**) Optical microscope images of single-layer, (**c**) double-layer, and (**d**) triple-layer MoS_2 thin films; (**e**) Single-layer, (**f**) double-layer, and (**g**) triple-layer MoS_2 Images of the corresponding atomic force microscope (AFM) for the marked area of the film

thermal method has the advantages of low reaction temperature and convenient adjustment of reaction conditions. There are many factors affecting the synthesis of ultra-thin nanostructures, including temperature, reaction time, and reactant ratio. This strategy can be used to prepare various compounds, such as MoS_2, TiO_2, ZnO, Co_3O_4, MnO_2, etc. For example, ultra-thin Rh nanoplates can be synthesized by solvent thermal method. The main steps are to first dissolve $Rh(acac)_3$ and PVP in a solvent mixed with benzyl alcohol and formaldehyde, stir vigorously for 1 h, then transfer to a reaction kettle lined with polytetrafluoroethylene, maintain at 180 °C for 8 h, and then cool to room temperature. The resulting black product is precipitated with acetone (10 mL), separated in a centrifuge, and washed with ethanol (10 mL) three times, and finally vacuum dried.

Wet chemical synthesis can be used to prepare two-dimensional non-layered metal oxides. As shown in Fig. 7.6a, a general solvent (ethanol/water/ethylene glycol) thermal synthesis method is used to prepare transition metal oxide nanoplates. Using surfactants to assemble reverse layered micelles, hydrate inorganic oligomers

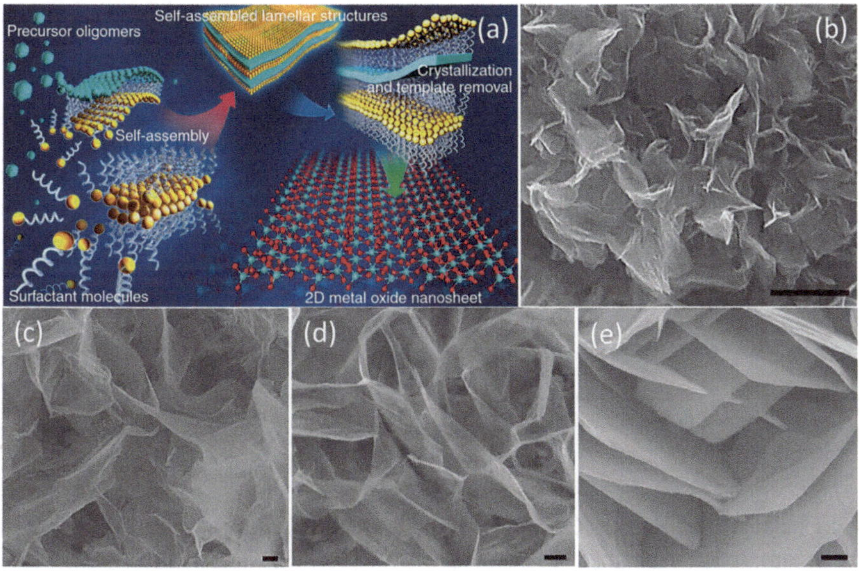

Fig. 7.6 (**a**) Schematic diagram of the self-assembly of two-dimensional metal oxide nanoplates; (**b–e**) SEM images of different nanoplates. (**b**) TiO_2; (**c**) ZnO; (**d**) Co_3O_4; (**e**) WO_3

are confined within the reverse layered micelles, and then promote the formation of layered inorganic oligomer aggregates. Condensation and crystallization will occur during the solvent thermal synthesis process, and after removing the surfactant template, well-crystallized ultra-thin two-dimensional transition metal oxide nanoplates can be obtained.

The advantage of the surfactant-guided solvent thermal synthesis method is that it is not limited to layered host materials with a limited number of layers, and it is universal for the synthesis of metal oxide nanoplates. It provides a flexible and universal method for synthesizing various two-dimensional metal oxides (Fig. 7.6b–e), such as TiO_2, ZnO, Co_3O_4 and WO_3, and can achieve large-scale production. Studies have shown that many types of two-dimensional nanostructures can be prepared by simple, low-cost wet chemical methods, such as SnO_2, ZnO, CuO, NiO, In_2O_3, etc. Compared with the above-mentioned exfoliation method, CVD method, the wet chemical synthesis method has the advantages of lower cost and better size controllability.

7.2.4 Atomic Layer Deposition (ALD)

Atomic Layer Deposition (ALD) allows for precise control over the thickness of ultra-thin metal oxides through self-limiting growth cycles achieved by alternating exposure to metal and oxygen precursors. Since the precursor reaction chemistry has been established, the initial nucleation step (which requires reaction with the substrate) is key to obtaining uniform and homogeneous thin films. The precise control over thickness allows ALD to manufacture sensing layers with lengths close to the semiconductor Debye length, maximizing sensor sensitivity. Two-dimensional metal disulfides, such as WS_2 and $MoSe_2$, can be synthesized using atomic layer deposition. For example, Kim and others have achieved the synthesis of large-area WS_2 nanosheets with precisely controlled layers of 1, 2, and 4, and the size of the nanosheets exceeds 4 inches. In addition, ALD is also an effective strategy for producing hybrid two-dimensional nanomaterials through graphene surface functionalization.

7.2.5 Selective Extraction Method

The selective extraction method is mainly used for the preparation of novel two-dimensional materials MXenes. The structure and characteristics of MXene have been introduced in Sect. 7.1.3. The preparation of MXene usually involves selective etching of layered bulk metal carbides, nitrides, or carbonitrides 3D MAX phases. The earliest reported method for etching to prepare MXene was using hydrofluoric acid. Due to the strong corrosiveness, biological toxicity, and environmental hazards of hydrofluic acid, researchers have developed new fluoride-free systems, using HCl, NaOH, or Lewis-acidic molten salts for selective etching to obtain MXenes. Because selective etching uses strong acids or bases, and the M layer itself is highly active, various functional groups are introduced during the preparation process. Therefore, the control of MXenes' physical and chemical properties can be achieved through the regulation of reactions during the preparation process to meet various needs in different fields.

7.2.6 Other Methods

In addition to the above methods for synthesizing two-dimensional materials, there are other methods, such as microwave method, electron beam evaporation method, localized chemical conversion method, etc., which also have their specific advantages in the preparation of some two-dimensional materials. For example, the microwave method is used to thermally reduce graphene oxide (GO) to produce micrographene nanosheets, the whole process only takes a few seconds. The microwave exfoliation method is used to produce only a few layers of N-doped graphene nanosheets. It has been proposed that under the presence of sodium amide, the high-energy state caused by dielectric heating under microwave radiation can directly convert graphite into N-doped graphene. Compared with the other exfoliation methods mentioned above, microwave exfoliation has the advantages of being simple and fast.

7.3 Two-Dimensional Material-Based Smart Sensor Devices

7.3.1 Mechanical Sensing

The mechanical-electrical conversion performance of two-dimensional materials is related to their geometric dimensions, as their sensing mechanisms vary at different scales. At the microscopic scale, when two-dimensional materials are subjected to stress or strain, their lattice, interlayer spacing, and electronic band structure will change. This change in their crystallographic behavior allows for the adjustment of electrical and optical properties, thereby responding to mechanical information. At the macroscopic scale, in addition to the influence of microscopic crystallographic behavior, the mechanical-electrical coupling performance of the material will also be affected by many factors such as macrostructure, chemical modification, interlayer sliding interface, substrate effect, size of the two-dimensional atomic layer, substrate morphology, etc. Therefore, based on various emerging two-dimensional materials, different sensing mechanisms, and emerging advanced preparation methods, the basic mechanics sensing technology of two-dimensional materials has developed rapidly. Mechanical sensors composed of two-dimensional materials have different sensing mechanisms, including piezoresistive, capacitive, piezoelectric, and frictional.

7.3.1.1 Piezoresistive Sensing

Most two-dimensional nanomaterials, such as graphene, MoS_2, WSe_2, $MoSe_2$, etc., have piezoresistive properties, which are conducive to the design application of strain or pressure sensing. The sensing mechanism of piezoresistive sensing varies for different structural designs. Currently, structural designs mainly include in-plane lateral designs based on nanoplates, stacking of nanoplates, and composites of heterogeneous materials. In-plane lateral design does not involve interplanar interactions and signal transmission. The planes are strongly bonded, and when the nanoplate is subjected to external stress, the plane will undergo strain or shear, leading to resistance changes. In Sect. 7.2, we introduced various preparation methods for two-dimensional nanomaterials. It is known that structural defects will inevitably be introduced during the preparation of materials, which will affect their physical and chemical properties. Due to the uncontrollability of defects, even for the same two-dimensional material, its piezoresistive properties are not exactly the same. The main reason is the changes in physical and chemical properties caused by the presence of defects or grain boundaries, so it is difficult to achieve consistency in sensing performance based on in-plane lateral sensing design of two-dimensional materials.

The stacking design of two-dimensional materials, that is, vertically stacking the crystal layers of two-dimensional materials, is mostly weakly bonded by van der Waals forces between the crystal layers. Therefore, for stacked sensor devices, there are not only in-plane strain changes, but also interlayer strain changes. For in-plane loads, the resistance changes due to the strain or shear of the crystal layer; for interplanar loads, the resistance changes due to changes in the interlayer spacing of the crystal layer. The piezoresistive properties of the vertical stacking structure also depend on the mechanical properties of each layer and the van der Waals force between the layers. When two-dimensional materials are subjected to strain levels greater than 1%, interfacial sliding between 2D layers may occur instead of lattice deformation. Therefore, by controlling the conductivity through sliding between adjacent crystal layers, highly sensitive mechanical sensor devices can be prepared. In addition, when two-dimensional materials contain isolated islands or penetrating cracks, their piezoresistive properties are mainly determined by the tunneling effect, thus having extremely high sensitivity. Generally speaking, the above sensing mechanisms do not exist alone, but are triggered by different stimuli to work together.

Introducing heterogeneous composite materials formed by other functional components into the preparation of two-dimensional material sensors, or utilizing the synergistic effects of different functional materials, can also construct more intelligent materials with stimulus responses. Currently, researchers have achieved the combination of 2D materials with materials of different forms to form heterogeneous structures, which are used to construct mechanical sensor devices, including binary material systems and ternary material systems, etc. As shown in Fig. 7.7, inspired by the layered structure of pearl shell, researchers constructed a Ti_3C_{2x}-AgNW-PDA/Ni^{2+} strain sensor, which has a "brick-mortar" architecture. By utilizing the functional synergy of different materials, 2D MXene/1D AgNW acts as the "brick", PDA acts as the "mortar", and their combination can control the generation of cracks, thereby improving its sensitivity. In addition, by incorporating conductive metal two-dimensional materials into a system with a flexible material framework, conductive flexible materials can be obtained. Compared with one-dimensional conductive fillers, two-dimensional fillers have a larger contact area, thus having higher strain and pressure sensitivity. Sensors based on piezoresistive effects usually have the advantages of easy manufacturing and low cost, and are widely used in cost-effective wearable electronic products.

7.3.1.2 Capacitive Principle Sensing

The huge specific surface area, rich porous structure, and flexible surface chemical modification of two-dimensional materials make them very suitable as ion sensing electrodes. In the construction of capacitive sensing, two-dimensional materials can be integrated into sensor devices as electrodes or dielectric layers. By utilizing the supercapacitor characteristics of the electric double layer on the elastic medium/electrode interface, the interface supercapacitor sensor devices based on two-dimensional materials can detect tiny mechanical deformations and have excellent performance, including ultra-high unit area capacitance, ultra-high sensitivity, excellent anti-interference ability, and the light and flexible structure of the device. For example, using MXene ($Ti_3C_2T_X$) to construct capacitive electrodes, when subjected to external pressure, the accessible surface area of MXene increases and the ion transport path shortens, the sensor shows ultra-high sensitivity and ultra-wide sensing range, which are 46,730 kPa^{-1} and 1.4 MPa respectively. Capacitive sensing

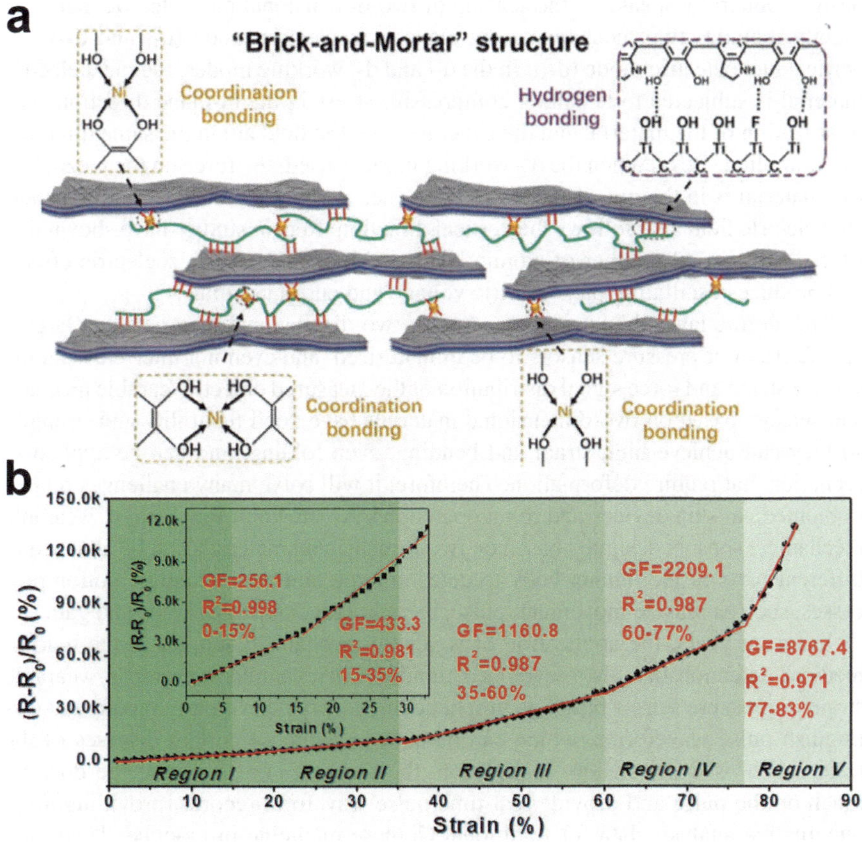

Fig. 7.7 Heterogeneous composite structure of $Ti_3C_2T_x$-AgNW-PDA/Ni^{2+} sensor based on two-dimensional crystals, and the corresponding electrical performance response under different strains

based on two-dimensional materials shows good application prospects in the fields of intelligent robots, electronic skin, wearable health monitoring, etc.

7.3.1.3 Piezoelectric and Triboelectric Principle Sensing

Different from the piezoresistive and capacitive principles, piezoelectric and triboelectric sensing are self-powered sensors that do not require an external energy source. Using materials with piezoelectric effects, when the material deforms under external force in a certain direction, positive and negative charges will appear on its two opposite surfaces. Theoretical and experimental studies have shown that MoS_2, $MoSe_2$, WS_2, WSe_2 and a series of two-dimensional materials are piezoelectric or ferroelectric materials. Due to the high strength, flexibility, and ultra-thin characteristics of two-dimensional nanomaterials, they have great advantages in sensor design. Generally speaking, the sensing of two-dimensional piezoelectric materials mainly includes two mechanisms: in-plane piezoelectric mode (d_{11} and d_{22}) and vertical piezoelectric mode (d_{33}). In the d_{11} and d_{22} working modes, the piezoelectric material is subjected to tensile or compressive stress in the in-plane direction. The polarization of the material and the generated electric field are in the same direction as the applied stress. When the d_{33} working mode is used, the force on the piezoelectric material is in the out-of-plane direction (i.e., vertical direction), and the generated electric field also follows the vertical direction. Recent studies have shown that MoS_2 with an odd number of atomic layers exhibits in-plane piezoelectric effects and produces oscillating piezoelectric voltage and current outputs.

The atomic layer thickness advantage of two-dimensional materials can largely enable strain or pressure sensors to be miniaturized, and even not interfere with the surface shape and force signal distribution of the measured object. Wearable mechanical sensors based on two-dimensional materials have good flexibility and strength, so they can achieve high strain and bending, even folding, and can be applied to scenarios that require deformation. Therefore, it will solve many challenges related to biomedical skin devices and robot operations. As shown in Fig. 7.8a, b, wearable mechanical sensors designed based on two-dimensional materials can be attached to different parts of the human body to detect a large number of human motion processes, such as muscle movement, pulse, breathing, facial expressions, and gait.

In recent years, the application of wearable mechanical sensors in the field of medical detection has also developed rapidly. For example, wearing a wearable strain or pressure sensor based on graphene on the wrist can clearly record and distinguish pulse waveforms, which can be used to diagnose certain diseases of the tester, such as hypertension. In addition, the sensor can also mimic the doctor's touch on the pulse and provide real-time pulse waveform records, providing more quantitative analysis data for traditional Chinese medicine in practice. Therefore, wearable strain/pressure sensors based on two-dimensional materials have broad application potential in home portable diagnostic and therapeutic devices.

7.3.2 *Biosensing*

Two-dimensional biosensors are typically devices that use two-dimensional materials as signal conversion elements to transmit biological signals. Thanks to the excellent characteristics of two-dimensional materials, compared to traditional biosensors, sensors made from two-dimensional materials have superior performance in terms of electron transfer efficiency, specific surface area, and doping sensitivity. With the introduction of emerging two-dimensional materials and the continuous development of sensor construction methods, the sensing performance of two-dimensional material-based sensors has been further improved, and the detectable biomolecules have been further expanded, which has now surpassed traditional biosensors. The excellent optical, electrical, and electrochemical properties of two-dimensional materials make them suitable as platforms or for developing sensitive electrochemical and optical probe biosensors to detect environmental pollutants and bioactive molecules.

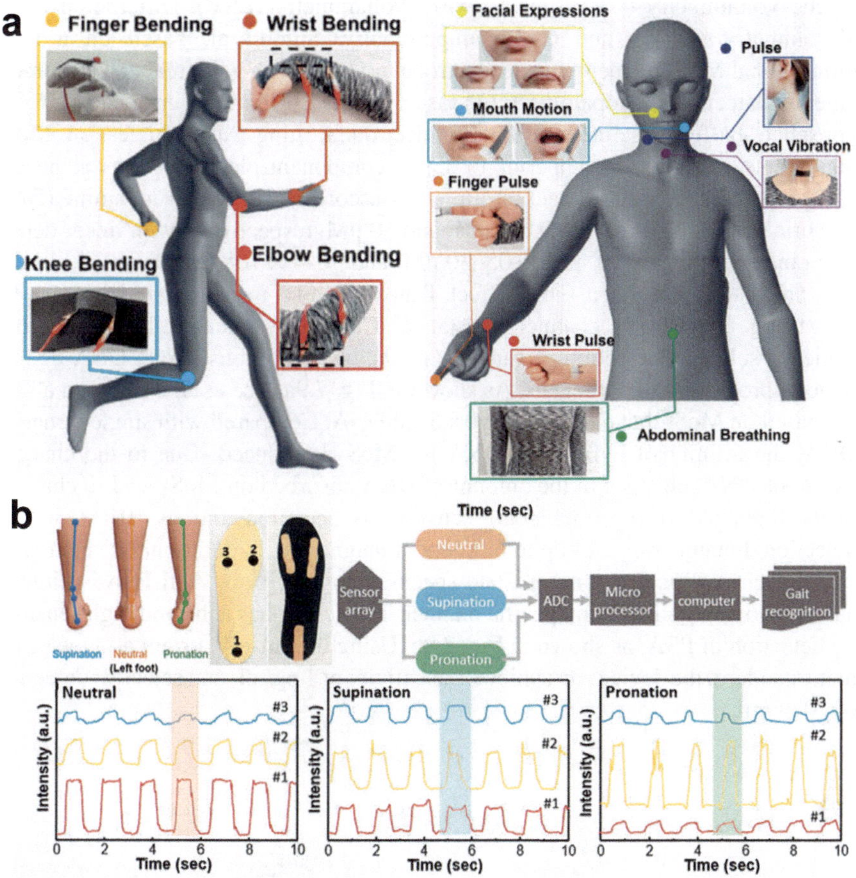

Fig. 7.8 Wearable applications and intelligent sensors assisted by machine learning. (a) Graphene textile strain sensor for human motion detection; (b) Graphene pressure sensor for various gait detections

7.3.2.1 Electrochemical Biosensors

In biosensing applications, electrochemical sensors are usually the first choice and are the most promising analytical devices. They mainly sense by detecting changes in current (Ampere method, Volt-Ampere method) or interfacial impedance (electrochemical impedance spectroscopy). They rely on biological recognition elements and have high selectivity for target analytes. In the design of sensors, two-dimensional materials can serve as both sensing elements and substrates, with metals and their oxides, carbon nanomaterials, and enzymes modified on their surfaces. The basic principle of biosensors is that the recognition action between fixed biomolecules and target analytes causes changes in the electrical characteristics of sensing materials or solutions, such as current, conductivity, potential, and ion strength. Current (or Volt-Ampere) detection includes various techniques such as

Cyclic Voltammetry (CV), Linear Sweep Voltammetry (LSV), Differential Pulse Voltammetry (DPV), and Chronoamperometry. Sun et al. based on a two-dimensional MoS_2 nanocomposite electrode modified with gold, realized the simultaneous detection of dopamine (DA), ascorbic acid (AA), and uric acid (UA) in phosphate-buffered saline (PBS). By electrodepositing Au particles on MoS_2 nanosheets, performance superior to single-component electrodes was achieved. The prepared sensor has detection limits for ascorbic acid (AA), dopamine (DA), and uric acid (UA) of 50 μM, 0.05 μM, and 50 μM, respectively, with linear detection ranges of 50–1 × 10^5 μM, 0.05–30 μM, and 50–4 × 10^4 μM. Potential biosensors are mainly based on Field Effect Transistor (FET) sensors, which sense by measuring changes in channel resistance. Currently, potential biosensing has achieved selective detection of various targets, including ions, target DNA double helices, proteins, and pathogens. As shown in Fig. 7.9a, Lee et al. developed a sensor based on MoS_2 that can detect hybridized DNA. Compared with single-stranded DNA, the affinity of hybridized DNA for MoS_2 is reduced. Due to the charged nature of DNA, changes in the amount of DNA adsorbed on MoS_2 lead to changes in the threshold voltage, achieving sensing. Its detection limit is 10 fM, with a detection dynamic range of up to 6 orders of magnitude. In addition, Lee et al. also studied sensors for detecting prostate-specific antigen (PSA). Anti-PSA is directly adsorbed on MoS_2 in a non-specific manner. The sensor has achieved high sensitivity detection of PSA, as shown in Fig. 7.9b. Using the cut-off current as a measurement standard, the device can achieve a sensitivity of 1 pg/mL and selective detection of 10 ng/mL.

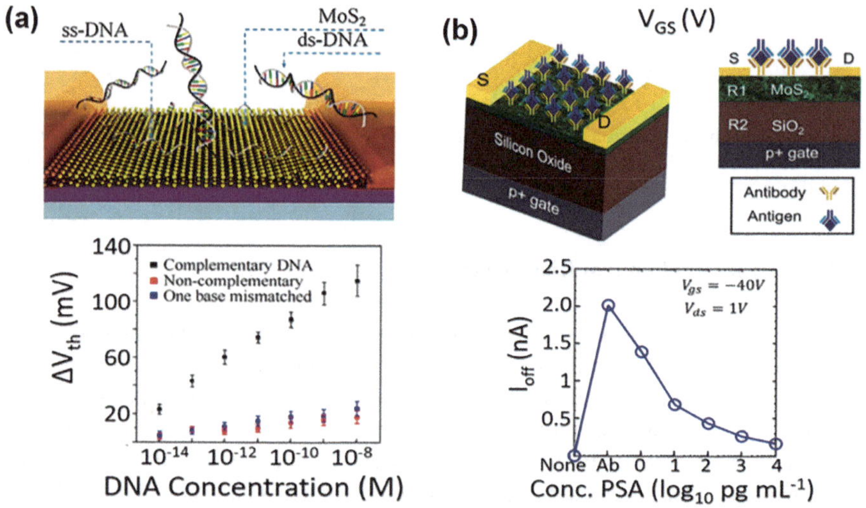

Fig. 7.9 (**a**) Top: MoS_2 based sensor for detecting hybrid DNA; Bottom: The relationship between the concentration of complementary, non-complementary, and hybrid DNA and the responsiveness of the device; (**b**) Top: MoS_2 based sensor for detecting PSA; Bottom: The relationship between the cut-off current and the concentration of PSA

7.3.2.2 Optical Sensors

Compared with traditional electronic sensors, optical biosensors have characteristics such as no electromagnetic interference, corrosion resistance, and high sensitivity, and their sensitivity can even detect single molecules. Common optical sensing technologies include Fluorescence Resonance Energy Transfer (FRET), Surface Plasmon Resonance (SPR), and Surface Enhanced Raman Spectroscopy (SERS). Sensors based on fluorescence quenching or enhancement, such as FRET, can detect changes in fluorescence to sense target analytes. FRET is a non-radiative energy transfer mechanism between two fluorophores through dipole-dipole coupling, and its performance mainly depends on three factors: the fluorescence donor, the fluorescence acceptor, and the distance between the donor and the acceptor. FRET is extremely sensitive to the distance between the fluorophores. To produce fluorescence, the excitation band of the acceptor must overlap with the emission band of the donor. At the same time, the distance between the excitation band and the emission band of the donor must be within the typical range of the nanoscale. Kong et al., Deng et al., and Huang et al. used MoS_2 quenchers to sensitively detect prostate-specific antigen, DNA methyltransferase activity, and DNA detection through hybrid chain reaction (HCR). By designing a substrate DNA that has affinity for MoS_2, the detection of DNA methyltransferase (MTase) activity was achieved. It consists of ssDNA and dsDNA, where ssDNA allows substrate DNA to be adsorbed on MoS_2 nanoplates through van der Waals interactions. Therefore, the fluorophore attached to the end of dsDNA will undergo fluorescence quenching. Kong et al. constructed a surface-adsorbed fluorescently modified single-strand (ss) DNA probe, which can detect prostate-specific antigen (PSA), as shown in Fig. 7.10a. When there is no PSA, the fluorescence is quenched, and when PSA is added, the fluorescence recovers. Since a PSA value greater than or equal to 4 ng/ml can be considered to indicate prostate cancer, this technology has practical application significance in the field of early disease diagnosis.

The surface plasmon resonance SPR sensor generates surface plasmon excitons that are particularly sensitive to changes in local refractive index. Sensing can be achieved by detecting changes in local refractive index caused by the adsorption of target analytes. This technology has now been commercialized. By integrating two-dimensional materials into SPR sensors, the sensitivity of SPR sensors can be improved. Zhou et al. developed a sensor for detecting E. coli using a fiber optic probe coated with rGO/AgNP/AuNP, as shown in Fig. 7.10b. The selectivity is improved by depositing the antimicrobial peptide Magainin I-C on the surface. The linear range of the sensor is from 10^3 to 5×10^7 CFU/mL, and the detection limit is 5×10^2 CFU/mL. This SPR detection strategy can be completed without any labels or secondary antibodies, making the detection system faster, lower cost, and more stable. The authors tested this by adding the test substance to tap water, apple juice, orange juice, and mixed fruit and vegetable juice, and the results showed that the recovery rate was between 88% and 110%, proving the potential application of this sensor in real life.

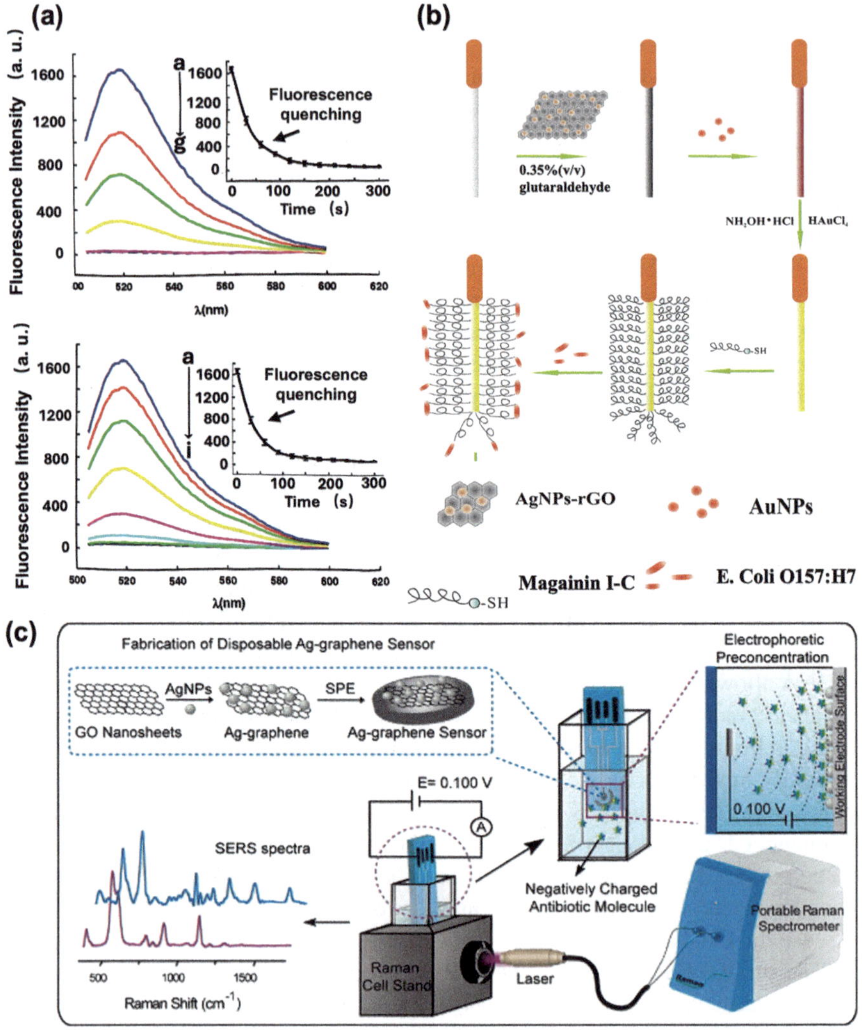

Fig. 7.10 (**a**) For a given concentration of PSA aptamer (PA) (50 nM), adding more MoS_2 nanosheets (0–40 μg/ml) will quench the fluorescence intensity. After increasing the PSA concentration (shown here 0–300 ng/ml), the fluorescence will recover; (**b**) Schematic diagram of the preparation of rGO/AgNP/AuNP sensor; (**c**) Schematic diagram of the heterojunction structure of WS_2 nanosheets and AuNP phase for SERS

SERS technology can detect analytes adsorbed on the surface of Raman enhancement substrate materials, and enhance the Raman scattering signal through the action of surface plasmons. As shown in Fig. 7.10c, Li and others developed a SERS sensor for detecting polar antibiotics by depositing graphene/silver composite material on a screen-printed electrode. Four different antibiotics (methotrexate (MT), ampicillin trihydrate (AT), 6-aminopenicillanic acid (6-AA) and penicillin G, Na

(PG)) were first ionized, where MT and 6-AA were positively ionized, while AT and PG were negatively ionized. The sensor was used to test river water samples with 100 nM antibiotics added, and the results showed that Raman peaks corresponding to the four different antibiotics could be distinguished.

7.3.3 Gas Sensing

2D materials exhibit excellent sensor-related characteristics, including adjustable high specific surface area, abundant surface active sites and functional groups, atomic-level thickness, and excellent electronic properties. Two-dimensional materials provide a large number of active sites and sufficient interaction area for adsorbing gas molecules, so effective detection can be achieved even when the concentration of the target analyte is extremely low, indicating its extremely high sensitivity and lower detection limit. In addition, abundant surface functional groups can achieve specific recognition of the target analyte, which will further improve its sensitivity and give the sensor selectivity. The main limiting factors for two-dimensional materials used for chemical gas sensing include: (1) they lack environmental stability and selectivity; (2) lack of effective methods for large-scale production for commercial sensor manufacturing. Since the stability of two-dimensional materials is difficult to guarantee, the surface area of single-layer two-dimensional materials will be significantly reduced, thereby reducing its active sites and the adsorption capacity of gas molecules. Directly preparing nanosheets vertically on the substrate using vapor deposition is one of the effective strategies that can solve this problem to some extent.

Due to the high sensitivity and ease of synthesis of metal oxides and graphene as gas sensors, they are the most common two-dimensional materials used for gas sensing. In recent years, metal disulfides, phosphorene, boron nitride (BN) and MXene have also been proven to have significant application potential in the field of gas sensing. However, their sensing mechanisms are not the same. For metal oxides, the sensing mechanism of their gas sensors is mainly due to the change in charge carriers caused by the oxidation or reduction reaction occurring on the surface of the metal oxide. Taking n-type SnO_2 as an example, the brief process of its sensing mechanism is: oxygen molecules are adsorbed on the surface of the semiconductor material, and chemically adsorbed to form O^{-2}, O^- and O^{2-} Surface matcrials, these materials extract electron density from the surface of SnO_2. When the sensor is exposed to a reducing gas (such as ethanol), a reaction occurs between the adsorbed oxygen species and the reducing molecules, resulting in changes in the sensor's conductivity or resistance. However, they still have shortcomings such as short lifespan, poor selectivity, high operating temperature, and their sensitivity is not sufficient to meet the needs of precise measurements. In addition, many other 2D metal oxides have also been used for chemical gas sensors, such as ZnO, SnO_2, CuO, NiO, In_2O_3, WO_3, and CO_3O_4, etc., showing good sensing characteristics for various analytes.

Unlike metal oxides, graphene-based gas sensing is based on a direct charge transfer mechanism, originating from the physical adsorption of gas molecules on the material. Theoretical and experimental results show that graphene and its derivatives (GO, rGO) have a large specific surface area, excellent conductivity, easy adsorption of gas molecules, and therefore have good gas sensitivity. Graphene has excellent electron mobility and a large specific surface area, occupying an important position in chemical gas sensors. Graphene materials as p-type semiconductors contain many holes and have an electron-pulling effect in the gas atmosphere. After the gas molecules are adsorbed by graphene, the gas molecules undergo weak hybridization, coupling with the surface electrons, and the Fermi level moves up and down slightly. Adsorbed gas molecules will change its Fermi level, thereby causing changes in graphene conductivity to achieve sensing. Therefore, graphene is particularly sensitive to the detection of small molecule gases adsorbed, and can monitor the surrounding chemical environment with ultra-high sensitivity.

Reduced graphene oxide (rGO) has some surface oxygen groups, such as epoxy, hydroxyl and carbonyl groups, which are believed to enhance the adsorption ability of gas molecules on its surface. At the same time, through the action of these functional groups, rGO can be combined with metal oxide nanoparticles to construct composite materials with functional synergy. Researchers have obtained a variety of new materials for gas sensing by combining graphene-derived materials with various metals and metal oxide nanoparticles. Studies have shown that compositing nanoparticles with graphene, the introduced defects and functional groups can effectively enhance the material's adsorption capacity for gases. The resulting composite material retains the high conductivity characteristics of graphene, and also has the high sensitivity of metal nanoparticles. In addition, the Schottky junction formed between graphene and metal oxides also enhances gas sensing performance, and its Schottky barrier height is strongly affected by the adsorbed chemicals. Based on this feature, Xia and others constructed a high-sensitivity gas sensor for detecting NO_2 based on the rGO/ZnO Schottky junction, as shown in Fig. 7.11a. This sensor is constructed by loading ZnO nanorods on the surface of rGO nanosheets, and a Schottky junction will form at the interface between rGO and ZnO, where the zero-gap rGO acts as a metal, and a Schottky barrier height and depletion layer will form at the interface. Compared with pure rGO, this composite material with NO_2 Having stronger interactions, therefore, it has a stronger adsorption detection response, as shown in Fig. 7.11b, c.

7.3.4 Photo-detector

A photo-detector is a sensor that converts photons into measurable electrical signals to detect light signals. Due to the unique electronic and optoelectronic properties of two-dimensional materials, they have significant advantages in the design and application of photo-detectors. After years of development, two-dimensional materials have shown excellent performance in this field, including covering an ultra-wide band from visible light to terahertz frequency (THz, 10^{12} Hz), ultra-high light responsiveness, polarization-sensitive light detection, high-speed light response, high spatial resolution

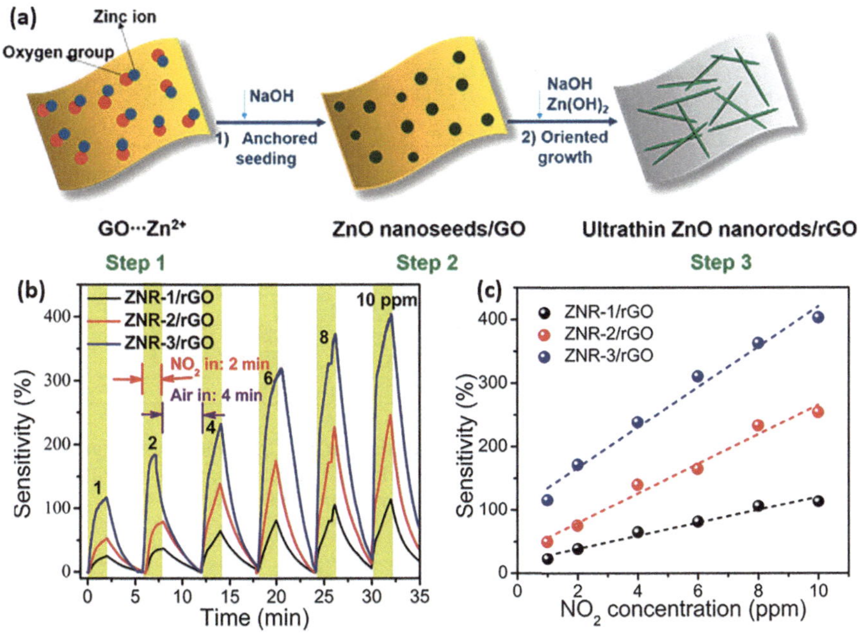

Fig. 7.11 (**a**) Schematic diagram of the preparation of ultra-thin zinc oxide nanorods/reduced graphene oxide nanocomposite; (**b**) Dynamic response curves and (**c**) sensitivity of sensors based on three types of ZnO nanorods/rGO nanocomposites to 1.0–10.0 ppm NO_2 at room temperature

imaging, etc. The crystal layer of two-dimensional materials is atomically thin, but it has strong interactions with light, including graphene, transition metal dichalcogenides (TMD) and black phosphorus (BP), etc. For example, the absorption of visible light by single-layer graphene (thickness about 0.335 nm) is 2.3%, and the maximum absorption of MoS_2 (thickness about 0.615 nm) in the visible light region exceeds 20%. However, to design a photo-detector, it is necessary to further increase its light absorption rate. The common method is to construct a heterostructure. By combining two-dimensional materials with other materials, such as flexible materials, silicon and its compounds, to form a heterostructure, performance synergy is achieved, thereby improving its light absorption performance. For example, the Graphene-WS_2-Graphene sandwich heterostructure constructed using graphene and WS_2 has a high photovoltaic external quantum efficiency of up to 30%, the ultra-high light response of Graphene-MoS_2 is up to about 10^{10} A W^{-1}, the ultra-high light gain of Graphene-MoS_2 is up to about 10^8, with ultra-high sensitivity D* of about 10^{15} Jones, the heterostructure WSe_2 -Graphene-MoS_2 has a low dark current, and the detection wavelength is extended to 2.4 μm. In addition, two-dimensional materials can achieve a wide response range of spectra, as shown in Fig. 7.12a, b, the bandgap ranges from graphene's zero bandgap, TMDs medium bandgap b-P narrow bandgap to insulating h-BN wide bandgap, its spectrum can respond to the range from visible light to terahertz.

High-sensitivity photo-detectors based on two-dimensional materials are expected to be used in human electronic eyes (Human E-Eye), high-quality

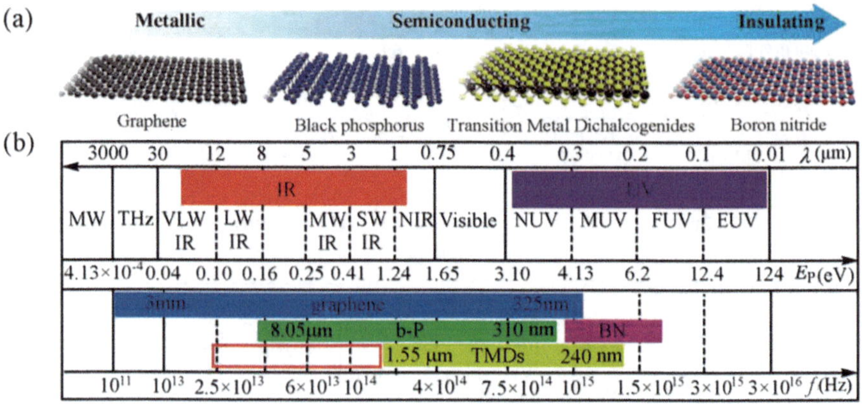

Fig. 7.12 (**a**) Bandgap of two-dimensional layered materials, from metallic-like graphene to semiconductor b-P and TMD, to insulator h-BN; (**b**) Photo-detectors based on two-dimensional materials cover a wide electromagnetic spectrum from terahertz to ultraviolet

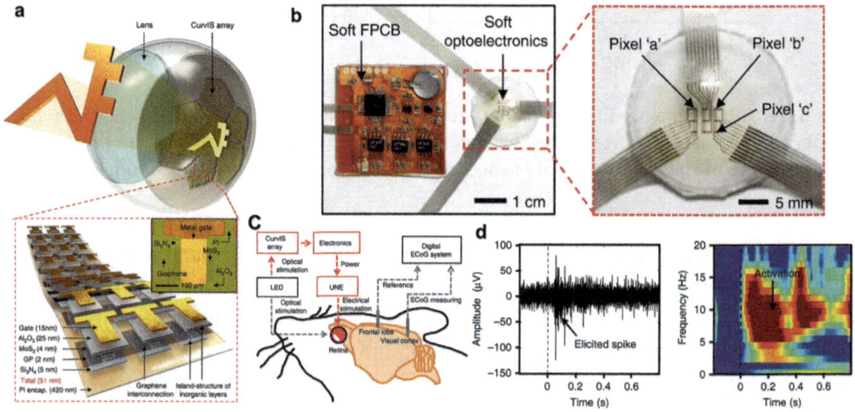

Fig. 7.13 (**a**) Schematic diagram and corresponding architecture schematic of the curved image sensor array based on MoS_2-graphene heterostructures. (**b**) Images of flexible electronic devices, CurvIS arrays, and ultra-thin neural interface electrodes used for light signal processing. (**c**) Schematic diagram of the rat retina stimulation experiment device. (**d**) Verification of rat cortical stimulation by detecting induced spikes (left) and changes in local field potentials (right)

imaging, and other applications. The human eye converts light signals into electrical signals through the retina, and then transmits these signals to the brain through nerves. If the process of converting light signals into electrical signals is obstructed, the human eye will not be able to perform its visual function. Using photodetectors, optical information can be converted into electrical signals, which can then be transmitted to the brain through an electrode array, thus realizing the visual function of the human eye. Choi and others have developed a hemispherical photodetector array based on MoS_2/graphene heterostructures (as shown in Fig. 7.13a), which is used to

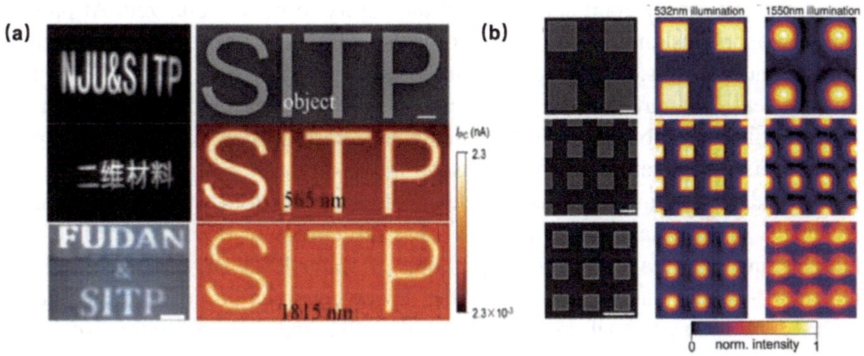

Fig. 7.14 (**a**) Photocurrent images based on p-g-n heterostructures, GaTe-Si p-n photodiodes, and graphene thermoelectric single-element detectors. (**b**) High-resolution and high-contrast images based on black phosphorus single-element detectors

convert light signals into electrical signals, and then transmitted to the optic nerve through the developed ultra-thin neural interface electrode (Fig. 7.13b). Through in vivo experiments on rats, researchers evaluated the stimulation of the retina by this soft optoelectronic device. Figure 7.13c is a schematic diagram of the in vivo animal experiment device, which combines the optoelectronic device and the neural recording system. When a pulse signal is applied to the device, the induced spikes and local field potentials (Fig. 7.13d) indicate that the electrical pulses will be transmitted to the rat's visual nerves. The authors first demonstrated the electronic eye technology based on two-dimensional materials, which can effectively stimulate nerves, but there is still a long way to go to achieve real visual functions.

In addition, photodetectors are expected to be used for high-quality imaging. W. D. Hu and others have prepared various imaging sensors based on 2D materials, including p-g-n heterostructures, graphene thermoelectric single-element detectors, and GaTe-Si p-n photodiodes, and obtained high spatial resolution images through these sensors, as shown in Fig. 7.14a. Engle's research shows that some two-dimensional materials can produce high-resolution (sub-micron) and high-contrast ($V > 0.9$) imaging, which can distinguish micron-level metal patterns. The photodetector based on two-dimensional multilayer black phosphorus they made can realize microscopic pattern imaging within the visible light and infrared spectrum, as shown in Fig. 7.14b, indicating that black phosphorus has broad potential in the field of multispectral imaging.

7.4 Chapter Summary

In summary, due to the atomic-level thickness of two-dimensional materials and their excellent physical and chemical properties, sensors based on emerging two-dimensional materials have advantages such as high sensitivity, fast response speed, wide sensing range, good stability, and the ability to be miniaturized. At present, smart sensors designed based on two-dimensional materials can realize sensing of various mechanisms, including mechanical sensing, gas sensing, biosensing, optical sensing, etc., and have shown great application potential in fields such as biomedicine, flexible wearables, and environmental monitoring. However, despite the great progress in the research of sensors based on two-dimensional materials in the past few decades, there are still great challenges to apply them to practical life. For the synthesis of materials, there are still great difficulties in defect control of two-dimensional materials and the preparation of wafer-level large-area controllable layer two-dimensional materials. For the device construction process, the integration of two-dimensional materials and devices, and the realization of multi-functional coordination of devices, there are still challenges. For the prepared sensors, most of them are still in the prototype concept verification stage, and there is still a gap from practical application. However, with the development of material synthesis technology and integration technology, the application of two-dimensional materials in the sensing field will further develop, bringing infinite possibilities for multi-functional intelligent sensing applications. They will be one of the very important candidates for the next generation of intelligent sensing technology, playing an important role in various application scenarios.

Bibliography

1. ZHENG L, WANG X, JIANG H, et al. Recent progress of flexible electronics by 2D transition metal dichalcogenides [J]. Nano Research, 2021, 15(3): 2413-32.
2. LIU X, MA T, PINNA N, et al. Two-Dimensional Nanostructured Materials for Gas Sensing [J]. Advanced Functional Materials, 2017, 27(37): 1702168.
3. CHOI W, CHOUDHARY N, HAN G H, et al. Recent development of two-dimensional transition metal dichalcogenides and their applications [J]. Materials Today, 2017, 20(3): 116-30.
4. PEI Y, CHEN R, XU H, et al. Recent progress about 2D metal dichalcogenides: Synthesis and application in photodetectors [J]. Nano Research, 2020, 14(6): 1819-39.
5. YANG T, JIANG X, HUANG Y, et al. Mechanical sensors based on two-dimensional materials: Sensing mechanisms, structural designs and wearable applications [J]. iScience, 2022, 25(1): 103728.
6. JIANG H, ZHENG L, LIU Z, et al. Two-dimensional materials: From mechanical properties to flexible mechanical sensors [J]. InfoMat, 2019, 2(6): 1077-94.
7. BOLOTSKY A, BUTLER D, DONG C, et al. Two-Dimensional Materials in Biosensing and Healthcare: From In Vitro Diagnostics to Optogenetics and Beyond [J]. ACS Nano, 2019, 13(9): 9781-810.
8. LEE C W, SUH J M, JANG H W. Chemical Sensors Based on Two-Dimensional (2D) Materials for Selective Detection of Ions and Molecules in Liquid [J]. Frontiers in Chemistry, 2019, 7(708):1-21.
9. LEI Z L, GUO B. 2D Material-Based Optical Biosensor: Status and Prospect [J]. Advanced Science, 2022, 9(4): e2102924.
10. MUNTEANU R E, MORENO P S, BRAMINI M, et al. 2D materials in electrochemical sensors for in vitro or in vivo use [J]. Analytical and Bioanalytical Chemistry, 2021, 413(3): 701-25.
11. TYAGI D, WANG H, HUANG W, et al. Recent advances in two-dimensional-material-based sensing technology toward health and environmental monitoring applications [J]. Nanoscale, 2020, 12(6): 3535-59.

12. ZHANG L, KHAN K, ZOU J, et al. Recent Advances in Emerging 2D Material-Based Gas Sensors: Potential in Disease Diagnosis [J]. Advanced Materials Interfaces, 2019, 6(22):1-27
13. CHEN W Y. 2D Materials for gas-sensing application [D]. 2022, Thesis, Purdue University.
14. LONG M, WANG P, FANG H, et al. Progress, Challenges, and Opportunities for 2D Material Based Photodetectors [J]. Advanced Functional Materials, 2018, 29(19): 1803807
15. Naguib, M., Barsoum, M. W., Gogotsi, Y., Ten Years of Progress in the Synthesis and Development of MXenes. Adv. Mater. 2021, 33, 2103393.
16. Shi X, Wang H, Xie X, Xue Q, Zhang J, Kang S, Wang C, Liang J, Chen Y. Bioinspired Ultrasensitive and Stretchable MXene-Based Strain Sensor via Nacre-Mimetic Microscale "Brick-and-Mortar" Architecture. ACS Nano. 2019,13(1):649-659.
17. Kong, RM., Ding, L., Wang, Z. et al. A novel aptamer-functionalized MoS2 nanosheet fluorescent biosensor for sensitive detection of prostate specific antigen. Anal Bioanal Chem, 2015, 407, 369–377
18. Xie H, Li Z, Cheng L, et al. Recent advances in the fabrication of 2D metal oxides[J]. Iscience, 2021: 103598.
19. Yang Z, Pang Y, Han X, et al. Graphene textile strain sensor with negative resistance variation for human motion detection[J]. ACS nano, 2018, 12(9): 9134-9141.
20. Pang Y, Zhang K, Yang Z, et al. Epidermis microstructure inspired graphene pressure sensor with random distributed spinosum for high sensitivity and large linearity[J]. ACS nano, 2018, 12(3): 2346-2354.
21. Li YT, Qu LL, Li DW, Song QX, Fathi F, Long YT. Rapid and sensitive in-situ detection of polar antibiotics in water using a disposable Ag–graphene sensor based on electrophoretic preconcentration and surface-enhanced Raman spectroscopy. Biosensors and Bioelectronics. 2013 May 15;43:94-100.
22. Sun H, Chao J, Zuo X, Su S, Liu X, Yuwen L, Fan C, Wang L. Gold nanoparticle-decorated MoS 2 nanosheets for simultaneous detection of ascorbic acid, dopamine and uric acid. Rsc Advances. 2014;4(52):27625-9.
23. Lee J, Dak P, Lee Y, Park H, Choi W, Alam MA, Kim S. Two-dimensional layered MoS2 biosensors enable highly sensitive detection of biomolecules. Scientific reports. 2014, 17;4(1): 1-7.
24. Lee DW, Lee J, Sohn IY, Kim BY, Son YM, Bark H, Jung J, Choi M, Kim TH, Lee C, Lee NE. Field-effect transistor with a chemically synthesized MoS2 sensing channel for label-free and highly sensitive electrical detection of DNA hybridization. Nano Research. 2015, 8(7): 2340-50.
25. Zhou C, Zou H, Li M, Sun C, Ren D, Li Y. Fiber optic surface plasmon resonance sensor for detection of E. coli O157: H7 based on antimicrobial peptides and AgNPs-rGO. Biosensors and Bioelectronics. 2018 Oct 15;117:347-53.
26. Dong, T., Simões, J., Yang, Z., Flexible Photodetector Based on 2D Materials: Processing, Architectures, and Applications. Adv. Mater. Interfaces 2020, 7, 1901657.
27. Long M, Liu E, Wang P, Gao A, Xia H, Luo W, Wang B, Zeng J, Fu Y, Xu K, Zhou W. Broadband photovoltaic detectors based on an atomically thin heterostructure. Nano Letters. 2016 Apr 13;16(4):2254-9.
28. Guo N, Hu W, Jiang T, Gong F, Luo W, Qiu W, Wang P, Liu L, Wu S, Liao L, Chen X. High-quality infrared imaging with graphene photodetectors at room temperature. Nanoscale. 2016;8(35):16065-72.
29. Yuan X, Tang L, Wang P, Chen Z, Zou Y, Su X, Zhang C, Liu Y, Wang W, Liu C, Chen F. Wafer-scale arrayed pn junctions based on few-layer epitaxial GaTe. Nano Research. 2015 Oct;8(10):3332-41.
30. Engel M, Steiner M, Avouris P. Black phosphorus photodetector for multispectral, high-resolution imaging. Nano letters. 2014 Nov 12;14(11):641

Chapter 8
Liquid Metal Micro-Nano Processing and Flexible Electronic Sensors

8.1 Basic Properties of Liquid Metal

Liquid Metal (LM) is a large category of emerging functional materials with unique physical and chemical properties such as low melting point, high electrical and thermal conductivity, high boiling point, and low toxicity. In recent years, with the development of flexible electronics and wearable sensor technology, the application of LM in flexible wearable devices for physiological signal monitoring has become a hot research topic, among which gallium (Ga) based alloys are the most widely used liquid metals. This chapter mainly summarizes the basic physical properties of Ga-based liquid metals, such as toxicity, oxidizability, wettability, electrical conductivity, and thermal conductivity. It introduces three material forms of liquid metal flexible electronic devices: original state liquid metal, liquid metal nanoparticles, and liquid metal silicon-based composite materials, and introduces the processing technology of flexible electronic devices based on these three forms of liquid metal materials. Ga-based liquid metal flexible electronic devices can monitor various forms of physiological signals, for example, capacitive/resistive pressure sensors based on Ga-based liquid metal can monitor pulse signals, flexible stretchable strain sensors can monitor limb movements, liquid metal electrodes can monitor blood glucose, etc. At present, liquid metal flexible wearable monitoring devices are still in the preliminary research stage, including the synthesis of liquid metal materials, processing and preparation methods, functional flexible electronic devices, etc., and still have a lot of room for development.

X. Ma et al., *Functional Micro/Nanomaterials for Intelligent Sensing Devices*,
https://doi.org/10.1007/978-981-96-4990-7_8

8.1.1 Gallium-Based Liquid Metal

Metals such as Hg, Fr, and Cs are also in a liquid state at room temperature, but due to their radioactivity, toxicity, etc., they are greatly restricted in applications, especially since they cannot be applied in fields such as biological signal monitoring. Unlike the above-mentioned liquid metals, Ga and Ga-based alloys are currently the most widely used liquid metals in flexible electronics manufacturing, mainly due to their physical characteristics such as low melting point, high electrical and thermal conductivity, low viscosity, almost non-toxic, and low saturated vapor. The physical characteristics of pure Ga, Ga-based liquid metal, and Hg alloys are summarized in Table 8.1. The pure Ga alloy with a melting point of 29.8 °C is usually solid at room temperature. By adding alloying elements such as In and Sn, the melting point can be further reduced (for example, the EGaIn alloy with 24.5 wt.% In element added, its melting point is 15.8 °C; the Galinstan alloy with 20.5 wt.% In element and 12.5 wt.% Sn element added, its melting point drops to −19 °C). Ga-based liquid metal has an extremely high boiling point and can remain liquid at temperatures up to 2000 °C. This ensures that Ga-based liquid metal is not easy to evaporate at room temperature, thus having higher safety. In addition, Ga-based liquid metal has significant supercooling characteristics, and its solidification temperature is usually much lower than its melting point. This allows the alloy to continue to remain liquid at temperatures far below the melting point, which is of great significance for improving the service of Ga-based liquid metal flexible electronics at room temperature or temperatures far below room temperature, and maintaining the stability of the physical state. However, the supercooling characteristic of liquid metal is unstable, and it is easily affected by the thermal history of the alloy, the cooling rate, and the purity of the material. Experiments have found that due to its higher melting point, pure Ga liquid metal may be at the temperature above 0 °C causes solidification, which leads to the failure of flexible devices made from pure Ga due to loss of flexibility. In contrast, EGaIn and Galinstan metals have lower freezing points (<−10 °C). In addition, the excellent metallic conductivity and thermal conductivity of Ga-based liquid metals play an important role in heat dissipation and power transmission in wearable flexible electronic products. The viscosity of Ga-based liquid metals is greatly affected by the oxide film. Without the oxide film, the

Table 8.1 Comparison of the physical properties of four liquid metals: pure Ga, EGaIn, Galinstan, and Hg

Parameter	Pure Ga	EGaIn	Galinstan	Hg
Melting point (°C)	29.8	15.5	−19.0	−38.8
Viscosity (kg m^{-1} s^{-1})	2.04×10^{-3}	1.99×10^{-3}	2.40×10^{-3}	1.526×10^{-3}
Electrical conductivity (S m^{-1})	3.4×10^{6}	3.3×10^{6}	3.1×10^{6}	1.04×10^{6}
Thermal conductivity (W m^{-1} K^{-1})	28.7 at 77 °C	26.4 at 37 °C	25.4 at 37 °C	8.5
Vapor pressure (Pa)	1 at 1037 °C	1.33×10^{-3} at 300 °C	$<1.33 \times 10^{-6}$ at 500 °C	1 at 42 °C

viscosity of the liquid metal will decrease, which is about twice that of water. Moreover, the viscosity of liquid metals can be adjusted by adding other alloying elements, such as the viscosity of EGaIn is 1.99×10^{-3} kg/m/s, which is lower than that of liquid pure Ga metal (2.04×10^{-3} kg m^{-1} s^{-1}). The decrease in viscosity may be due to the lower viscosity of liquid In alloy (about 1.69×10^{-3} kg m^{-1} s^{-1}).

8.1.2 Toxicity

The vapor pressure of Hg at 42 °C is 1 Pa, and the vapor pressures of EGaIn and Galinstan are 1.33×10^{-3} Pa (300 °C) and $< 1.33 \times 10^{-6}$ Pa (500 °C), respectively. Compared with the higher vapor pressure of mercury metal and its easy volatilization at room temperature, the vapor pressure of Ga-based liquid metal is extremely low and its biocompatibility is better, which means that there is no need to worry about inhalation when operating Ga-based liquid metal at room temperature. Like most metals, gallium does not dissolve in water, but it can dissolve in blood in the form of Ga^{2+} and metabolize, so it is generally believed in the academic community that the biological application of Ga-based liquid metal is relatively safe. Liquid metal can be directly printed on human skin as an electrode for electric field therapy. Compared with traditional electrodes, the fluid properties of liquid metal increase the contact area between the electrode and the skin, thereby enhancing the therapeutic effect without obvious side effects. In addition, EGaIn/calcium alginate hydrogel can be used as a candidate drug for intravascular embolization and tumor embolization treatment. Ga-based liquid metal electrodes have no adverse effects on the growth of artificially cultured neurons, and cultured neurons can effectively simulate the function of target neurons. For wearable flexible electronic products, liquid metal is usually encapsulated in biocompatible elastic silicones such as polydimethylsiloxane (PDMS) and Ecoflex, and indirect contact with the skin further ensures the safety of wearable liquid metal flexible electronics. However, current research on the toxicity of liquid metals is not systematic, and more stringent toxicity studies are needed to understand the impact of liquid metals on human health. Liquid metal flexible devices usually need to use large pieces of liquid metal, and the packaging accuracy needs to be strictly controlled to prevent liquid metal from entering the body.

8.1.3 Oxidation and Wettability

Similar to other metals such as aluminum, magnesium, and copper, when the oxygen concentration is above the ppm level, a Ga_2O_3 oxide layer will immediately form on the surface of the liquid metal, and the thickness of this oxide layer is related to the oxygen concentration (about 0.7 nm under vacuum conditions), about 3 nm under environmental conditions). The surface oxide film has a certain elasticity and stability under the critical yield surface stress (~0.5–0.6 N/m) (as shown in

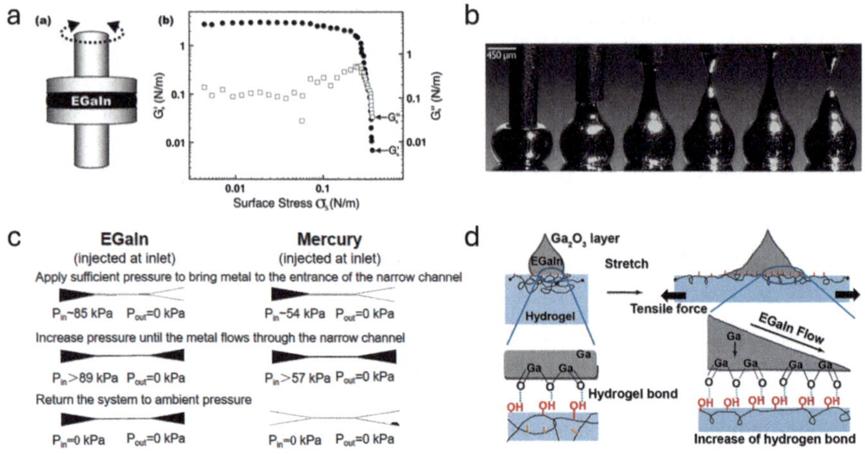

Fig. 8.1 (**a**) Yield stress of Ga_2O_3 oxide film on the surface of liquid metal; (**b**) A set of images of cone-shaped liquid metal droplets formed with the help of the oxide film; (**c**) Compared with mercury metal, liquid metal can be injected into the channel and maintain its shape due to the action of the surface oxide film; (**d**) Due to the interaction between Ga_2O_3 and –OH groups, liquid metal can wet the newly generated surface after stretching the hydrogel

Fig. 8.1a), and only when the stress is higher than the yield strength, the oxide layer will rupture and the internal liquid metal will flow out. In addition, the oxide film plays an important role in the preparation of flexible electronic products, for example, it can stabilize the droplets of liquid metal into non-spherical shapes (as shown in Fig. 8.1b), and maintain the structure of the liquid metal injected into the microchannels after the applied pressure is withdrawn (as shown in Fig. 8.1c). By directly stirring to increase the content of Ga_2O_3, the wettability of liquid metal on various flexible substrate surfaces such as PDMS and PVC can be enhanced. Surface modification of flexible substrates is also a common method to improve the wettability of liquid metal. Due to the interaction between –OH and Ga_2O_3 layers, chemically cross-linked hydrogels composed of aliphatic alkyl chains with many hydroxyl (–OH) groups can promote the extension of liquid metal wettability on their surfaces (as shown in Fig. 8.1d). In addition, the wettability of liquid metal can also be improved by removing Ga_2O_3 through acid-base treatment, allowing the liquid metal to react with metals such as Au/Cu to enhance its wettability on the substrate.

8.2 Preparation Methods of Liquid Metal Micro/Nanoparticles

The preparation methods of liquid metal micro/nanoparticles (LMMPs, LMNPs) can generally be divided into: template method, jet method, microfluidic method, slicing method (shearing method), ultrasonic method, and physical vapor deposition

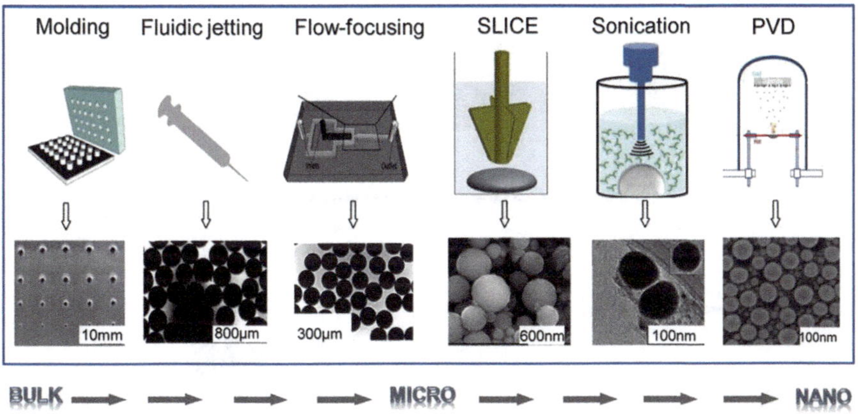

Fig. 8.2 Typical preparation methods of liquid metal micro/nanoparticles

(PVD) method, as shown in Fig. 8.2. Among them, the template method, jet method, microfluidic method, slicing method (shearing method), and ultrasonic method belong to the top-down preparation methods, and the PVD method belongs to the bottom-up preparation method.

8.2.1 Template Method

The template method is often used to prepare LMMPs with diameters ranging from several hundred micrometers to a few millimeters. Taking the preparation of EGaIn microspheres by the template method as an example (as shown in Fig. 8.3), a laser cutting machine is used to draw a columnar pattern on the acrylic sheet, and then PDMS is poured and solidified on the acrylic sheet to form a cover with many voids. The dispersed liquid metal is then used to spread and fill the voids (as shown in Fig. 8.3a). Because the thin oxide layer on the surface of the liquid metal counteracts the surface tension, it is difficult to obtain regular spherical particles. Hydrochloric acid vapor can be used to fill the cover to remove the surface oxide layer and obtain regular spheres (as shown in Fig. 8.3b). The disadvantage of the template method is that it is time-consuming and laborious to produce on a large scale, and it is difficult to expand the preparation of various diameters of LMMPs.

8.2.2 Jetting Method

The jetting method uses a combination of a syringe and a petri dish to conveniently prepare LMMPs at room temperature, as shown in Fig. 8.4. By injecting LM into a water solution containing a surfactant through a syringe (as shown in Fig. 8.4a), a

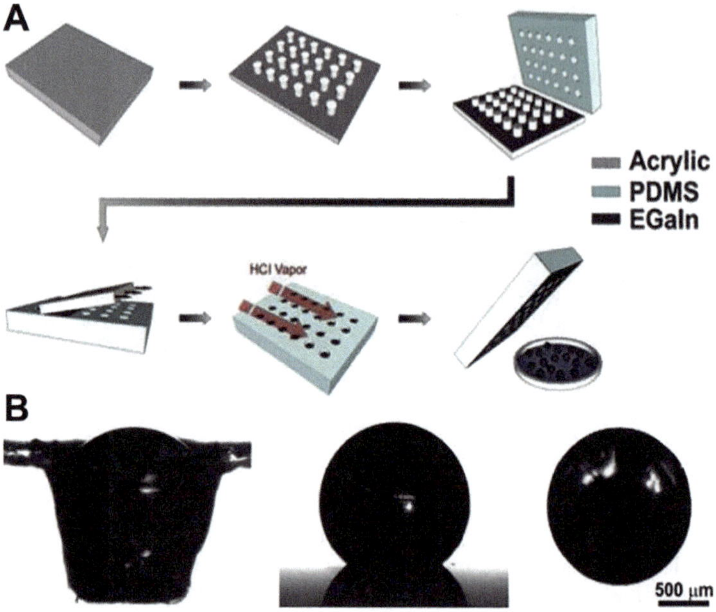

Fig. 8.3 Preparation of liquid metal by the template method (**a**) Process of preparation by the template method; (**b**) Formation process of particles prepared by the template method

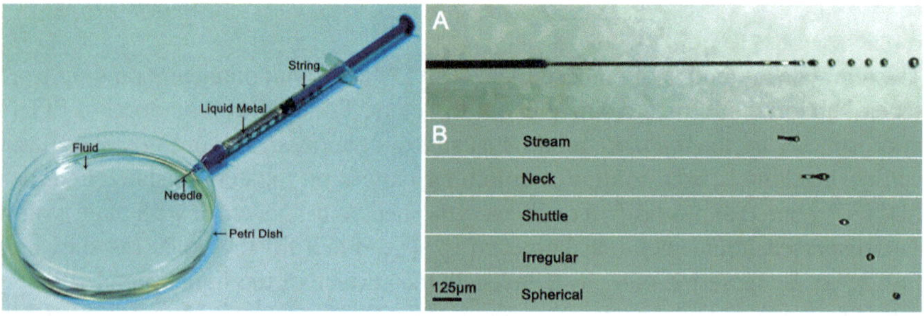

Fig. 8.4 Preparation of liquid metal by the jetting method (**a**) Equipment of the jetting method; (**b**) Droplet manufacturing process under a high-speed camera

continuous thin stream will appear. LM will break into various shapes, including neck-shaped, shuttle-shaped, irregular-shaped, and spherical LMMPs, with an average diameter of about 313 ± 34 μm, as shown in Fig. 8.4b.

In addition, the particle size of LMMPs can be controlled by changing the needle hole diameter and jet speed. Smaller needles and higher jet speeds both contribute to the formation of denser LMMPs. If a 60 μm needle is used, the average particle size produced is 590 ± 71 μm; if the needle diameter is reduced to 41 μm, the average droplet diameter can reach 173 ± 18 μm.

The jetting method is simple and efficient, and is currently the main preparation method for LM micro-motors. However, its disadvantage is that after large-scale preparation, the dispersion of particle size is poor.

8.2.3 Microfluidic Method

The microfluidic method can be understood as a more complex and refined jetting method. By designing a microfluidic flow focusing device, LM and a fluid that is incompatible with LM (glycerin or water) are forced to pass through the microchannel. When the two meet in the micropore, the continuous phase fluid (glycerin or water) will generate enough shear force to decompose the dispersed phase LM into LMMPs, finally forming a mixed system of LMMPs and glycerin or water, and flowing out of the device. The specific steps and devices are shown in Fig. 8.5a. Based on this principle, Gol and others designed a microfluidic transfer system (as shown in Fig. 8.5b), which realized the dynamic transfer of LMMPs to sodium hydroxide solution. When the confluence of LMMPs, NaOH, and glycerin occurs, LMMPs will gradually cross the interface from the high viscosity flow (glycerin) into the low viscosity flow (NaOH).

The size of particles prepared by microfluidic methods usually depends on three factors: interfacial tension, inertial force, and shear force. However, these factors are closely related to the device itself. Once the device is fixed, it is difficult to adjust the size of the prepared microdroplets by changing the experimental conditions. For this reason, Tang and others have developed a microfluidic device that can simultaneously use electrochemistry and electrowetting to change the surface tension of LM, thereby adjusting the diameter distribution and formation frequency of LM droplets. As the voltage increases (0–10 V), the diameter of the LM droplets smoothly decreases (185–85 μm), and the droplet yield sharply increases. To further

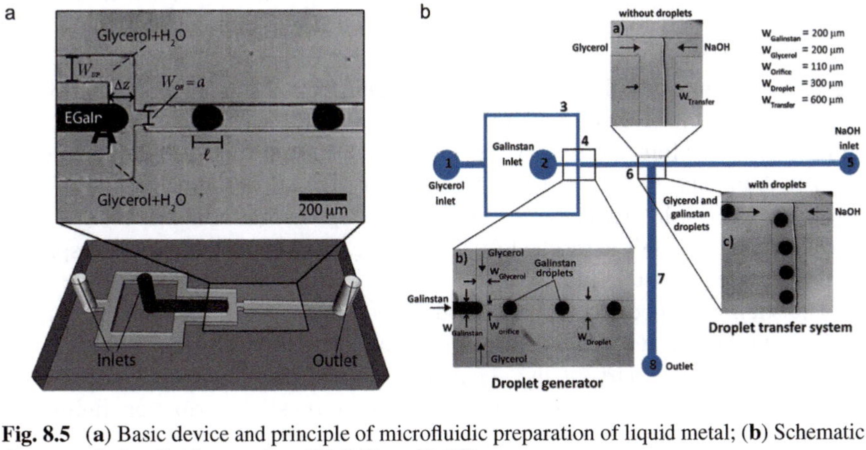

Fig. 8.5 (**a**) Basic device and principle of microfluidic preparation of liquid metal; (**b**) Schematic diagram of microfluidic transfer of LMMPs to NaOH

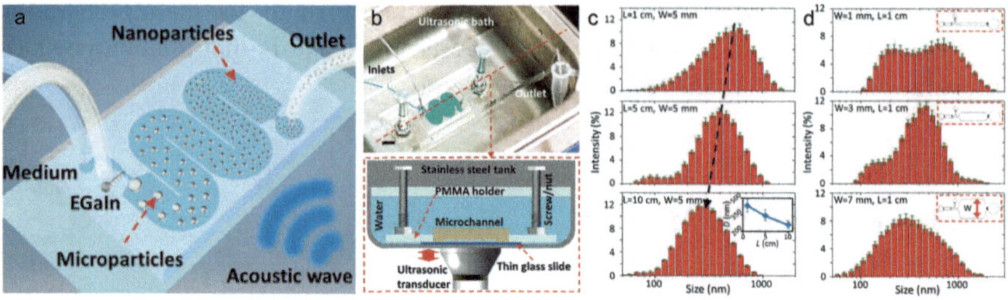

Fig. 8.6 Device and working mechanism of ultrasonic microfluidics (**a**) Schematic diagram of the decomposition of the LMNPs production platform; (**b**) The assembled system is placed in an ultrasonic bath; (**c**) Cross-sectional schematic of the entire system; (**d**) Droplet size generated under different ultrasonic powers

obtain nanoscale LMNPs, Tang and others continued to combine the microfluidic method with the ultrasonic method (see Sect. 8.2.5), using the high shear force caused by ultrasonic waves to further reduce the particle size of the produced particles, as shown in Fig. 8.6a–c. The device is designed as a serpentine channel, with T-shaped joints at the inlet and outlet (50 μm high, 500 μm wide), and the average size of the particles can be adjusted by changing the width and length of the channel, for example, when the channel length increases from 10 mm to 100 mm, the average particle size of LMNPs decreases from 450 nm to 400 nm (as shown in Fig. 8.6d).

8.2.4 Slicing Method (Shearing Method)

The slicing method, also known as the shearing method, is a direct synthesis method for soft micro/nanoparticles, and is an extension of traditional emulsification technology. A thin inert coating is added to the LM surface to inhibit solidification, and then it is placed in a stirrer for stirring (as shown in Fig. 8.7a). The shear force produced by this method helps to form LMPs ranging from 6.4 nm to more than 10 μm. During the stirring process, the forces acting on the droplets include shear force (γ), gravity (F_g), drag force (F_d), centrifugal force (F_c), and buoyancy (F_b), among which the γ is dominant. As the speed increases during the stirring process, the initially static droplets will gradually be stretched into a cylindrical shape. When the Rayleigh-Plateau limit is reached ($r_{droplet} > 1.5\ r_{cylinder}$), the cylindrical LM splits to form LMPs (as shown in Fig. 8.7b).

The slicing method (shearing method) provides a tunable, low-cost, green, and convenient method to manufacture micro/nanoparticles using fluid shear force. However, the asymmetry of the shear force results in the slicing method being unable to directly prepare nanoparticles with a uniform size distribution, limiting its application in the field of periodic structures.

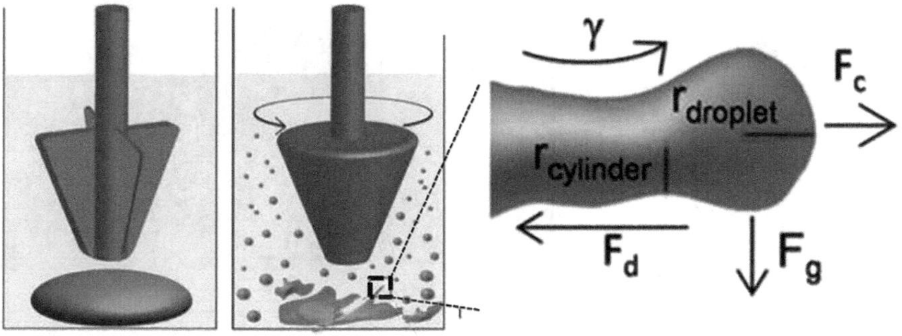

Fig. 8.7 Preparation of LMPs by shearing method (**a**) Schematic diagram of the shearing method process; (**b**) Structural schematic diagram of LMPs prepared by the shearing method

8.2.5 Ultrasonic Method

The ultrasonic method is currently the most common method for preparing LMNPs. After LM is mixed with water or organic solvent, the ultrasonic probe is introduced into this system. The ultrasonic energy generated by the probe will cause cavitation in the water or organic solvent, inducing a large amount of LM to transform into LMMPs, and finally disperse into LMNPs (as shown in Fig. 8.8a). The particle size can be controlled by controlling the ultrasonic output power, ultrasonic treatment time, and changes in environmental temperature. For example, ultrasonication at lower temperatures usually induces the production of smaller particles. The average particle size of particles synthesized at 50 °C (60 nm) is about 1.7 times that of particles produced at 20 °C (35 nm) (as shown in Fig. 8.8b–d). In addition, with the increase of ultrasonic time and output power, the average diameter of the obtained particles and the time consumed will decrease. The smallest particle size of LMNPs that can be obtained by the ultrasonic method reported at present is 35 nm.

It is worth noting that, due to the oxidation layer on the surface of LMNPs gradually falling off under the action of ultrasonic energy, the Ga droplet surface exposed with the reduction of particle size will be rapidly oxidized, forming a new Ga_2O_3 layer. The repetition of this process will greatly promote the transformation of Ga to Ga_2O_3, thus forming nanoscale flake Ga-based oxide, rather than micro/nanoparticles. Therefore, it is particularly important to control the oxidation layer and add surfactants to ensure the stable dispersion of particles. Usually, the strength of the LM surface oxidation layer is enough to cope with the particle aggregation driven by surface tension, ensuring the stable dispersion of LMNPs in a neutral solution. However, if the oxide film continuously dissolves or peels off, the balance between dispersed particles will be quickly broken, causing particle aggregation or fusion. For example, the increase or decrease of pH value, and the application of mechanical force in a neutral solution will cause this phenomenon, only with the help of surfactants, particles can remain stable in an acidic or alkaline environment for a long time. Surfactants should be added to the liquid environment that is insoluble in

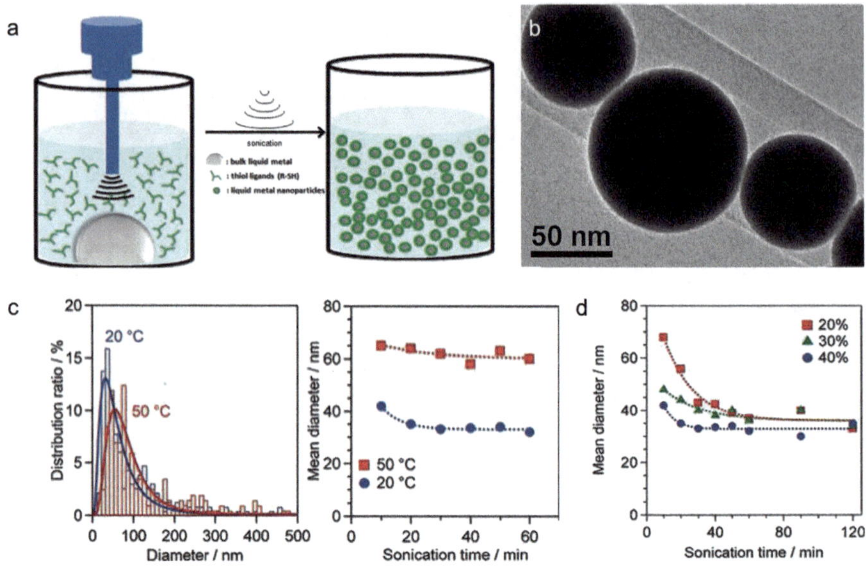

Fig. 8.8 Preparation of LMNPs by ultrasonic method (**a**) Schematic diagram of ultrasonic treatment; (**b**) SEM image of EGaIn nanodroplets; (**c**) Relationship between ultrasonic time and average particle size under different ultrasonic temperatures; (**d**) Relationship between ultrasonic time and average size under different ultrasonic powers

LM before ultrasonic treatment, alcohol, 1-dodecanethiol, thiol-based surfactants, chitosan, etc. are all optional, which can not only effectively stabilize LMNPs, but also further improve the yield of nanoparticles.

8.2.6 Physical Vapor Deposition (PVD)

PVD is an emerging bottom-up method for preparing LMNPs. Taking the preparation of EGaIn nanoparticles as an example, bulk EGaIn is placed in a tungsten boat and evaporated into atoms under resistance heating. These atoms diffuse to each other and collide, forming stable atomic nuclei on the cold target substrate (silicon, glass, or mica) (as shown in Fig. 8.9a, b). Before 100 s of deposition, the average particle size is about 25 nm (almost the smallest average particle size) (as shown in Fig. 8.9c); after the deposition time increases to 500 s, the EGaIn particles fuse, and the particle size grows to more than 100 nm (as shown in Fig. 8.9d), and obvious particle size separation occurs. The high vapor pressure and low boiling point of EGaIn make it difficult to add surfactants during the PVD process, and the stability of the obtained micro and nano particles needs to be studied.

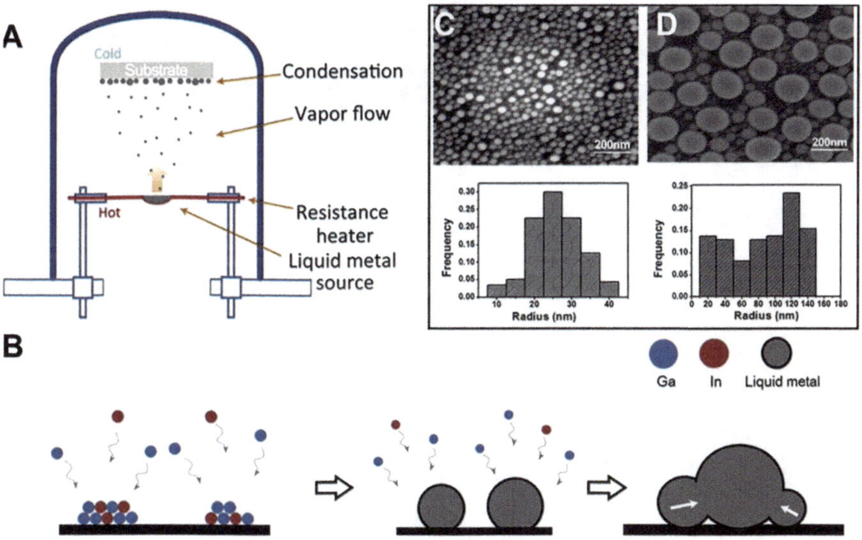

Fig. 8.9 Preparation of LMNPs by PVD (**a**) Schematic diagram of the device for preparing EGaln LMNPs by PVD technology; (**b**) Process of producing EGaln LMNPs by PVD. Ga atoms and In atom alloys (left) grow (middle) and finally merge into large particles (right); (**c**) Average particle size before 100 s of deposition; (**d**) Average particle size after 500 s of deposition

8.3 Methods of Preparing Liquid Metal Flexible Devices

Liquid metal, being a liquid itself, theoretically has no limit to its elongation, and its conductivity is relatively high, making it the most suitable choice for flexible electronic devices. However, its high surface tension and non-wetting with many common materials limit its patterning. To solve this problem, researchers have developed many new preparation methods. According to the form of processing, these preparation methods can be divided into four categories: injection, photolithography, additive, and subtractive methods.

8.3.1 Injection Method

The injection method generally involves injecting liquid metal into a pre-prepared channel using a syringe or pump, as shown in Fig. 8.10a. The required circuit shape is pre-designed, and the liquid metal is injected into the channel to obtain the required device. Its advantage is that it can directly use the prepared channels, the injection operation is simple, and it is suitable for large-scale production. The difficulties of the injection method mainly lie in two points: First, the surface tension of the liquid metal is high, and there is a spontaneously formed oxide film on the

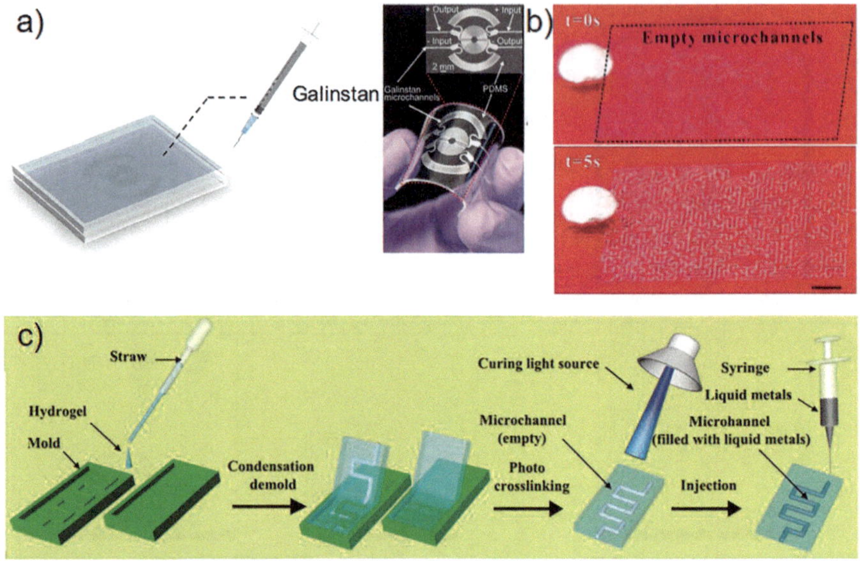

Fig. 8.10 (**a**) Pressure sensor prepared by injection method; (**b**) Vacuum-assisted filling of liquid metal; (**c**) Mold prepared by photolithography

surface. The injection of liquid metal often requires extremely high pressure. Research shows that injecting liquid metal into a 150 nm quartz tube requires a pressure of 10 MPa, which requires high-performance equipment. Therefore, most research only considers injecting liquid metal into channels of several hundred micrometers; The second difficulty of the injection method is to prepare suitable channels. The fluidity of liquid metal is very good. When the strength of the encapsulated flexible material is not enough, the injection method can easily cause channel damage, especially when the channel design spacing is too small, it is easy to cause liquid metal leakage, leading to device preparation failure. At the same time, whether the liquid metal can completely fill the channel during injection also requires adjustment of process parameters.

Early liquid metal devices were mostly prepared by the injection method, a typical example is injecting liquid metal into hollow fibers. These fibers are usually prepared by the reverse mold method, and materials can be selected from various flexible polymers, such as Etienne Palleau and others who used Reverlink polymer to prepare a hollow fiber with an inner diameter of 500 μm, and directly injected EGaIn with a syringe to construct a stretchable conductive fiber. Because the inner diameter of the channel is large, the syringe can provide enough pressure. However, when we need more precise and complex devices, this large-size channel is no longer applicable. Therefore, researchers have developed more injection-based preparation methods. The first is the vacuum-assisted channel filling method for blind holes and small channels. As shown in Fig. 8.10b, when using a syringe to directly pour liquid metal, excessive pressure may cause channel damage, and blind holes in the channel are also difficult to fill. Vacuum assistance can fill blind holes and protect the channels to a certain extent. The second is the preparation of molds. When preparing small-sized channels, PDMS can be poured into a mold with a pre-made circuit shape. After the PDMS is solidified, it is demolded. Then, another semi-solidified PDMS film is combined with it. After complete solidification, the required channel can be obtained, as shown in Fig. 8.10c. The mold preparation process can choose machining, 3D printing, or photolithography according to the resolution. It should be noted that whether choosing PDMS or other elastic polymer materials, there may be situations where the polymer material adheres to the mold and the channel shape is destroyed during the demolding process, especially when the channel size reaches tens of micrometers. In addition, the filling of uncured polymers into the channel due to capillary action also needs special attention.

The methods mentioned above are used to prepare two-dimensional devices. Currently, in order to improve the integration of devices, three-dimensional devices are gradually attracting attention. The preparation of three-dimensional channels through 3D printing, and then the construction of three-dimensional electronic devices by injecting liquid metal, has also become a research hotspot. Thanks to the excellent fluidity of liquid metal, the injection method is more convenient for preparing three-dimensional flexible electronic devices.

8.3.2 Photolithography

Photolithography, as the name suggests, is a method of making liquid metal devices based on photolithography. Photolithography is a method of using light of a specific wavelength to change the solubility of a polymer film on a substrate, and then removing the unnecessary areas. It should be noted that photolithography is an indirect method of preparing liquid metal devices, so many methods can use photolithography to prepare the required templates, and there are many methods for patterning after preparing the template. For example, combining screen printing and masking to prepare patterns; using photolithography to prepare masks for transferring liquid metal; using templates to prepare microchannels and injecting liquid metal; using photolithography to make PDMS channels, and transferring liquid metal to the channels by imprinting to achieve patterning. Here we only discuss the method of using photolithography to prepare transfer molds.

The advantage of photolithography is its high precision. When the precision requirements of the required device reach the level of several hundred nanometers, other methods basically cannot meet the requirements, and researchers can only rely on photolithography to prepare templates. The unique fluid properties of liquid metal allow us to transfer specific patterns to the required substrate by the method of mold imprinting. The basic steps of preparing liquid metal patterns by photolithography include: (1) Spin-coating a PAA (polyacrylic acid) sacrificial layer on a silicon wafer, and then depositing a layer of poly-p-xylylene-C as a protective layer by PVD to protect the bottom sacrificial layer; (2) Using electron beam or photolithography to prepare a pattern of poly(methyl methacrylate) (PMMA) on these two layers; (3) Evaporating a layer of Ti/Au to improve the wettability of the substrate to liquid metal, with a thickness of about 5/30 nm, then using a PDMS stamp to transfer the liquid metal to the surface, this pressing process may need to be repeated 2–3 times to ensure that the liquid metal can fill the surface; (4) Stripping PMMA with acetone, leaving only the required liquid metal pattern; (5) Encapsulating the liquid metal pattern with PDMS, dissolving the PAA layer with water, and finally removing the protective layer with oxygen plasma etching; (6) Encapsulating the exposed area with PDMS, as shown in Fig. 8.11.

The polymer materials used in the above methods can also be replaced by others, the key is to be able to remove them by one method. At the same time, some researchers mix liquid metal particles with photoresists to make specific ink, and then pattern it through photolithography. The main steps include: (1) Using ultrasound to disperse large pieces of liquid metal into nanoparticles in the photoresist; (2) Spin-coating the photoresist mixed with liquid metal particles on a silicon wafer; (3) Directly completing patterning with photolithography; (4) After developing, use a roller to squeeze to make the pattern conductive. Due to the presence of the oxide film, the directly photolithographed circuit is non-conductive, and it is necessary to break the oxide film by mechanical squeezing. This process is easy to causes damage to the pattern, and therefore its precision is not as high as the aforementioned method, with a feature line width of 10 μm.

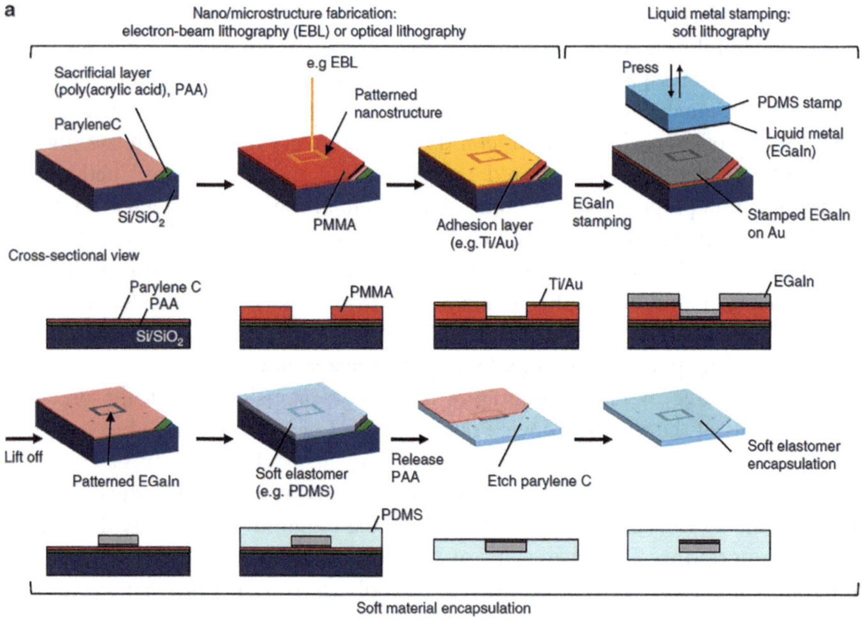

Fig. 8.11 Preparation of liquid metal devices by photolithography

The use of photolithography to create highly precise and accurately arranged liquid metal patterns is essentially unachievable by other methods. However, to obtain a precise and accurate pattern, the process is complex, the cost is high, and the requirements for the equipment are also high. Taking the exposure patterning process in photolithography as an example, not only does it require the design of reasonable patterns and the use of expensive masks, but the exposure process parameters also need to be adjusted repeatedly over a long time. If the pattern needs to be changed frequently, this will also significantly increase the cost. Therefore, the use of photolithography is best established in scenarios where the pattern has already been designed and only batch production is required subsequently.

8.3.3 Additive Method

The additive method is a means of patterning by depositing liquid metal in a designated area. Many related methods have been developed by researchers, including printing, stamping, thermal evaporation, and 3D printing. These methods have all evolved from mature printing methods of the past. The uniqueness of gallium-based liquid metals lies in the fact that the outer oxide film is wettable with many materials, while the internal liquid metal is the opposite. Therefore, researchers need to reasonably control the oxide film of the liquid metal to ensure that the obtained liquid metal circuit can stably bond with the substrate.

There are relatively many means of preparing liquid metal electronic devices by printing. It can effectively combine drawing software and printing platforms, customize the construction of the circuits required by users, and achieve "what you see is what you get". The technical means of printing include direct writing, mechanical printing, and mask printing. Direct writing is a method of directly squeezing liquid metal onto the substrate with a pen or nozzle. The oxide film on the surface of the liquid metal can adhere to the substrate material. Due to the fluid properties of the liquid metal itself, we can use it as a printing material like ink, as shown in Fig. 8.12a. This method is easy to operate, but the accuracy is low, mainly because the amount of liquid metal squeezed out is too large, and it is difficult to control the line width when spreading on the substrate. If the diameter of the nozzle or pen tip is reduced, it is difficult to squeeze out the liquid metal. To further improve the printing accuracy, researchers have designed many liquid metal printing instruments for mechanical printing, as shown in Fig. 8.12b. Technologies such as air pump pressurization, inkjet printing, and electro-wetting printing can be used for patterning liquid metal, mainly by applying pressure or shear force to break the oxide film, forcing the liquid metal to adhere to the substrate. Due to the high surface tension of the liquid metal itself, it is not easy to flow out of the nozzle. Researchers have also developed various types of liquid metal inks. The ink is generally made by mechanically stirring or ultrasonically dispersing the liquid metal into a solution, becoming a mixture of liquid metal micro/nanoparticles, dispersants, and solvents. The proportion of ink is regulated according to the requirements of the instrument and the printed circuit to

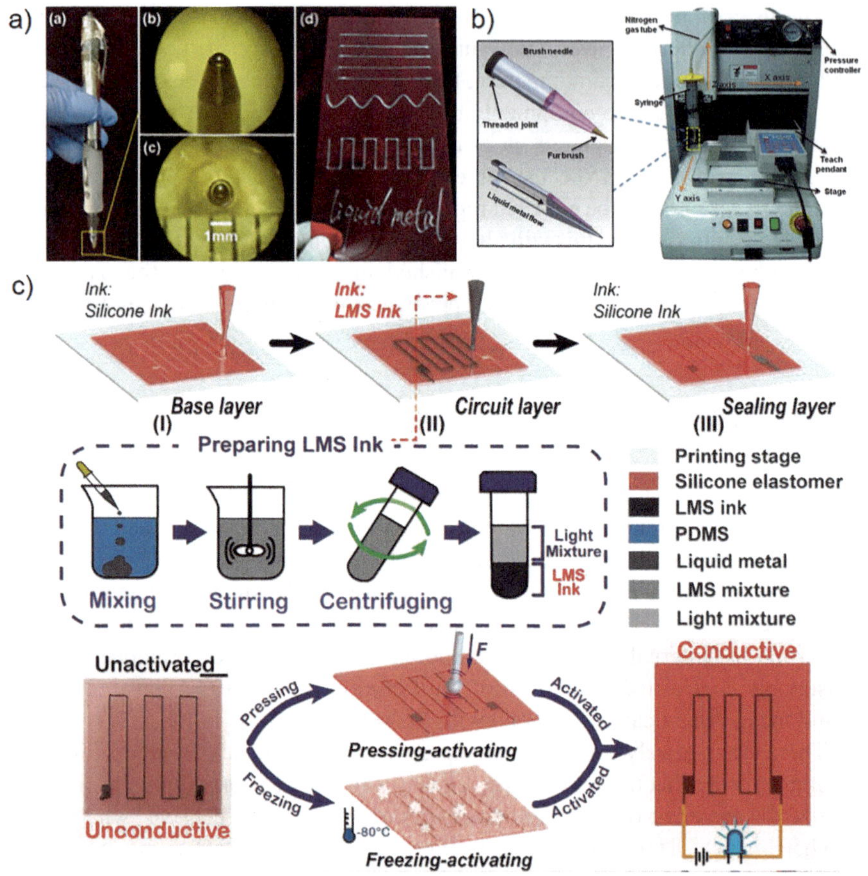

Fig. 8.12 (**a**) Liquid metal circuit prepared by direct writing; (**b**) Direct printing of liquid metal; (**c**) Preparation, printing, and mechanical sintering process of liquid metal ink

obtain the appropriate pattern. The method of printing with liquid metal ink can be seen in Fig. 8.12c. The liquid metal ink prepared by this method can easily pass through small apertures, overcoming the problem brought by high surface tension. Since the dispersed small particles are covered with a large area of oxide film, the printed ink circuit often does not conduct electricity. However, this oxide film layer is very thin and can be broken by mechanical extrusion, stretching, laser sintering, and self-sintering, and the liquid metal can flow out to form a conductive circuit. Some researchers also print liquid metal circuits based on mask templates, usually using 3D printing, laser processing, or photolithography to obtain templates, stick them on the printing substrate, and obtain liquid metal patterns by spraying or coating. This method is relatively simple to operate, but high-precision printing requires high-precision line processing on the template.

In addition to the various methods developed based on printing, other methods also have their characteristics, and we need to choose reasonably according to the use scenario. For example, when constructing liquid metal patterns on complex surfaces, polymethyl methacrylate (PMA) and other adhesives can be coated on the material surface to improve wettability, so that the liquid metal can adhere to the designated position, as shown in Fig. 8.13a. Of course, water transfer can also be used to construct liquid metal electronic devices on rough surfaces and curved surfaces. The method of preparing liquid metal patterns by the stamping method can be seen in Fig. 8.13b, where a stamp is used to transfer the liquid metal pattern to the designated PDMS substrate. This method is suitable for some scenarios where the precision requirements are not high and the pattern volume of printed electronic devices is extremely large. The technical key of this method lies in the wettability control of the substrate and the stamp. Ideally, the wettability of the stamp to liquid metal is less than that of the substrate to liquid metal. In this process, the liquid metal is prone to incomplete adhesion with the stamp or substrate. To solve this situation, selective wetting treatment can be performed on the material, such as pre-vapor deposition of small patterns made of gold on the stamp. Because the wettability of liquid metal and gold is good, it can adhere completely to the stamp. Of course, other chemical modification or physical modification methods can also be used to adjust the wettability. The method of preparing liquid metal patterns by thermal evaporation is to deposit liquid metal in a designated area by thermal evaporation. To obtain the required pattern, selective wetting methods are often needed, such as screen printing gold patterns on the PDMS substrate in advance, so that the liquid metal can only stay on the surface of the gold. The surface oxide film of liquid metal can limit its flow and also provide the basis for maintaining its shape. This allows us to directly construct liquid metal-related devices using 3D printing. The method of 3D printing liquid metal circuits developed by Young-Geun Park and others, as shown in Fig. 8.13c, can change the spatial arrangement of liquid metal circuits by driving the nozzle to move with a motor. The authors pointed out that the speed at which the nozzle is lifted needs to be controlled between 0.001 and 0.1 mm/s to ensure that the adhesion of the liquid metal is less than the fracture energy of the oxide film, so that the circuit will not break during the 3D printing process. Liquid

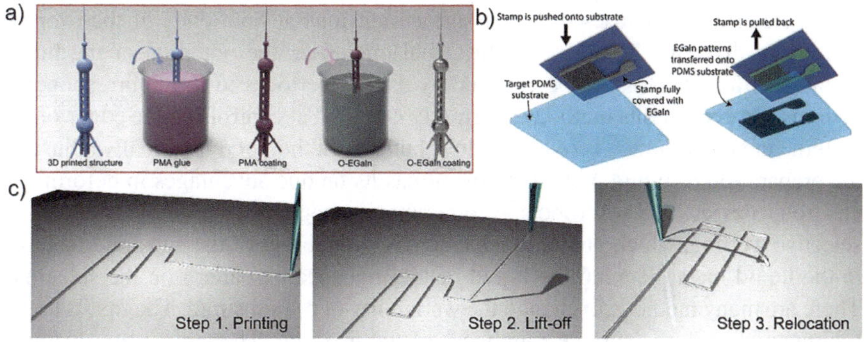

Fig. 8.13 (**a**) Liquid metal pattern obtained by PMA modified surface; (**b**) Liquid metal device prepared by stamping method; (**c**) 3D printed liquid metal circuit

metal 3D printing has broad prospects in constructing flexible antennas, highly sensitive sensors, and other three-dimensional devices. Of course, the strength of the liquid metal itself often cannot support overly complex three-dimensional structures, which also makes the three-dimensional structure unable to be designed as freely as patterns on a two-dimensional plane. If micrometer-level linewidths are needed, it often relies on pre-made polymer tubes to build complex structures using injection methods.

8.3.4 Subtractive Method

The subtractive method is to first coat the liquid metal in the designated area, and then remove the excess liquid metal by some method, leaving only the predetermined pattern. The commonly used subtractive methods include: laser ablation, electrowetting, and selective wetting. Laser ablation, as shown in Fig. 8.14a, obtains extremely fine circuits by causing excess liquid metal to evaporate through local high temperatures induced by lasers. Pan et al. obtained a linewidth of up to 4.5 μm using a Nd: YAG laser. Laser ablation to prepare liquid metal circuits is an effective means to obtain precise patterns, but it requires expensive light sources and supporting equipment. Moreover, appropriate laser power, scanning speed, and substrate materials should be determined through repeated experiments, otherwise, the laser can cause damage to the circuit and substrate materials. Electrowetting is a method of forming patterns by adjusting the surface potential of the liquid metal, making some areas of the liquid metal non-wetting with the substrate, as shown in Fig. 8.14b. The theory of electric wetting was established by Lippmann based on mercury. He believed that the double layer of charge accumulated by the electrochemical reaction of droplets and electrolytes caused a capacitive effect, leading to changes in the energy of the entire system and causing changes in surface tension. This principle applies to electrolytes and liquid metals exposed to air. When a positive voltage is applied to the liquid metal, it will spread on the substrate, and conversely, the liquid metal will retract into a spherical shape. In the electrolyte, the uneven distribution of the electric field causes different surface tensions at both ends of the droplet, which can drive the liquid metal. In actual operation, the potential of the liquid metal in different areas can be changed by using electrodes to form corresponding patterns. Of course, this method is currently not good at controlling the edges of the pattern, and the accuracy is relatively low, but as a method of dynamically adjusting the preparation of liquid metal patterns, it has its unique advantages in deformable electronic devices and soft robots. Another major category of subtractive methods is selective wetting, which generally changes the wettability of the substrate material to the liquid metal, leaving the liquid metal on a specific pattern on the substrate. There are many methods to change the wettability of the substrate. The first is to use a laser to change the surface roughness of the material. The wettability of liquid metal on a smooth surface is better, and its wettability on a rough surface drops significantly (Cassie state). This can first use a laser to make corresponding patterns

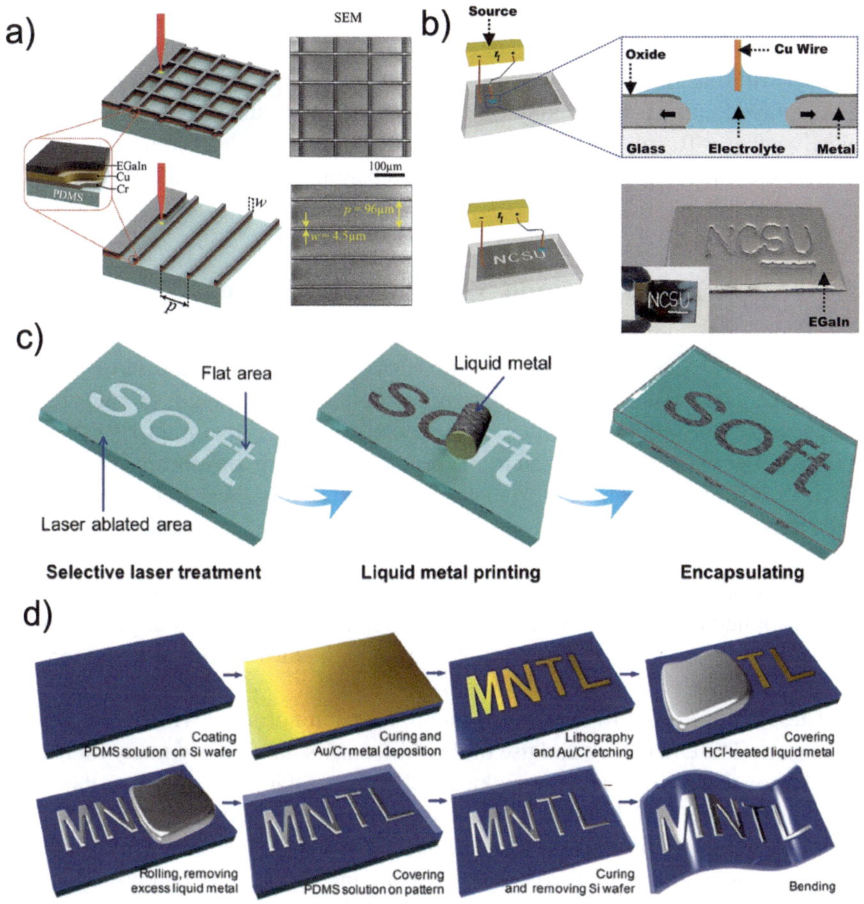

Fig. 8.14 (**a**) Laser ablation to construct liquid metal patterns; (**b**) Electric wetting to construct liquid metal patterns; (**c**) Laser modification of the substrate material surface; (**d**) Surface gold plating selective wetting

on the substrate material, and then spread the liquid metal, as shown in Fig. 8.14c. The second is to evaporate other metals, generally by masking and spraying gold on the substrate to obtain corresponding patterns, as shown in Fig. 8.14d. Because gallium-based liquid metal has better wettability with gold, it will only stay in the gold-sprayed area, so the required electronic devices can be prepared. In addition, chemical modification methods can be used to control the surface of the substrate or liquid metal, adding hydroxyl and other better wettability functional groups. Subtractive methods for preparing liquid metal devices are classified into many types, their operations are relatively simple, and they can already use some mature processes and instruments to obtain the required patterns, such as lasers, evaporation, etc., which further reduces the cost of developing new instruments. However, except for laser ablation to make lines, other methods involve controlling the wettability of liquid metals. The fluid properties of liquid metals, coupled with the complex physical and chemical states of the material surfaces, together cause its contact with the substrate material to be not completely controllable. Even on the gold-sprayed surface, there may be cases where the liquid metal does not completely adhere. To solve this problem, it is necessary to further explore the surface state of the liquid metal and optimize process parameters based on this.

8.4 Application of Liquid Metal Flexible Sensors

At present, wearable flexible electronics inspired by human skin have attracted widespread attention due to their potential applications in health monitoring and diagnosis, robots, and prosthetics, and are considered an important branch of the next generation of intelligent electronic products. Flexible sensors are composed of flexible substrates and conductive materials. In recent years, researchers have developed various excellent flexible substrate materials, such as platinum-catalyzed silicone (Ecoflex 00–30), flexible materials such as PDMS and polyurethane are used. These flexible materials, due to their inherent good scalability and biocompatibility, can be applied to various flexible strain sensors, meeting the needs of people for the development of flexible sensors towards miniaturization, integration, and intelligence.

These types of electronic devices cannot function normally without the support of sensors. The emergence and use of flexible sensors have solved many limitations of traditional sensors, such as wrinkles and breakage failures caused by the deformation of rigid materials, burdens, and inconveniences caused to people when worn, etc. Among them, LM plays a huge role due to its ability to deform while maintaining high conductivity. Compared with mercury, the low toxicity and no vapor pressure of Ga-based liquid metal are the key factors for its preparation and use at room temperature. Based on theirs unique properties, this section will introduce and summarize the applications and latest developments of LM flexible sensors in pressure, strain, and other different fields.

8.4.1 Liquid Metal Pressure Sensor

The main performance measurement indicators of flexible pressure sensors are sensitivity, force range, response time, etc. They must be able to work while deforming, so flexible pressure sensors based on liquid metal have great application prospects and development space. LM pressure sensors are of great significance in medical applications, human-machine interfaces, real-time monitoring of physiological signals, etc., and have been widely studied. They can produce flexible pressure sensors that meet various needs of people by designing based on different conversion principles or structures. At present, pressure sensors can be roughly divided into

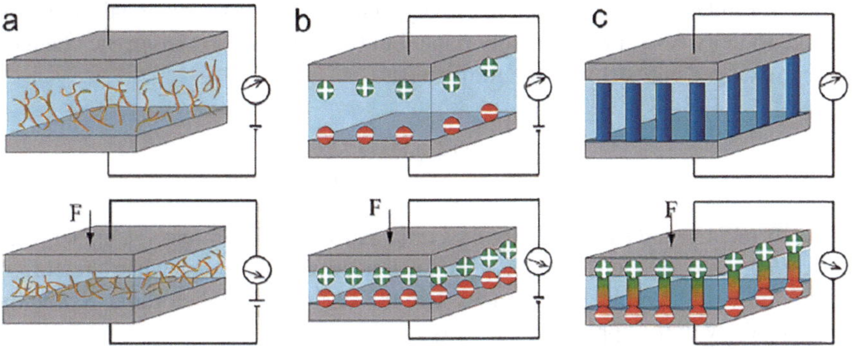

Fig. 8.15 Classification and mechanism of flexible pressure sensors: (**a**) resistive; (**b**) capacitive; (**c**) piezoelectric

resistive, capacitive, piezoelectric, etc., as shown in Fig. 8.15. On this basis, people have designed and developed corresponding LM flexible sensors according to different principles.

8.4.1.1 Piezoelectric Sensor

The piezoelectric pressure sensor is based on the piezoelectric effect. Because the commonly used materials for the piezoelectric effect are piezoelectric polymers and inorganic materials rather than LM, such as poly(vinylidene fluoride), ZnO, GaN, InN, ZnS, and CdS, etc., it is not detailed in the following text. Only its mechanism is introduced here. Under external pressure stimulation, dipole moments are generated in anisotropic crystal materials, and macro potentials are generated in the crystal. Because the voltage output of the piezoelectric sensor is a pulse signal, it is suitable for measuring dynamic pressure but not suitable for static measurement. The piezoelectric coefficient is a physical quantity to evaluate the energy conversion rate of piezoelectric materials. The larger the value, the higher the energy conversion rate.

8.4.1.2 Resistive Sensor

The principle of the resistive sensor is to convert the pressure signal into a change in resistance. The advantages of resistive sensors are easy to manufacture, simple structure, low cost, and easy signal collection. It is also currently the most widely used type of flexible tactile sensor, but its disadvantages are also obvious, that is, the signal-to-noise ratio is low. The expression of conductor resistance is as follows:

$$R = \frac{\rho l}{A} \tag{8.1}$$

In the formula, ρ, l, and A are resistivity, length, and cross-sectional area. The change in body resistance is mainly related to two factors: (1) Geometric changes, (2) For changes in resistivity, the formula is:

$$\frac{\Delta R}{R} = \left(1+2v\right)\varepsilon + \frac{\Delta \rho}{\rho} \tag{8.2}$$

In the formula, υ and ε are the Poisson's ratio and strain, respectively. Contact resistance (R_c) refers to the resistance generated at the contact point when the current passes through it, and its size changes with the size of the applied force, R_c and the relationship with the applied force (F) is:

$$R_c \propto F^{-1/2} \tag{8.3}$$

The resistance change of the pressure sensor mainly comes from two aspects: (1) the change of the body resistance of the material itself; (2) the change of contact resistance. Kim et al. prepared a mechanical sensor based on changing the shape of the liquid metal microchannel to change the resistance of the liquid metal itself, which integrates the elastic substrate, liquid metal, and microchannel, has a stable signal response under cyclic load, and has a high sensitivity (0.158 kPa^{-1}), as shown in Fig. 8.16a. In terms of using contact resistance changes, Peng et al. compounded liquid metal with hydrogel. When external pressure is applied, the contact area

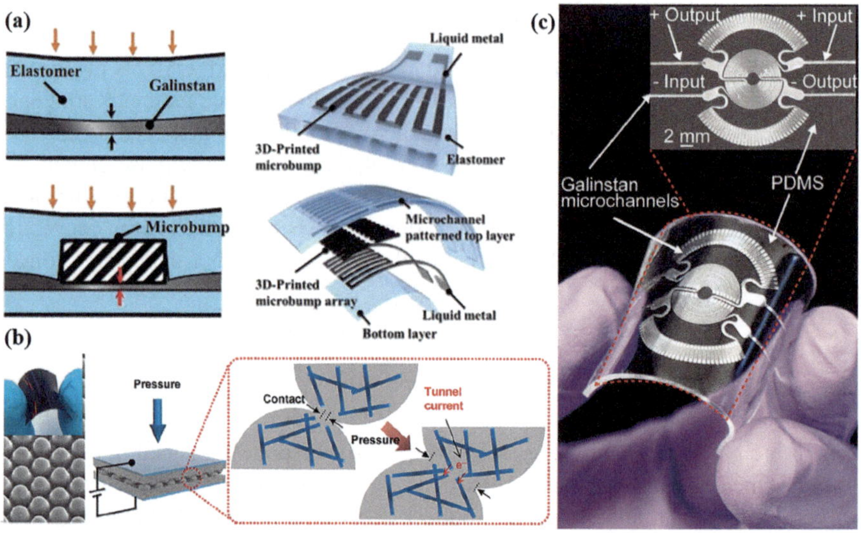

Fig. 8.16 (**a**) The effect of rigid micro-convex blocks on pressure sensitivity and the schematic diagram of the pressure sensor based on LM integrated with 3D printed rigid micro-convex blocks; (**b**) Schematic diagram of a pressure sensor that changes the contact resistance of the electrode due to changes in microstructure; (**c**) Highly sensitive pressure sensor obtained based on the principle of equivalent Wheatstone bridge circuit

between the liquid metal in the composite material increases, reducing the contact resistance, and thereby causing resistance changes. Similarly, based on the principle that the change of microstructure leads to the change of electrode contact resistance, combined with active materials such as carbon nanotubes (CNT), gold, etc., and the preparation of interlocking structure pressure sensors with PDMS substrate microstructure can improve sensitivity and response time, as shown in Fig. 8.16b.

To enrich the application scenarios of pressure sensors and improve the sensitivity of mechanical sensors, Gao et al. designed and demonstrated a highly sensitive sensor as shown in Fig. 8.16c, which can resolve pressure changes below 50 Pa in less than 90 ms. This sensor uses embedded Galinstan microchannels and fully utilizes the tangential and radial strain fields with the equivalent Wheatstone bridge circuit, making its sensitivity as high as 0.0835 kPa^{-1}, which can be used in flexible sensing gloves to provide comprehensive tactile feedback for human hands when touching or holding objects.

8.4.1.3 Capacitive Sensors

Capacitance is a basic physical quantity that represents the storage capacity of a capacitor. Capacitive sensors are based on the principle of parallel plate capacitors, as shown in Fig. 8.15b. It is mainly composed of electrodes and dielectric layers, and its capacitance equation is:

$$C = \varepsilon_0 \varepsilon_s A / d \tag{8.4}$$

In the formula, ε_0 and ε_s are the vacuum dielectric constant and the dielectric constant of the dielectric layer, respectively; *A* represents the overlapping area between the upper and lower electrodes; *d* represents the vertical distance between the two electrodes.

The principle of change in capacitive pressure sensors is: when vertical pressure is applied to the electrode, *d*, *ε* change, causing a change in capacitance, thereby achieving the measurement of force. The capacitive pressure sensor designed by Ali and others is shown in Fig. 8.17a. When pressure is applied to the sensor, due to the

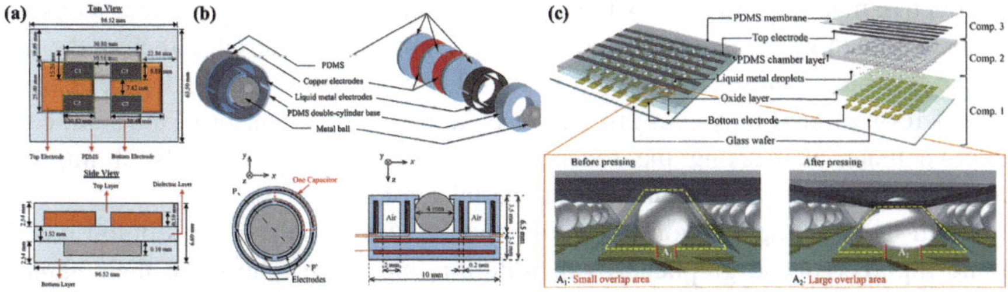

Fig. 8.17 (**a**) Schematic diagram of a capacitive pressure sensor; (**b**) LM as the electrode of the capacitive 3-axis mechanical sensor schematic diagram; (**c**) Schematic diagram of the truncated pyramid-shaped capacitive pressure sensor

change in the thickness of the dielectric layer, the distance between the top and bottom electrodes will decrease. The capacitance of the manufactured pressure sensor is similar to the capacitance of a parallel plate capacitor in configuration, so it changes proportionally with the applied pressure. If there is a structure of air gaps in the dielectric layer in the capacitor, when the dielectric layer is compressed, the volume of air decreases. Because the dielectric constant of air is less than the dielectric constant of the elastomer, the capacitance will also increase. Based on this, the LM can be designed as the electrode of the capacitive 3-axis mechanical sensor, its schematic diagram is shown in Fig. 8.17b. This 3-axis force sensor can scan the surface to measure shear forces and reaction forces from 0 N to 5 N, which has great prospects in tumor detection applications.

To further improve the sensitivity of capacitive pressure sensors (usually less than 10 pF/N), Won and others proposed and designed a new truncated pyramid-shaped sensor, which has a large dynamic range, high sensitivity, and relatively good spatial resolution (Fig. 8.17c). The sensor is caused by LM The capacitance change caused by the overlap area between the droplet and the flat bottom electrode improves its dynamic range and increases the sensitivity to 147 pF/N.

Overall, capacitive pressure sensors have higher sensitivity and faster response speed compared to other types, and can be used for the measurement of small forces, but they are also easily affected by environmental interference causing capacitance fluctuations, resulting in measurement errors. The requirement for a stable working environment is high. Therefore, the future research direction is how to overcome these weaknesses and play the advantages of capacitive pressure sensors.

8.4.2 Liquid Metal Strain Sensor

Traditional strain sensors are mainly composed of rigid devices. Due to their high cost, poor ductility, and high rigidity, they reduce the comfort of human wear, and some sensors can no longer meet the needs of people for sensors in modern society. Emerging flexible strain sensors can solve these problems. When a flexible substrate is used but a solid conductive material is used to prepare a flexible strain sensor, although it has good sensitivity, there will be a large friction force between the solid conductive material and the flexible substrate, which will not only produce a hysteresis phenomenon but also consume the solid conductive material, affecting the life of the flexible strain sensor. Even if such a flexible strain sensor undergoes excessive deformation, it may cause irreversible damage to the interior of the solid conductive material, causing the sensor to fail completely, which also limits the application range of solid conductive material sensors.

With the wide application of liquid conductive materials such as liquid metal, ionic liquid, reduced graphene oxide solution, etc., researchers use their natural ductility and fluidity to adapt to a wide range of deformations. Compared with solid conductive materials, liquid conductive materials can reduce the friction force with the flexible substrate. However, it should be noted that the preparation process of many liquid-type flexible strain sensors is relatively complicated, the preparation efficiency is low, not only need to prepare liquid channels, but also need to inject liquid conductive materials into the liquid channels and seal them.

Similar to flexible pressure sensors, flexible strain sensors can be mainly divided into resistive and capacitive types based on their sensing signals. Among them, resistive ones are more common, so this section focuses on resistive flexible strain sensors.

8.4.2.1 Resistive Type

When a resistive flexible strain sensor is subjected to external force, it deforms, causing the length or cross-sectional area of the conductive material on the surface or inside the elastic body to change, thereby changing the resistance of the flexible strain sensor. After the external force disappears, the elastic body returns to its initial state, and the length or cross-sectional area of the conductive material also returns to its original state, ultimately causing the sensor's resistance to return to its initial resistance. This type of flexible strain sensor uses resistance signals as sensing signals to achieve the purpose of sensing.

A typical example is the use of LM-filled fiber optic or microchannel strain sensors. The LM elongates in the direction of strain and narrows perpendicular to the direction of strain, resulting in increased resistance, which complies with the equation:

$$R = R_0 \left(1+\varepsilon\right)^2 \tag{8.5}$$

where R_0 is the initial resistance; ε is the strain; R is the resistance under the applied strain ε. Inspired by the ability of blood vessels to sense temperature changes, Wu et al. designed a high-performance fiber optic strain sensor filled with LM, which can adjust the sensitivity by controlling the size of the tubular channel. It shows nearly no hysteresis (0.11%), low detection limit (0.3% strain), ultra-low resistance (0.344 Ω), and good repeatability over 3500 cycles (as shown in Fig. 8.18a). Due to its good biocompatibility, it has high application value in the development of implantable biomedical fiber sensors.

One of the main applications of resistive strain sensors in flexible wearable electronic devices is to capture changes in joint angles on the sagittal plane. They can be integrated into wired/wireless gloves (as shown in Fig. 8.18b), and the changes in gestures can be judged by real-time monitoring of output voltage or resistance changes to detect joint movement. Similarly, Jeong et al. developed an advanced wireless near-field communication (NFC) LM strain sensor (as shown in Fig. 8.18c), which can be directly installed on the joint to monitor human body movement in real time. In this sensor, the relationship between the change in resistance and strain satisfies:

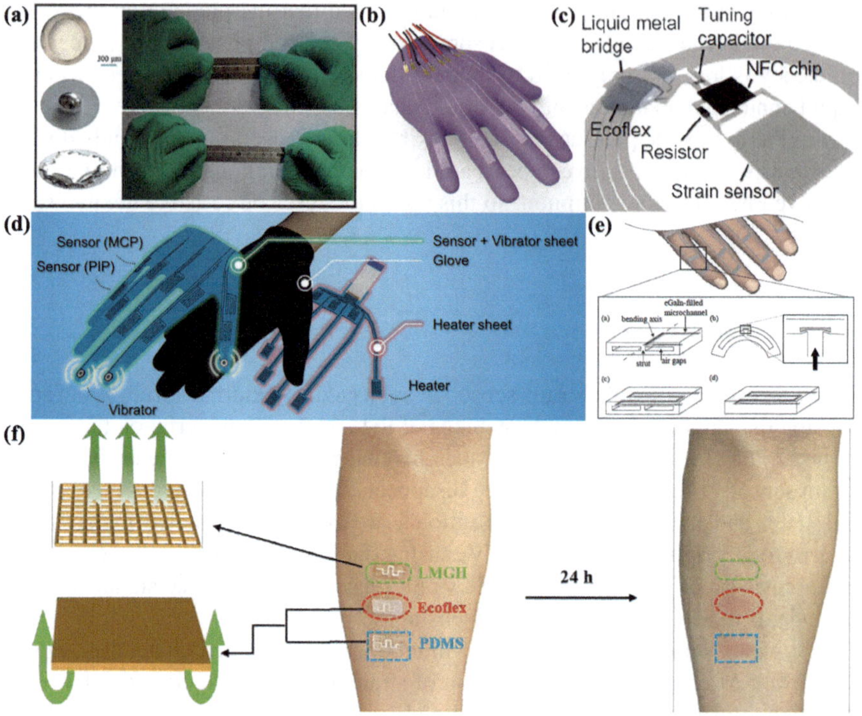

Fig. 8.18 (**a**) Cross-section of LM fiber; (**b**) Schematic diagram of a glove integrated with flexible strain sensors; (**c**) Schematic diagram of the LM strain sensor for NFC; (**d**) Schematic diagram of a VR sensor glove for human-computer interaction; (**e**) Flexible strain sensor based on GelMA hydrogel; (**f**) Schematic diagram of a flexible strain sensor with a hollow structure

$$\frac{\Delta R}{R_0} = \frac{R}{R_0} - 1 = \left(\frac{l}{l_0}\right)^2 - 1 = \varepsilon^2 + 2\varepsilon \tag{8.6}$$

where R_0 is the initial resistance; ε is the strain; ΔR is the resistance changed by the strain; l_0 is the initial length; l is the length after applying the strain.

Obviously, the above LM strain sensor cannot simultaneously sense the movement of multiple points, including local strain and global deformation. For this reason, Oh and others have developed a sensing glove integrated with ten sensors and three vibrators to measure finger movement and provide vibration tactile feedback (as shown in Fig. 8.18d). It can be used in virtual reality (VR) sensing applications for human-computer interaction in the future. In addition to integrating multiple sensors, designing an LM strain sensor with multiple cross-sections can achieve short-distance, multi-point strain measurement.

The outstanding advantage of the LM strain sensor compared to traditional strain sensors or solid-state conductive material sensors is that its resistance will not change significantly after thousands of cycles, which means that the LM strain sensor has high stability and durability. However, it should be noted that the sensitivity or strain coefficient (GF) of sensors that usually use microchannel injection LM to achieve strain sensing function is very low (see Table 8.2).

$$GF = \frac{1}{\varepsilon}\left[\lambda\left(1+\varepsilon\right)^2 - 1\right] \tag{8.7}$$

In the formula, λ is the deformation factor of the microchannel. Its average strain coefficient GF cannot meet the requirements of commercial wearable strain sensors with GF greater than 100. To improve the sensitivity of the LM strain sensor, Kramer and others introduced a curvature sensor with a hollow structure, as shown in Fig. 8.18e. The sensor can both elongate and apply pressure to the embedded microchannel during bending, thereby enhancing the resistance change. The low-sensitivity LM strain sensor is greatly limited in terms of sensing accuracy and small strain sensing. If you want to further improve the sensitivity of the LM strain sensor, you can consider using channel structure design or elastic body elasticity adjustment strain redistribution to improve sensitivity.

Further, considering the poor breathability of the substrate of general flexible strain sensors, it is difficult for people to wear them in the same position for a long time. Yuan and others developed a new type of hydrogel substrate GelMA, where LM can be well sealed in the microchannel without leakage. Combining the advantages of LMs and GelMA hydrogels, LMGE has good biocompatibility, excellent air and ion permeability, and stable electrical response to strain, as shown in Fig. 8.18f. In addition, thanks to the characteristics of GelMA hydrogel, it can monitor people's behavior changes in air or water, even assess the production of secretions, and monitor human sweat secretion in real time during exercise through LMGE, which has broad application prospects.

Table 8.2 Comparison of sensor characteristics based on the principle of resistance change caused by microchannel deformation

Conductive material	Gage factor	Detction limit (%)	Degree of hysteresis (%)	Working range (%)
Graphene	*>500*	–	–	$0 < \varepsilon < 10$
GFs	> 10	–	–	$0 < \varepsilon < 35$
Au	1.09	0.5	–	$0 < \varepsilon < 60$
Ag NP	2.05	0.8	21	$0 < \varepsilon < 25$
Ag NW	2	2	38	$0 < \varepsilon < 100$
GWF	1	2.3	5	$0 < \varepsilon < 10$
EGaAu	*1*	–	–	$0 < \varepsilon < 4$
EGaln	0.66	2.9	–	$0 < \varepsilon < 100$
EGaInSn	2.2	0.3	0.11	$0 < \varepsilon < 140$

In terms of mass production, Kim and others used Direct Ink Writing technology to ensure relative consistency of the microchannels under the premise of emphasizing efficiency, ensuring that multiple sensors produced under the same manufacturing conditions have similar behavior.

8.4.2.2 Capacitive

Capacitive flexible strain sensors achieve sensing by changing the capacitance. When external changes occur, the dielectric constant, distance, or facing area of the parallel plates in the flexible strain sensor will change accordingly, and the capacitance will change accordingly. Lu and others used a method of flexible anisotropic conductors sealing LM circuits to prevent leakage, and proposed and designed a sensor based on observing capacitance changes to obtain strain values. This sensor consists of two overlapping EGaIn films, which are used as capacitive electrodes, and are insulated by a thin layer of PDMS. When placed on the wrist, joint movement will cause the electrodes and dielectric to deform and change the capacitance (as shown in Fig. 8.19a). Cooper and others injected liquid metal into the hollow and highly stretchable elastic fiber core, wound the resulting fiber into a spiral, and made a capacitive sensor for torsion, strain, and touch, as shown in Fig. 8.19b. It can measure up to 800 rad·m^{-1} of torque, the intertwined fibers can sense strain, the stretch length and capacitance are linearly related, and it has good cycle repeatability.

In addition to the above, bulk LM, LMNPs ink, and LM composites can all be used to prepare liquid metal strain sensors, most of which are also based on the resistance principle to measure strain changes. For liquid metal ink and liquid metal composite strain sensors, external forces will change the microstructure of the material and cause resistance changes.

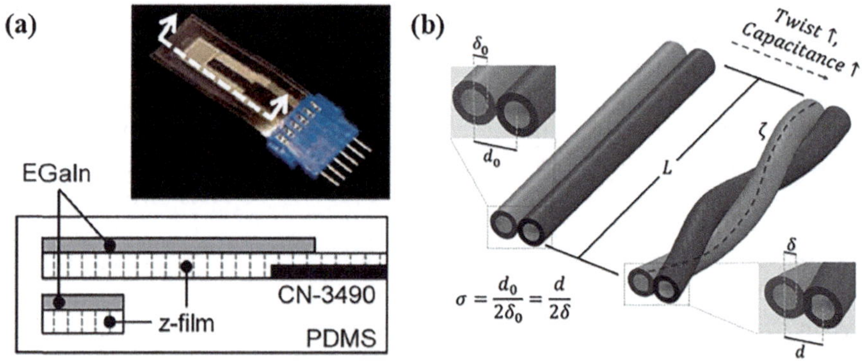

Fig. 8.19 (**a**) Real and schematic diagram of capacitive flexible strain sensor; (**b**) Schematic diagram of LM-filled spiral flexible strain sensor

8.4.3 Other Applications of Liquid Metal Flexible Sensors

Flexible sensors based on liquid metals not only shine in the field of wearable flexible electronic products, but researchers have also successfully developed their applications in electrical components based on their excellent conductivity, such as fully flexible capacitors, inductors, resistors, etc. In addition, applications in friction nanogenerators (TENG), antennas, and auxiliary circuits all demonstrate that flexible sensors prepared with Ga-based liquid metals have wide applicability and strong practicality. As people deepen their understanding of the physical properties of Ga-based liquid metals, the flexible sensors produced will bring infinite possibilities to people's lives.

8.4.3.1 Application in Electrical Components

Capacitor

Capacitors connect to alternating current and act as a circuit breaker when direct current (DC) is applied. They are widely used in coupling, DC isolation, bypass, filtering, tuning, energy conversion, automatic control, and other fields. Compared with traditional (rigid) electronic devices, flexible circuit components can maintain electronic functions even when stretched to several times their natural length. As shown in Fig. 8.20a, the flexible capacitor developed by Fassler and others, composed of liquid-phase gallium-indium-tin alloy (Galinstan) microchannels embedded in a soft silicone elastomer, monotonically changes the capacitance of the embedded liquid channels as the surrounding elastomer stretches. Based on this, Kim and others further developed a fully flexible electronic passive component based on EGaIn that includes parallel plate capacitors, as shown in Fig. 8.20b. The capacitance can be changed by adjusting the area of the parallel plates, the distance between the parallel plates, and the dielectric constant of the flexible material. Similarly, this can also be applied to interdigital capacitors (as shown in Fig. 8.20c).

It should be noted that most of the LM flexible electrical sensors currently prepared are realized by injecting LM into microchannels. However, the gallium-based LM oxide skin in the atmosphere will adhere to the surface of the microchannel, thereby hindering the transport of gallium-based LM in the channel. Konishi and others have studied this and found that the shape of the channel is the key to the smooth transport of Galinstan, regardless of the channel material. It was also found that when liquid metal is transported in a microchannel, if it passes between electrodes, it will change the local capacitance. Therefore, the flow of liquid metal can be detected using capacitance, and conversely, the transport of liquid metal can also adjust and control the capacitance, as shown in Fig. 8.20d.

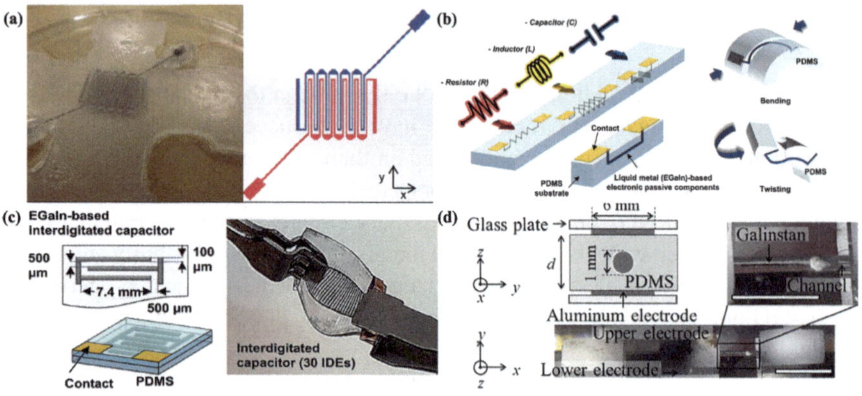

Fig. 8.20 (**a**) Real and schematic images of the flexible capacitor; (**b**) Schematic of the fully flexible electronic passive component based on EGaIn; (**c**) Schematic and real images of the interdigital flexible capacitor; (**d**) Serial capacitor device based on LM

Inductor

Inductors connect to alternating current and hinder direct current. The work of the aforementioned Fassler and others on flexible electrical components also includes research on flexible inductors. As the surrounding elastomer stretches, the inductance of the embedded liquid channel monotonically changes. Do and others have studied a miniature soft electromagnetic actuator made of organosilicon polymer, liquid metal, and magnetic powder. The core component of the electromagnetic sensor is the 3D the spiral coil conductor (as shown in Fig. 8.21a) can provide high thermal conductivity and high current. According to a recent study, the inductance is related to the diameter, length, and number of turns of the helical tube. When the inductance is stretched axially, the inductance increases with the increase in the length of the helical tube, and when bent, the inductance decreases with the increase in the radius of curvature. This spiral structure inductor has been successfully applied to the posture detection of serpentine soft robots (as shown in Fig. 8.21b). In addition, because wearable integrated electronic devices may undergo severe deformation in applications, it is necessary to carefully design the inductor to ensure the normal operation of the electronic device.

Resistors

Compared with rigid resistors made of copper-nickel alloy, other flexible conductive materials such as ionic liquids and hydrogels, Ga-based liquid metals have higher conductivity. According to the definition formula of resistance:

$$R = \rho \cdot \frac{l}{s} \tag{8.8}$$

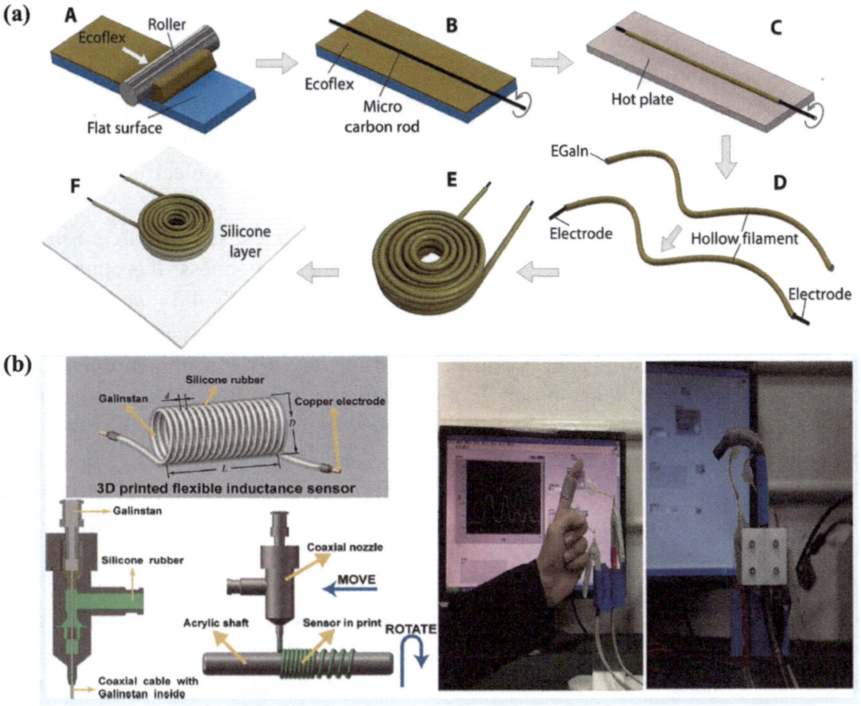

Fig. 8.21 (**a**) Schematic diagram of the manufacturing process of the flexible 3D spiral coil inductor; (**b**) Schematic diagram and physical diagram of the stretchable inductor based on LM

where R is the resistance; ρ is the resistivity; l is the length of the resistor; s is the cross-sectional area of the resistor. The resistance of Ga-based liquid metal can be adjusted by controlling the length and cross-sectional area of the channel. Jin et al. changed the channel width while keeping other geometric parameters constant to explore the relationship between the resistance, temperature, Joule heat, and color change of the flexible thermochromic elastomer. The ability of liquid metal to deform combines geometric changes with Joule heating, thus achieving adjustable thermomechanical color sensing for touch and strain. In more complex circuits, the deformation of the metal can redistribute electrical energy to the far end of the network, thereby converting analog tactile "input" into digital chromaticity "output". Applying pressure deforms the liquid metal inside the channel, thereby changing the local resistance. Combined with the flexible thermochromic elastomer, it can convert analog tactile "input" into digital chromaticity "output", providing the possibility for the "visualization" of pressure. The liquid metal channel will change in size when subjected to external force, causing a change in resistance, which is also the principle used by most liquid metal force/strain sensors.

8.4.3.2 Application in Triboelectric Nanogenerators (TENG)

Wireless wearable flexible electronic devices that do not require an external power source are very important for improving the convenience and comfort of wearing, and the most critical is to achieve self-supply of energy from the integrated electronics themselves. TENG can convert mechanical energy into electrical energy to power wearable electronic devices.

In the application of TENG, LM has many advantages, such as increasing the contact area between electrodes. Affected by surface roughness, the contact area between the two layers of the friction nanogenerator made of solid materials cannot reach 100%, which greatly reduces the total charge density that can be transferred, thereby reducing the total energy conversion efficiency. To solve this problem, Tang et al. developed a mercury-based friction nanogenerator that allows full contact between metal and dielectric. The friction nanogenerator they studied has an instantaneous energy conversion efficiency of 70.6% and a high output charge density of 430 C/m^2, which is four to five times the charge density of solid film electrodes, as shown in Fig. 8.22a. Similarly, Yang et al. designed a new stretchable LM-TENG, which can be stretched to 300% of its initial size without failure (as shown in Fig. 8.22b). Dong et al. fully utilized the flexibility of LM and applied it to the new spinning process, using a thermal stretching process to manufacture microstructured stretchable friction elastomer electric fibers (as shown in Fig. 8.22c), combining the microtextured surface with multiple liquid metal electrodes, its efficiency is comparable to the flat TENG system. Even under large deformation, this fiber shows high electrical output, it can withstand up to 560% strain without failure, and can be used to weave into machine-washable deformable textiles. The charge density of

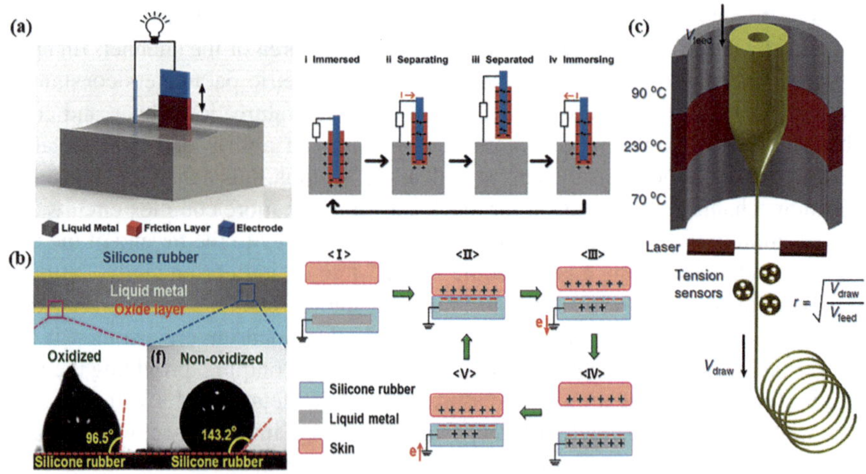

Fig. 8.22 (**a**) Schematic diagram of the working principle of LM-based TENG; (**b**) Cross-sectional diagram of TENG and schematic diagram of the generator; (**c**) Schematic diagram of long fiber thermal stretching process production

Ga-based TENG is about 68.4–84.5 C/m^2, which is lower than that of Hg-based TENG. This is related to the instantaneous formation of Ga_2O_3 oxide layer on the surface of Ga-based alloy. If we want to further improve the performance of Ga-based TENG in the future, how to deal with the surface oxidation of Ga-based LM is a problem that needs to be solved urgently.

8.4.3.3 Application in Antennas

With the increase of wearable smart terminal devices, antenna designs with flexible compliance characteristics are receiving more and more attention. Materials such as carbon nanotubes, copper, and Ga-based liquid metals are used in the manufacture of flexible antennas. Especially for Ga-based LM, due to its good conductivity and deformability, it is considered one of the candidate materials for flexible antennas. The spectral characteristics of the antenna are a function of its shape, therefore, the ability to physically deform and manipulate the shape of liquid metal provides unique possibilities for tunable and reconfigurable antennas. Liquid metal injected into microchannels has been used to manufacture various types of antennas, including dipoles (as shown in Fig. 8.23a), loops (as shown in Fig. 8.23b), patch antennas (as shown in Fig. 8.23c), RF antennas (as shown in Fig. 8.23d), and reconfigurable antennas, etc., and there are more examples using mercury. Flexible dipoles can change their shape by mechanical stretching, thereby changing spectral characteristics (for example, resonance frequency, see Fig. 8.23a), this feature can be used for wireless strain sensing, where the antenna itself is the sensor. But in some cases, the change in frequency during deformation is undesirable, this problem can be solved by aligning the important part of the antenna perpendicular to the direction of strain.

It should be noted that the electrical conductivity of gallium-based liquid metal is lower than that of copper, which is a widely used antenna material. For certain antenna geometries, such as dipoles, the radiation efficiency of antennas made from gallium-based LM and copper does not decrease, and the efficiency may exceed 90%. However, in some geometries, the efficiency of gallium-based LM is lower than that of copper, for example, the radiation efficiency of patch antennas is 60%. Although this efficiency is still acceptable, it indicates that if we want to take advantage of the deformability of gallium-based LM sensors, we should focus on the antenna's stretchability, reconfigurability, or patterning in new ways.

8.4.3.4 Applications in Auxiliary Circuits

In addition to the various flexible sensors made from liquid metal mentioned above, if we take advantage of its good conductivity and plasticity, liquid metal can also play an important role in various flexible sensors as an auxiliary circuit. At this time, LM no longer acts as an information collector, but serves as a "signal transmission" wire or "carrier", such as the auxiliary circuits, bionic electronic eyes, heart rate monitoring, and respiratory monitoring that will be introduced next.

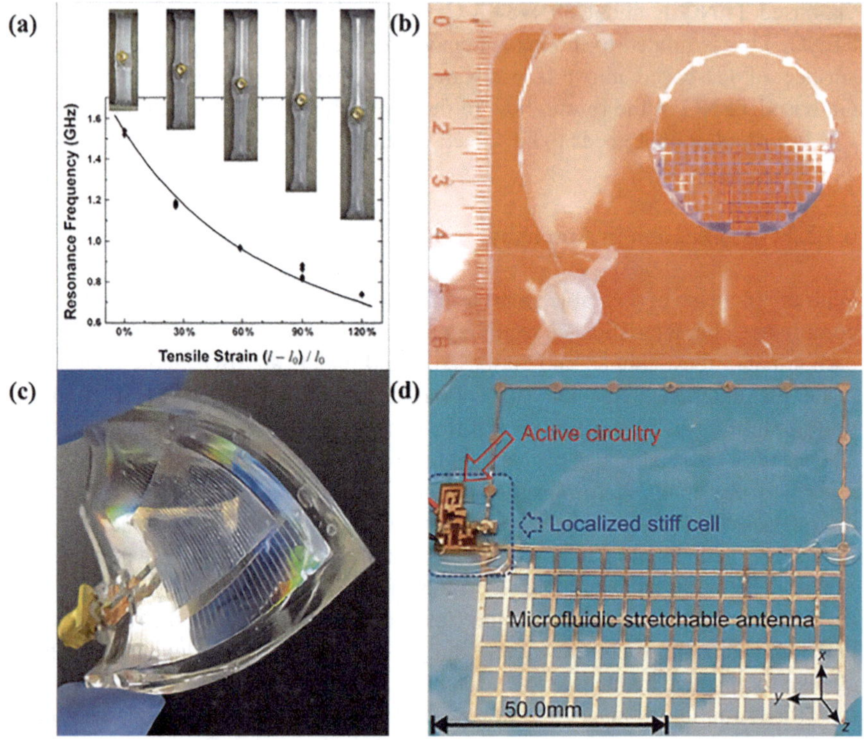

Fig. 8.23 Application of LM flexible sensor in the field of antennas: (**a**) Dipole; (**b**) Flexible loop; (**c**) Patch antenna; (**d**) RF antenna

Pan and others use laser-based microfabrication technology to produce soft, elastic (strain limit >100%) conductive films (thin layer resistance = 2.95 Ω sq^{-1}; resistivity = 1.77 × 10^{-6} Ω m), the biggest feature of which is the use of visually imperceptible ultra-fine liquid metal wires to achieve optical transparency on the circuit, as shown in Fig. 8.24a. These films consist of grid arrays on transparent elastomers, not only have high transmittance (>85% at 550 nm), but are also visually invisible under typical lighting conditions and reading distances, thanks to the laser direct writing technology that makes the LM line width and spacing as small as 4.5 μm and 100 μm respectively. At the same time, the grid wiring has sufficient conductivity to realize digital functions, and can be applied to emerging technologies in transparent displays and sensor electronic products in the future. Similarly, Zhang and others used projection micro-stereolithography 3D printing technology combined with liquid metal filling to manufacture bionic compound eye sensors (as shown in Fig. 8.24b).

Takaya and others took advantage of the high resistance stability of LM after multiple deformations to develop a highly deformable electrocardiograph (as shown in Fig. 8.24c) for diagnosing heart diseases. To achieve the large deformation function of the system electrodes, heterogeneous connectors composed of liquid metal

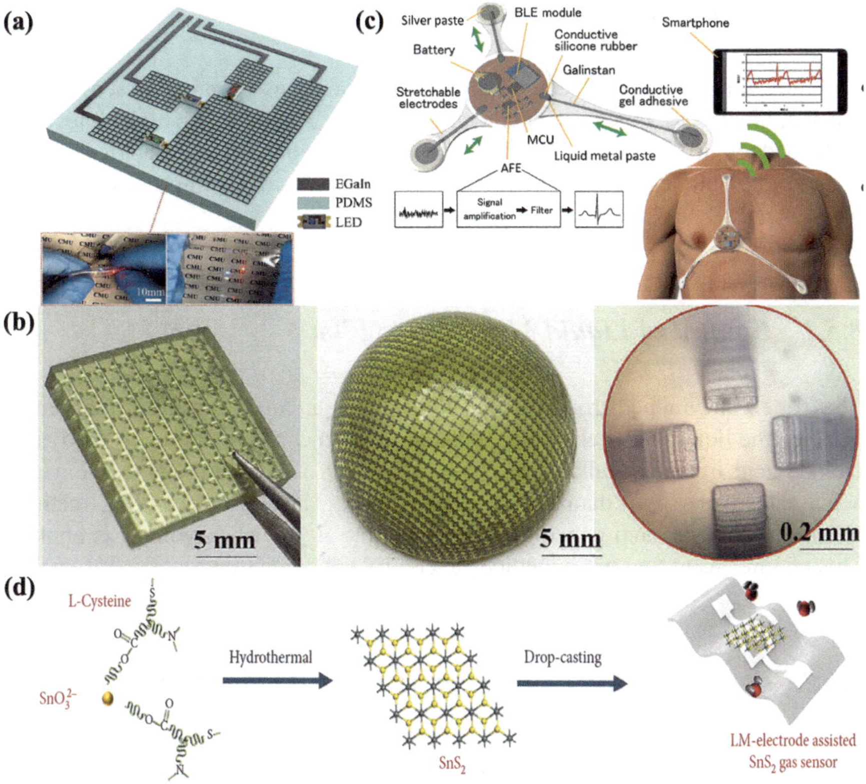

Fig. 8.24 (**a**) Schematic and physical picture of the optically transparent conductive thin film; (**b**) Biomimetic compound eye sensor; (**c**) Highly deformable electrocardiograph schematic; (**e**) Gas sensor preparation schematic

paste and carbon-based conductive rubber were used. These electrodes showed quite high resistance stability under large deformation and multiple stretching use environments. The device developed can achieve a strain of 200% after 100 stretching tests, and the resistance change is only 6%. This device balances resistance stability and interface adaptability, providing an important reference for the stretchable medical device field.

In addition to serving as a wire, LM can also serve as a "carrier" for other sensors. For example, Huang and others studied a flexible electrode based on liquid metal combined with SnS_2 nanomaterial (as shown in Fig. 8.24d), which serves as a wearable gas sensing device. The SnS_2 nanomaterial loaded on the LM flexible electrode can react with NO and H_2O, using this principle, can determine the respiratory pattern of the human body. Based on this, the author has added Bluetooth functionality for remote respiratory monitoring and diagnosis, achieving real-time monitoring of human respiratory and exercise patterns.

8.5 Chapter Summary

Currently, liquid metal has been widely used in strain sensing, pressure sensing, gas sensing, bioelectric signal sensing, and other fields. Thanks to the high elongation, high conductivity, and fluidity of liquid metal, the mechanical and electrical performance of flexible sensors with liquid metal has been effectively improved. However, liquid metal sensors still face many challenges:

8.5.1 Control of Liquid Metal Oxide Film

The oxide film wrapped around the outside of the liquid metal can affect the wetting between the liquid metal and other substrates. The means to remove the oxide film include acid treatment, alkali treatment, and electro-wetting methods. However, acids and alkalis may damage electronic devices, and electro-wetting methods require electrolytes and generate a large amount of gas when voltage is applied. This requires us to find more methods to control the oxide film.

8.5.2 Stability of the Connection Between Liquid Metal and Devices

When liquid metal is connected to rigid devices, the stability of the circuit needs special attention, especially metals that react with liquid metal, such as aluminum. In addition, the connection between the rigid device and the liquid metal is prone to failure during deformation, causing fluctuations in the electrical signal. Although researchers have proposed a series of structural designs to solve this problem, designing a new material interface is still the best solution.

8.5.3 Resolution of Liquid Metal Circuits

With the help of photolithography, the resolution of liquid metal circuits can reach about 500 nm, but there is still a gap compared with traditional silicon-based electronic devices. Future flexible electronic devices need to further reduce the size to improve integration, which puts forward new requirements for precision equipment, the growth of liquid metal oxide films, and phase transition research during the molding process.

8.5.4 *Biocompatibility of Liquid Metal*

At present, many literatures have proven that liquid metal is biocompatible, but the safe amount of liquid metal used in the human body still needs to be studied. There are few reports on long-term use experiments on the human body, and these issues require more in vivo and in vitro experiments to solve. In addition, as a liquid material at room temperature, liquid metal can easily flow and even enter the human body, which requires finding suitable packaging strategies when applying liquid metal to biosensors.

In summary, liquid metal has been proven to be an ideal material for the preparation of flexible sensors. In the future, more research should be carried out on the doping of liquid metal, the control of oxide films, interface properties, phase transition processes, and biocompatibility.

Bibliography

1. DICKEY M D. Stretchable and Soft Electronics using Liquid Metals. Advanced Materials[J]. 2017, 29 (27).
2. ZRNIC D, SWATIK D S. In the resistivity and surface tension of the eutectic alloy of gallium and indium. Journal of the Less-Common Metals[J]. 1969, 18 (1):67-68.
3. DUGGIN M J The thermal conductivity of liquid metal. Physics Letters[J]. 1969, 29 (8):470-471.
4. Yu S, KAVIANY M. Electrical, thermal, and species transport properties of liquid eutectic Ga-In and Ga-In-Sn from first principles. Journal of Chemical Physics[J]. 2014, 140 (6):064303.
5. KOSTER J N. Directional Solidification and Melting of Eutectic Galn. Crystal Research and Technology[J]. 1999, 34 (9):1129-1140.
6. Lu Y, Hu Q, Lin Y, et al. Transformable liquid-metal nanomedicine. Nature Communications [J]. 2015, 6:10066.
7. CICCO A, Di F A, Local Correlations in Liquid and Supercooled Gallium Probed by X-Ray Absorption Spectroscopy. Europhysics Letters (EPL)[J]. 1994, 27 (5):407-412.
8. Sostman H E. Melting point of gallium as a temperature calibration standard. Review of Scientific Instruments[J]. 1977, 48 (2):127-130.
9. Duggin, M. The thermal conductivity of liquid gallium. Physics Letters A[J]. 1969, 29 (8):470-471.
10. Yu S, KAVIANY, MASSOUD. Electrical, thermal, and species transport properties of liquid eutectic Ga-In and Ga-In-Sn from first principles. The Journal of Chemical Physics[J]. 2014, 140 (6):064303.
11. CHEN Q L, WU K J, HE C H. Thermal conductivity of ionic liquids at atmospheric pressure: database, analysis, and prediction using a topological index method. Industrial & Engineering Chemistry Research[J]. 2014, 53 (17):7224-7232.
12. ZRNIC D, SWATIK, D S. On the resistivity and surface tension of the eutectic alloy of gallium and indium. Journal of the Less-Common Metals [J]. 1969, 18 (1):67-68.
13. ZECH, O S, ALEXANDER B, RICHARD, et al. The conductivity of imidazolium-based ionic liquids from (248 to 468) KB Variation of the anion. Journal of Chemical Engineering Data[J]. 2010, 55 (5):1774-1778.

14. CAHILL, J A, KIRSHENBAUM A D. The Density of liquid copper from its melting point (1356 k.) to 2500 k. and an estimate of its critical constants1, 2. The Journal of Physical Chemistry[J]. 1962, 66 (6):1080-1082.
15. HARRISON D A, YAN D, BLAIRS S. The surface tension of liquid copper. The Journal of Chemical Thermodynamics[J]. 1977, 9 (12):1111-1119.
16. HARDY, S C. The surface tension of liquid gallium. Journal of Crystal Growth[J]. 1985, 71 (3):602-606.
17. BRILLO J, Egry, Ivan, Density determination of liquid copper, nickel, and their alloys. International Journal of Thermophysics[J]. 2003, 24 (4):1155-1170.
18. DICKEY M D, CHIECHI R C, LARSEN R J, et al. Eutectic gallium-indium (EGaIn): a liquid metal alloy for the formation of stable structures in microchannels at room temperature. Advanced Functional Materials[J]. 2008, 18 (7):1097-1104.
19. KOLBECK C, LEHMANN J, LOVELOCK K, et al. Density and surface tension of ionic liquids. The Journal of Physical Chemistry B[J]. 2010, 114 (51):17025-17036.
20. CHENTSOV V, SHEVCHENKO V, MOZGOVOI A, et al. Density and surface tension of heavy liquid-metal coolants: Gallium and indium. Inorganic Materials: Applied Research[J]. 2011, 2 (5):468-473.
21. LIU T, SEN P, KIM C J. Characterization of nontoxic liquid-metal alloy galinstan for applications in microdevices. Journal of Microelectromechanical Systems[J]. 2011, 21 (2):443-450.
22. GEDIK E, KURT H, RECEBLI Z. CFD simulation of magnetohydrodynamic flow of a liquid-metal galinstan fluid in circular pipes. Fluid Dynamics and Materials Processing[J]. 2013, 9 (1):23-33.
23. XU Q, OUDALOV N, GUO Q T, et al. Effect of oxidation on the Mechanical properties of liquid gallium and eutectic gallium-indium. Physics of fluids[J]. 2012, 24 (6):063101.
24. LIU S Q, SWEATMAN, K, MCDONALD S, et al. Ga-Based Alloys in Microelectronic Interconnects: A Review. Materials [J]. 2018, 11 (8):1384.
25. Li J, GUO C, WANG, Z S, et al. Electrical stimulation towards melanoma therapy via liquid metal printed electronics on skin. Clinical and Translational Medicine[J]. 2016, 5 (1).
26. FAN L L, DUAN, M H, XIE, Z C, et al. Injectable and Radiopaque Liquid Metal/Calcium Alginate Hydrogels for Endovascular Embolization and Tumor Embolotherapy. Small[J]. 2019, 16 (2).
27. HALLFORS N, KHAN A, DICKEY M D, et al. Integration of pre-aligned liquid metal electrodes for neural stimulation within a user-friendly microfluidic platform. Lab Chip[J]. 2013, 13 (4):522-526.
28. LIU S, SUN X, KEMME N, et al. Can liquid metal flow in microchannels made of its own oxide skin? Microfluidics and Nanofluidics[J]. 2016, 20 (1).
29. REGAN M J, TOSTMANN H, PERSHAN P S, et al. X-ray study of the oxidation of liquid-gallium surfaces[J]. 1997, 55 (16): 10786-10790.
30. CADEMARTIRI L, THUO M M, IJHUIS, C A., et al., Electrical Resistance of AgTS-$(CH_2)_{n-1}CH_3$//Ga_2O_3/EGaIn Tunneling Junctions. The Journal of Physical Chemistry C[J]. 2012, 116 (20):10848-10860.
31. DICKEY M D, Emerging applications of liquid metals featuring surface oxides. ACS Applied Materials & Interfaces [J]. 2014, 6 (21):18369-79.
32. CHIECHI R C, WEISS E A, DICKEY M D, et al. Eutectic Gallium–Indium (EGaIn): A Moldable Liquid Metal for Electrical Characterization of Self-Assembled Monolayers. Angewandte Chemie International Edition[J]. 2008, 47 (1):142-144.
33. PARK J E, KANG H S, KOO M, et al. Autonomous Surface Reconciliation of a Liquid-Metal Conductor Micropatterned on a Deformable Hydrogel. Advanced Materials[J]. 2020, 32 (37):e2002178.
34. DICKEY M D, CHIECHI R C, LARSEN R J, et al. Eutectic Gallium-Indium (EGaIn): A Liquid Metal Alloy for the Formation of Stable Structures in Microchannels at Room Temperature. Advanced Functional Materials[J]. 2008, 18 (7):1097-1104.

35. MOHAMMED M G, XENAKIS A, DICKEY M D, Production of Liquid Metal Spheres by Molding. Metals[J]. 2014, 4 (4):465-476.
36. YU Y, WANG Q, YI L T, et al. Channelless Fabrication for Large-Scale Preparation of Room Temperature Liquid Metal Droplets. Advanced Engineering Materials[J]. 2014, 16 (2):255-262
37. PAIBOON N, SURASSMO S, RUKTANONCHAI, U R, et al. Hydrodynamic control of droplet formation in narrowing jet and tip streaming regime using microfluidic flow-focusing. International Journal of Multiphase Flow[J]. 2022, 150:104013.
38. TEVIS I D, NEWCOMB L B, THUO M, Synthesis of Liquid Core–Shell Particles and Solid Patchy Multicomponent Particles by Shearing Liquids Into Complex Particles (SLICE). Langmuir[J]. 2014, 30 (47):14308-14313.
39. GAO Z F, Z L L, FU W L, et al. Feeding Alginate-Coated Liquid Metal Nanodroplets to Silkworms for Highly Stretchable Silk Fibers. Nanomaterials[J]. 2022, 12 (7):1177.
40. KUMAR R, GHAI V, SAHANI A K. A Surface Modification Approach to Overcome Wetting Behavior of Gallium-Based Liquid Metal Droplets. IEEE Transactions on Nanotechnology[J]. 2022, 21:158-162.
41. Sheng L, He Z Z, Yao, Y Y, et al. Transient State Machine Enabled from the Colliding and Coalescence of a Swarm of Autonomously Running Liquid Metal Motors. Small[J]. 2015, 11 (39):5253-5261.
42. Yuan B, Tan S C, Zhou, Y X, et al. Self-powered macroscopic Brownian motion of spontaneously running liquid metal motors. Science Bulletin[J]. 2015, 60 (13):1203-1210.
43. Thelen J, Dickey M D, Ward T. A study of the production and reversible stability of EGaIn liquid metal microspheres using flow focusing. Lab on a Chip[J]. 2012, 12 (20):3961-3967.
44. Gol B, Tovar-Lopez F J, Kurdzinski M E, et al. Continuous transfer of liquid metal droplets across a fluid–fluid interface within an integrated microfluidic chip. Lab on a Chip[J]. 2015, 15 (11):2476-2485.
45. Tang S Y, Joshipura I D, Lin Y L, et al. Liquid-Metal Microdroplets Formed Dynamically with Electrical Control of Size and Rate. Advanced Materials[J]. 2016, 28 (4):604-609.
46. Tang S Y, Qiao R, Yan S, et al. Microfluidic Mass Production of Stabilized and Stealthy Liquid Metal Nanoparticles. Small[J]. 2018, 14 (21):e1800118.
47. Cheeseman S, Elbourne A, Gangadoo S, et al. Interactions between Liquid Metal Droplets and Bacterial, Fungal, and Mammalian Cells. Advanced Materials Interfaces[J]. 2022, 9 (7).
48. Huang X M, Xu T H, Shen A, et al. Engineering Polymers via Understanding the Effect of Anchoring Groups for Highly Stable Liquid Metal Nanoparticles ACS Applied Nano Materials[J]. 2022.
49. Oloye O, Riches J D, O'Mullane A P. Liquid metal assisted sonocatalytic degradation of organic azo dyes to solid carbon particles. Chemical Communications[J]. 2021, 57 (73):9296-9299.
50. Liu Y, Zhang W, Wang H. Synthesis and application of core–shell liquid metal particles: a perspective of surface engineering. Materials Horizons[J]. 2021, 8 (1):56-77.
51. Ren L, Zhuang J C, Casillas G, et al. Nanodroplets for Stretchable Superconducting Circuits. Advanced Functional Materials[J]. 2016, 26 (44):8111-8118.
52. Yamaguchi A, Mashima Y, Iyoda T. Reversible Size Control of Liquid-Metal Nanoparticles under Ultrasonication. Angewandte Chemie International Edition [J]. 2015, 54 (43):12809-13.
53. Wang Q, Yu Y, Liu J. Preparations, Characteristics and Applications of the Functional Liquid Metal Materials. Advanced Engineering Materials[J]. 2018, 20 (5).
54. Wang D W, Wang X H, Rao W. Precise Regulation of Ga-Based Liquid Metal Oxidation. Accounts of Materials Research[J]. 2021, 2 (11):1093-1103.
55. Kurtjak M, Vukomanović M, Kramer L, et al. Biocompatible nano-gallium/hydroxyapatite nanocomposite with antimicrobial activity. Journal of Materials Science: Materials in Medicine[J]. 2016, 27 (11):170

56. Hohman J N, Kim M, Wadsworth G A, et al. Directing Substrate Morphology via Self-Assembly: Ligand-Mediated Scission of Gallium–Indium Microspheres to the Nanoscale. Nano Letters[J]. 2011, 11 (12):5104-5110.
57. Finkenauer L R, Lu Q Y, Hakem I F, et al. Analysis of the Efficiency of Surfactant-Mediated Stabilization Reactions of EGaIn Nanodroplets. Langmuir[J]. 2017, 33 (38):9703-9710.
58. Mou L, Xia Y, Jiang X. Liquid metal-polymer conductor-based wireless, battery-free epidermal patch. Biosens Bioelectron[J]. 2022, 197:113765.
59. ZHANG M K, YAO S Y, RAO W, et al. Transformable soft liquid metal micro/nanomaterials. Materials Science and Engineering: R: Reports[J]. 2019, 138:1-35.
60. ZHAO W, BISCHOF J L, HUTASOIT J, et al. Single-fluxon controlled resistance switching in centimeter-long superconducting gallium-indium eutectic nanowires. Nano Letters[J]. 2015, 15 (1):153-8.
61. PALLEAU E, REECE S, DESAI S C, et al. Self-healing stretchable wires for reconfigurable circuit wiring and 3D microfluidics. Advanced Materials [J]. 2013, 25 (11):1589-92.
62. GUO W T, ADRIAN J T, GAÑÁN-CALVO A M, et al. Pressure-Driven Filling of Closed-End Microchannel: Realization of Comb-Shaped Transducers for Acoustofluidics. Physical Review Applied[J]. 2018, 10 (5).
63. LIN Y, GORDON O, KHAN M R, et al. Vacuum filling of complex microchannels with liquid metal. Lab on a Chip[J]. 2017, 17 (18):3043-3050.
64. GAO Y, OTA H, SCHALER E W, et al. Wearable Microfluidic Diaphragm Pressure Sensor for Health and Tactile Touch Monitoring. Advanced Materials[J]. 2017, 29 (39).
65. WANG J, YANG S, DING P, et al. Omnidirectional Printing of Soft Elastomer for Liquid-State Stretchable Electronics. ACS Applied Materials & Interfaces [J]. 2019, 11 (20):18590-18598.
66. YUAN X, WU P, GAO Q, et al. Multifunctionally wearable monitoring with gelatin hydrogel electronics of liquid metals. Materials Horizons[J]. 2022, 9 (3):961-972.
67. LAZARUS N, BEDAIR S S, KIERZEWSKI I M, Ultrafine Pitch Stencil Printing of Liquid Metal Alloys. ACS Applied Materials & Interfaces [J]. 2017, 9 (2):1178-1182.
68. Park C W, Moon Y G, Seong H, et al. Photolithography-Based Patterning of Liquid Metal Interconnects for Monolithically Integrated Stretchable Circuits. ACS Applied Materials & Interfaces [J]. 2016, 8 (24):15459-65.
69. JIN Y, LIN Y, KIANI A, et al. Materials tactile logic via innervated soft thermochromic elastomers. Nature Communications [J]. 2019, 10 (1):4187.
70. KIM M G, BROWN D K, BRAND O. Nanofabrication for all-soft and high-density electronic devices based on liquid metal. Nature Communications [J]. 2020, 11 (1):1002.
71. ABBASI R, MAYYAS M, GHASEMIAN M B, et al. Photolithography–enabled direct patterning of liquid metals. Journal of Materials Chemistry C[J]. 2020, 8 (23):7805-7811.
72. Wang X, Liu J. Recent Advancements in Liquid Metal Flexible Printed Electronics: Properties, Technologies, and Applications. Micromachines (Basel)[J]. 2016, 7 (12).
73. ZHENG Y, ZHANG Q, LIU J. Pervasive liquid metal based direct writing electronics with roller-ball pen. AIP Advances[J]. 2013, 3 (11).
74. ZHENG Y, HE Z Z, GAO Y X, et al. Direct Desktop Printed-Circuits-on-Paper Flexible Electronics. Scientific Reports[J]. 2013, 3 (1).
75. KIM D, YOO J H, LEE Y, et al. In Gallium-based liquid metal inject printing, In Proceedings of the 2014 IEEE 27th International Conference on Micro Electro Mechanical Systems (MEMS)[C]. San Francisco, San Francisco, 2014; pp 967-970.
76. WATSON A M, COOK A B, TABOR C E, Electrowetting-Assisted Selective Printing of Liquid Metal. Advanced Engineering Materials[J]. 2019, 21 (10).
77. NEUMANN T V, DICKEY M D. Liquid Metal Direct Write and 3D Printing: A Review. Advanced Materials Technologies[J]. 2020, 5 (9).
78. LI Y, FENG S, CAO S, et al. Printable Liquid Metal Microparticle Ink for Ultrastretchable Electronics. ACS Applied Materials & Interfaces [J]. 2020, 12 (45):50852-50859.
79. MOHAMMED M G, KRAMER R. All-Printed Flexible and Stretchable Electronics. Advanced Materials[J]. 2017, 29 (19).

80. Zhou L Y, Fu J Z, Gao Q, et al. All-Printed Flexible and Stretchable Electronics with Pressing or Freezing Activatable Liquid-Metal–Silicone Inks. Advanced Functional Materials[J]. 2019, 30 (3).
81. LIU S, REED S N, HIGGINS M J, et al. Oxide rupture-induced conductivity in liquid metal nanoparticles by laser and thermal sintering. Nanoscale[J]. 2019, 11 (38):17615-17629.
82. LIU S, KIM S Y, HENRY K E, et al. Printed and Laser-Activated Liquid Metal-Elastomer Conductors Enabled by Ethanol/PDMS/Liquid Metal Double Emulsions. ACS Applied Materials & Interfaces [J]. 2021, 13 (24):28729-28736.
83. WU P C, ZHOU L Y, LV S, et al. Self-sintering liquid metal ink with LAPONITE® for flexible electronics. Journal of Materials Chemistry C[J]. 2021, 9 (9):3070-3080.
84. WANG M, MA C, UZABAKIRIHO P C, et al. Stencil Printing of Liquid Metal upon Electrospun Nanofibers Enables High-Performance Flexible Electronics. ACS Nano[J]. 2021, 15 (12):19364-19376.
85. GUO R, ZHEN Y, HUANG X, et al. Spatially selective adhesion enabled transfer printing of liquid metal for 3D electronic circuits. Applied Materials Today[J]. 2021, 25.
86. JIANG J, ZHANG S, WANG B, et al. Hydroprinted Liquid-Alloy-Based Morphing Electronics for Fast-Growing/Tender Plants: From Physiology Monitoring to Habit Manipulation. Small[J]. 2020, 16 (39):e2003833.
87. YALCINTAS E P, OZUTEMIZ K B, CETINKAYA T, et al. Soft Electronics Manufacturing Using Microcontact Printing. Advanced Functional Materials[J]. 2019, 29 (51).
88. LI X, LI M, XU J, et al. Evaporation-induced sintering of liquid metal droplets with biological nanofibrils for flexible conductivity and responsive actuation. Nature Communications [J]. 2019, 10 (1):3514.
89. ZHU H Y, WANG S L, ZHANG M H, et al. Fully solution processed liquid metal features as highly conductive and ultrastretchable conductors. NPJ Flexible Electronics[J]. 2021, 5 (1).
90. PARK Y G, AN H S, KIM J Y, et al. High-resolution, reconfigurable printing of liquid metals with three-dimensional structures. Science Advances[J]. 2019, 5 (6):eaaw2844.
91. PAN C, KUMAR K, LI J, et al. Visually Imperceptible Liquid-Metal Circuits for Transparent, Stretchable Electronics with Direct Laser Writing. Advanced Materials[J]. 2018, 30 (12):e1706937.
92. LIN Y L, LADD C, WANG S Y, et al. Drawing liquid metal wires at room temperature. Extreme Mechanics Letters[J]. 2016, 7:55-63.
93. FU J H, LIU T Y, CUI Y T, et al. Interfacial Engineering of Room Temperature Liquid Metals. Advanced Materials Interfaces[J]. 2021, 8 (6).
94. KHAN M R, TRLICA C, DICKEY M D. Recapillarity: Electrochemically Controlled Capillary Withdrawal of a Liquid Metal Alloy from Microchannels. Advanced Functional Materials[J]. 2015, 25 (5):671-678.
95. Yu Z W, Yun F F, Wang X L, A novel liquid metal patterning technique: voltage induced non-contact electrochemical lithography at room temperature. Materials Horizons[J]. 2018, 5 (1):36-40.
96. ZHANG J Z, YONG J L, ZHANG C J, et al., Liquid Metal-Based Reconfigurable and Repairable Electronics Designed by a Femtosecond Laser. ACS Applied Electronic Materials[J]. 2020, 2 (8):2685-2691.
97. LI G Y, WU X, LEE D W. Selectively plated stretchable liquid metal wires for transparent electronics. Sensors and Actuators B: Chemical[J]. 2015, 221:1114-1119.
98. WU Y H, XING S T, ZHENG R M, et al. Interface design for enhancing the wettability of liquid metal to polyacrylate for intrinsically soft electronics. Journal of Materials Chemistry C[J]. 2018, 6 (25):6755-6763.
99. LIN W G, QIU W K, TUERSUN Y, et al. Ultrastrong Spontaneous Surface Wetting of Room Temperature Liquid Metal on Treated Metal Surface. Advanced Materials Interfaces[J]. 2021, 8 (18).
100. WANG Y, GONG S, WANG S J, et al. Volume-invariant ionic liquid microbands as highly durable wearable biomedical sensors. Materials Horizons[J]. 2016, 3 (3):208-213.

101. ALI M M, NARAKATHU B B, EMAMIAN S, et al. Eutectic Ga-In Liquid Metal Based Flexible Capacitive Pressure Sensor. 2016 IEEE Sensors[J]. 2016.
102. YANG X F, WANG Y S, SUN H, et al. A flexible ionic liquid-polyurethane sponge capacitive pressure sensor. Sensors and Actuators a-Physical[J]. 2019, 285:67-72.
103. AMJADI M, KYUNG K U, PARK I, et al. Stretchable, Skin-Mountable, and Wearable Strain Sensors and Their Potential Applications: A Review. Advanced Functional Materials[J]. 2016, 26 (11):1678-1698.
104. WAN Y B, WANG Y, GUO C F. Recent progresses on flexible tactile sensors. Materials Today Physics[J]. 2017, 1:61-73.
105. ROLNICK H. Tension coefficient of resistance of metals. Physical Review[J]. 1930, 36 (3):0506-0512.
106. KIM K, CHOI J, JEONG Y, et al. Highly Sensitive and Wearable Liquid Metal-Based Pressure Sensor for Health Monitoring Applications: Integration of a 3D-Printed Microbump Array with the Microchannel. Advanced Healthcare Materials[J]. 2019, 8 (22).
107. PENG H, XIN Y M, XU J, et al. Ultra-stretchable hydrogels with reactive liquid metals as asymmetric force-sensors. Materials Horizons[J]. 2019, 6 (3):618-625.
108. PARK J, LEE Y, HONG J, et al. Giant Tunneling Piezoresistance of Composite Elastomers with Interlocked Microdome Arrays for Ultrasensitive and Multimodal Electronic Skins. ACS Nano[J]. 2014, 8 (5):4689-4697.
109. SU B, GONG S, MA Z, et al. Mimosa-Inspired Design of a Flexible Pressure Sensor with Touch Sensitivity. Small[J]. 2015, 11 (16):1886-1891.
110. GAO Y J, OTA H, SCHALER E W, et al. Wearable Microfluidic Diaphragm Pressure Sensor for Health and Tactile Touch Monitoring. Advanced Materials[J]. 2017, 29 (39).
111. ALI S, MADDIPATLA D, NARAKATHU B B, et al. Flexible Capacitive Pressure Sensor Based on PDMS Substrate and Ga-In Liquid Metal. IEEE Sensors Journal[J]. 2019, 19 (1):97-104.
112. NAGATOMO T, MIKI N. Three-dimensional Electrodes formation using liquid metal in micro channels for 3-axis Capacitive Force Sensor. 2017 39th Annual International Conference of the IEEE Engineering in Medicine and Biology Society (Embc)[J]. 2017:1579-1582.
113. NAKADEGAWA T, ISHIZUKA H, MIKI N. Three-axis scanning force sensor with liquid metal electrodes. Sensors and Actuators a-Physical[J]. 2017, 264:260-267.
114. WON D J, BAEK S, KIM H, et al. Arrayed-type touch sensor using micro liquid metal droplets with large dynamic range and high sensitivity. Sensors and Actuators a-Physical[J]. 2015, 235:151-157.
115. AMJADI M, PICHITPAJONGKIT A, LEE S, et al. Highly Stretchable and Sensitive Strain Sensor Based on Silver Nanowire-Elastomer Nanocomposite. ACS Nano[J]. 2014, 8 (5):5154-5163.
116. LIU H, LI M X, OUYANG C, et al. Biofriendly, Stretchable, and Reusable Hydrogel Electronics as Wearable Force Sensors. Small[J]. 2018, 14 (36).
117. LU T, WISSMAN J, RUTHIKA, et al. Soft Anisotropic Conductors as Electric Vias for Ga-Based Liquid Metal Circuits. ACS Applied Materials & Interfaces[J]. 2015, 7 (48):26923-26929.
118. MENGUC Y, PARK Y L, PEI H, et al. Wearable soft sensing suit for human gait measurement. International Journal of Robotics Research[J]. 2014, 33 (14):1748-1764.
119. YU Y R, GUO J H, SUN L Y, et al. Microfluidic Generation of Microsprings with Ionic Liquid Encapsulation for Flexible Electronics. Research[J]. 2019, 2019.
120. XU M X, QI J J, LI F, et al. Highly stretchable strain sensors with reduced graphene oxide sensing liquids for wearable electronics. Nanoscale[J]. 2018, 10 (11):5264-5271.
121. RUSSO S, RANZANI T, LIU H B, et al. Soft and stretchable sensor using biocompatible electrodes and liquid for medical applications. Soft Robotics[J]. 2015, 2 (4):146-154.
122. WU Y H, ZHEN R M, LIU H Z, et al. Liquid metal fiber composed of a tubular channel as a high-performance strain sensor. Journal of Materials Chemistry C[J]. 2017, 5 (47):12483-12491.

123. ZHOU L Y, FU J Z, GAO Q, et al. All-Printed Flexible and Stretchable Electronics with Pressing or Freezing Activatable Liquid-Metal-Silicone Inks. Advanced Functional Materials[J]. 2020, 30 (3).
124. YOON Y, KIM S, KIM D, et al. Four Degrees-of-Freedom Direct Writing of Liquid Metal Patterns on Uneven Surfaces. Advanced Materials Technologies[J]. 2019, 4 (2).
125. JEONG Y R, KIM J, XIE Z Q, et al., A skin-attachable, stretchable integrated system based on liquid GaInSn for wireless human motion monitoring with multi-site sensing capabilities. NPG Asia Materials[J]. 2017, 9.
126. OH J, KIM S, LEE S, et al. A Liquid Metal Based Multimodal Sensor and Haptic Feedback Device for Thermal and Tactile Sensation Generation in Virtual Reality. Advanced Functional Materials[J]. 2021, 31 (39).
127. KRAMER R K, MAJIDI C, SAHAI R, et al. Soft Curvature Sensors for Joint Angle Proprioception. 2011 IEEE/RSJ International Conference on Intelligent Robots and Systems[J]. 2011.
128. YUAN X, WU P, GAO Q, et al. Multifunctionally wearable monitoring with gelatin hydrogel electronics of liquid metals. Materials Horizons[J]. 2022.
129. KIM S, OH J, JEONG D, et al. Consistent and Reproducible Direct Ink Writing of Eutectic Gallium-Indium for High-Quality Soft Sensors. Soft Robotics[J]. 2018, 5 (5):601-612.
130. COOPER C B, ARUTSELVAN K, LIU Y, et al. Stretchable Capacitive Sensors of Torsion, Strain, and Touch Using Double Helix Liquid Metal Fibers. Advanced Functional Materials[J]. 2017, 27 (20).
131. TANG L X, CHENG S Y, ZHANG L Y, et al. Printable Metal-Polymer Conductors for Highly Stretchable Bio-Devices. iScience[J]. 2018, 4:302.
132. LI X K, LI M J, ZONG L, et al. Liquid Metal Droplets Wrapped with Polysaccharide Microgel as Biocompatible Aqueous Ink for Flexible Conductive Devices. Advanced Functional Materials[J]. 2018, 28 (39).
133. YUN G L, TANG S Y, SUN S S, et al. Liquid metal-filled magnetorheological elastomer with positive piezoconductivity. Nature Communications[J]. 2019, 10.
134. KONISHI S, KAKEHI Y, MORI F, et al. Smooth transportation of liquid metal droplets in a microchannel as detected by a serially arranged capacitive device. Scientific Reports [J]. 2021, 11 (1):7048.
135. KIM M G, ALROWAIS H, PAVLIDIS S, et al. Size-Scalable and High-Density Liquid-Metal-Based Soft Electronic Passive Components and Circuits Using Soft Lithography. Advanced Functional Materials[J]. 2017, 27 (3).
136. FASSLER A, MAJIDI C. Soft-matter capacitors and inductors for hyperelastic strain sensing and stretchable electronics. Smart Materials and Structures[J]. 2013, 22 (5).
137. DO T N, PHAN H, NGUYEN T Q, et al. Miniature Soft Electromagnetic Actuators for Robotic Applications. Advanced Functional Materials[J]. 2018, 28 (18).
138. ZHOU L Y, GAO Q, ZHAN J F, et al. Three-Dimensional Printed Wearable Sensors with Liquid Metals for Detecting the Pose of Snakelike Soft Robots. ACS Applied Materials & Interfaces[J]. 2018, 10 (27):23208-23217.
139. JIN Y, LIN Y L, KIANI A, et al. Materials tactile logic via innervated soft thermochromic elastomers. Nature Communications[J]. 2019, 10.
140. Pu X J, Guo H Y, Chen J, et al. Eye motion triggered self-powered mechnosensational communication system using triboelectric nanogenerator. Science Advances[J]. 2017, 3 (7):e1700694.
141. TANG W, JIANG T, FAN F R, et al. Liquid-metal electrode for high-performance triboelectric nanogenerator at an instantaneous energy conversion efficiency of 70.6%. Advanced Functional Materials[J]. 2015, 25 (24):3718-3725.
142. YANG Y Q, SUN N, WEN Z, et al. Liquid-Metal-Based Super-Stretchable and Structure-Designable Triboelectric Nanogenerator for Wearable Electronics. ACS Nano[J]. 2018, 12 (2):2027.

143. DONG C Q, LEBER A, DAS G T, et al. High-efficiency super-elastic liquid metal based triboelectric fibers and textiles. Nature Communications[J]. 2020, 11 (1).
144. ZHOU Y J, BAYRAM Y, DU F, et al. Polymer-Carbon Nanotube Sheets for Conformal Load Bearing Antennas. IEEE Transactions on Antennas and Propagation[J]. 2010, 58 (7):2169-2175.
145. NIKOLAOU S, PONCHAK G E, PAPAPOLYMEROU J, et al. Conformal double exponentially tapered slot antenna (DETSA) on LCP for UWB applications. IEEE Transactions on Antennas and Propagation[J]. 2006, 54 (6):1663-1669.
146. COSKER M, LIZZI L, FERRERO F, et al. Realization of 3-D Flexible Antennas Using Liquid Metal and Additive Printing Technologies. IEEE Antennas and Wireless Propagation Letters[J]. 2017, 16:971-974.
147. KUBO M, LI X F, KIM C, et al. Stretchable Microfluidic Radiofrequency Antennas. Advanced Materials[J]. 2010, 22 (25):2749.
148. SO J H, THELEN J, QUSBA A, et al. Reversibly Deformable and Mechanically Tunable Fluidic Antennas. Advanced Functional Materials[J]. 2009, 19 (22):3632-3637.
149. CHENG S, RYDBERG A, HJORT K, et al. Liquid metal stretchable unbalanced loop antenna. Applied Physics Letters[J]. 2009, 94 (14).
150. HAYES G J, SO J H, QUSBA A, et al. Flexible Liquid Metal Alloy (EGaIn) Microstrip Patch Antenna. IEEE Transactions on Antennas and Propagation[J]. 2012, 60 (5):2151-2156.
151. CHENG S, WU Z G. Microfluidic stretchable RF electronics. Lab on a Chip[J]. 2010, 10 (23):3227-3234.
152. HUANG Y A, WANG Y Z, XIAO L, et al. Microfluidic serpentine antennas with designed mechanical tunability. Lab on a Chip[J]. 2014, 14 (21):4205-4212.
153. PAN C F, KUMAR K, LI J Z, et al. Visually Imperceptible Liquid-Metal Circuits for Transparent, Stretchable Electronics with Direct Laser Writing. Advanced Materials[J]. 2018, 30 (12).
154. ZHANG K P, LIAO Y F, QIU B, et al. 3D Printed Embedded Metamaterials. Small[J]. 2021, 17 (50):e2103262.
155. TAKAYA M, MATSUDA R, INAMORI G, et al. Transformable Electrocardiograph Using Robust Liquid-Solid Heteroconnector. ACS Sensors[J]. 2021, 6 (1):212-219.
156. HUANG Y F, YANG F, LIU S H, et al. Liquid Metal-Based Epidermal Flexible Sensor for Wireless Breath Monitoring and Diagnosis Enabled by Highly Sensitive SnS_2 Nanosheets. Research[J]. 2021, 2021:1-13.

Chapter 9
Intelligent Sensor Devices Based on Active Micro-Nano Probes of Micro-Nano Motors

9.1 Micro-Nano Motors

9.1.1 Background Introduction of Micro-Nano Motors

The 1970s American science fiction film "Fantastic Voyage" tells the story of five doctors who were shrunk to one-millionth of their size and injected into a patient's blood vessels to perform surgery. Since then, people have begun to imagine entering the human body with micro-nano scale robots to solve some bottleneck problems in the medical field. In fact, there are many natural micro-nano motors in nature. For example: functional proteins in cells can move along the microtubules in the cells and transport protein molecules by consuming adenosine triphosphate (ATP), ensuring the normal operation of various functions in the cells, these functional proteins with mobility are natural micro-nano robots; E. coli can propel itself to obtain food and avoid adverse factors through flagellar movement; animal sperm cells can also achieve combination with egg cells through flagellar movement and their own chemotaxis characteristics. Inspired by these natural biological robots, scientists have researched and designed micro-nano structures that have autonomous driving ability and can achieve specific tasks by converting the energy of the surrounding environment, which we usually call micro-nano motors (or micro-nano robots). Micro-nano motors exhibit controllable movement ability at the micro-nano scale, and have important applications in drug delivery, minimally invasive surgery, in vivo sampling, single-cell operation, environmental governance, biological sensing detection and other fields, breaking new ground in the medical field of minimally invasive and highly specific treatment methods.

X. Ma et al., *Functional Micro/Nanomaterials for Intelligent Sensing Devices*,
https://doi.org/10.1007/978-981-96-4990-7_9

9.1.2 Driving Mechanism of Micro-Nano Motors

At the micro-nano scale, the rapid reduction in size leads to a rapid decrease in the Reynolds number, with viscous resistance dominating and inertia force negligible. This means that the driving methods and mechanisms of machines in the macro world (high Reynolds number) are no longer applicable at the micro-nano scale. In the 1970s, Purcell gave a lecture called "Life in a Low Reynolds Number Environment", proposing many pioneering theories and initiating nearly half a century of research on the locomotion mechanism of microorganisms. The process of scallop opening and closing is symmetrical in time, and if under low Reynolds number conditions, scallop opening and closing will not produce net movement. Research shows that in order to swim in a low Reynolds number environment, microorganisms need to break the symmetry of movement in time, which is the so-called scallop theorem. This is also the guideline followed in the design of different self-propelled micro-nano motors. Next, we will briefly introduce the main driving mechanisms of micro-nano motors, which are roughly divided into chemical drive, external field drive, and biomass drive (Fig. 9.1).

9.1.2.1 Chemical Drive

Bubble Drive

Bubble-driven micro-nano motors can achieve spontaneous directional movement by breaking the symmetry of bubble release direction. Specifically, it uses the catalyst material loaded on the micro-nano motor to react with the substrate (such as hydrogen peroxide) in the solution to produce bubbles, and the separation or bursting of the bubbles generates a recoil action to drive the motor. Because the rare metal platinum (Pt) has good catalytic activity of decomposing H_2O_2, it has become the most widely used inorganic catalyst for the preparation of micro-nano motors of various structures. Considering the scarcity and high cost of Pt, researchers also use other chemical reactions to replace Pt catalysts to achieve the autonomous driving of micro-nano motors. For example, based on the reaction of metal with water or acid to produce hydrogen bubbles, Mg/Al-based micro-nano motors have been manufactured, which are used in the digestive tract where stomach acid is present. In addition, due to the good catalytic performance and biocompatibility of enzymes, hydrogen peroxide enzymes can also be used to replace Pt to achieve hydrogen peroxide catalyzed bubble driving.

Self-propulsion Drive

Self-propulsion drive mainly includes self-electrophoresis drive and self-diffusiophoresis drive. (1) Self-electrophoresis drive: Redox reactions occur spatially separated, electrons are transferred inside the micro-nano motor, and protons

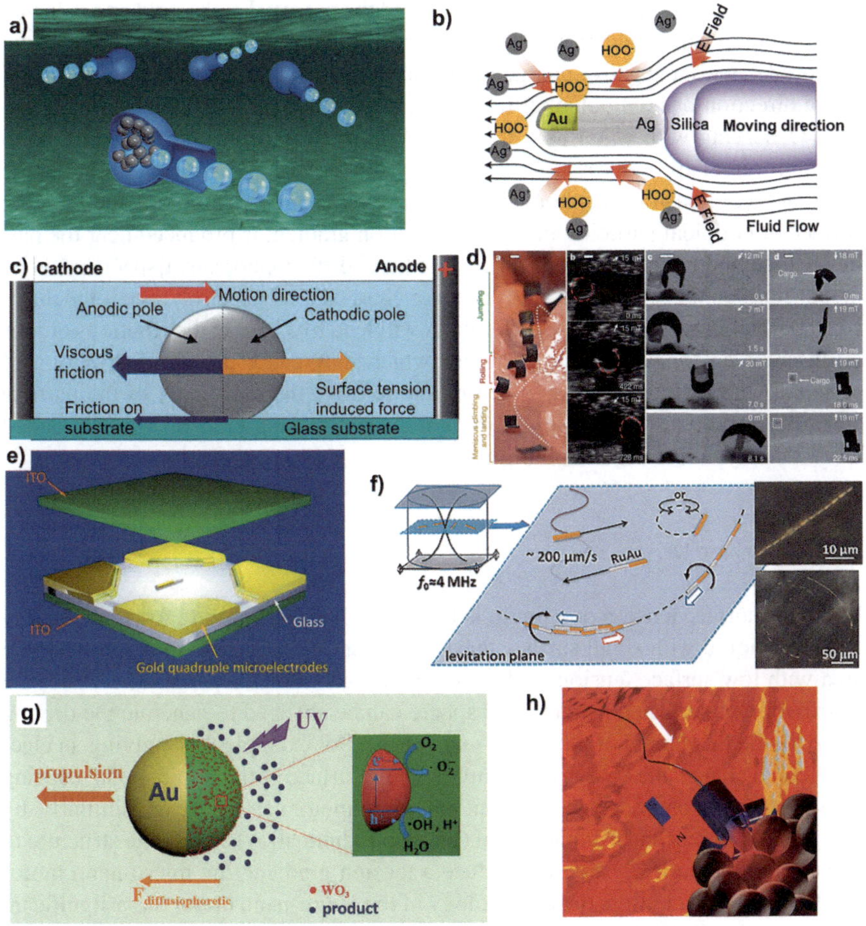

Fig. 9.1 Different driving methods of micro-nano motors. (**a**) Bubble-driven vase micro-nano motor. (**b**) Schematic diagram of self-diffusiophoresis micro-nano motor drive mechanism. (**c**) Schematic diagram of the force on gallium-indium alloy liquid metal in alkaline solution. (**d**) Flexible variable magnetic robot crawling in the stomach. (**e**) Electrically driven rod-shaped micro-nano motor realizes three-dimensional movement under an electric field. (**f**) Au-Ru metal chain performs axial movement along the axis in the ultrasonic field. (**g**) Schematic diagram of light-driven micro-nano motor drive. (**h**) Schematic diagram of sperm cell hybrid micro-nano motor movement

are generated and enriched at the anode to form a local electric field. For example: on the surface of the gold-platinum bimetallic rod, the electrochemical oxidation half-reaction of hydrogen peroxide occurs at the platinum end, and the electrochemical reduction half-reaction of hydrogen peroxide occurs at the gold end. Due to the asymmetric chemical reaction on the surface of the metal rod, an excess of hydrogen ions is produced at the platinum end, and more hydrogen ions are consumed at the gold end, forming a concentration gradient of hydrogen ions from the platinum end to the gold end. Since hydrogen ions carry a positive charge, the potential near the platinum end of the metal rod is higher, and the potential at the gold end is lower, forming an electric field from the platinum end to the gold end. Since the metal microrod carries a negative charge on the surface in aqueous solution, it will move in the direction of high potential in the electric field formed by the chemical reaction. Observing with water as a reference system, it will be found that the gold-platinum bimetallic microrod spontaneously moves in H_2O_2 with platinum as the front end. (2) Self-diffusiophoresis drive: When a chemical reaction occurs on the surface of colloidal particles and a concentration gradient is produced near the particles, the resulting particle motion is called self-diffusiophoresis. Especially when the reaction on the particle surface produces charged ions (electrolytes), if the diffusion coefficients of the ions are different, it will lead to different distributions of ions at different positions near the particles, which may cause uneven distribution of potential, and then generate an electric field to drive the movement of the particle and nearby particles. The particle motion caused by this kind of effect is called electrolyte-type self-diffusiophoresis.

Surface Tension Drive

When the surface tension of a liquid is large, it will exert a pulling force on the surrounding liquid with small surface tension, prompting the liquid to move from the area with low surface tension to the area with high surface tension. Therefore, the distribution of surface tension on the sphere can be changed to generate the driving force for the movement of the micro-nano motor. For example, by applying an electric field to change the charge distribution on the surface of the liquid metal, causing uneven distribution of surface tension, the micro motor can be driven; similarly, by using an aluminum sheet-liquid metal (gallium indium alloy) composite structure to react with sodium hydroxide to produce a tension gradient, the micro-nano motor can be driven; in addition, the morphology of the micro-nano motor has a significant impact on the distribution of surface charges, and the surface tension gradient can be formed by changing the distribution of surface charges to drive the micro-nano motor; there is also a design of microfluidic channels to introduce acid and alkali solutions on both sides of the liquid metal, resulting in uneven charge distribution, creating a tension gradient, and then driving the micro-nano motor.

9.1.2.2 External Field Drive

Magnetic Field Drive

Magnetic field driving is one of the most common driving methods for externally powered micro-nano motors. In nature, there are organisms with self-driving capabilities, such as bacteria and sperm cells, which can move through flagellar motion. Research has found this to be a highly efficient mode of movement. Therefore, researchers mimic this flagellar driving method to prepare helical magnetic micro-nano motors, applying an adjustable rotating magnetic field externally to achieve the purpose of driving the micro-nano motor. Magnetic field-driven micro-nano motors are mainly divided into: (1) Under a gradient magnetic field, magnetic particles or rod-shaped micro-nano motors are subjected to magnetic field forces and generate movement, which can be used for on-demand navigation to target locations and other application scenarios. (2) By rotating the magnetic field, the helical micro-nano motor is driven to move forward along the axis. A rotating magnetic field is generated by a Helmholtz coil, and the magnetic micro-nano motor inside the cell moves controllably within the cell under the action of the rotating magnetic field. (3) The reaction force generated by the interaction between the micro-nano motor and the substrate drives the micro-nano motor to move. Peanut-shaped magnetic micro-nano motors, under the conical rotating magnetic field generated by the electromagnetic coil group, can achieve rotational movement through interaction with the surface substrate. Compared to previously reported two-dimensional movements, this movement has the advantage of being able to overcome obstacles, improving its adaptability in different terrain environments, and better application in complex internal environments.

Electric Field Drive

Electric field driving is achieved by applying an external electric field, using electrophoresis or electroosmosis, to drive the micro-nano motor. A controllable electric field is formed in three-dimensional space by adding conductive glass on the Z-axis and metal electrodes on the X–Y plane. By changing the distribution of the electric field on different axes, the direction of the micro-nano motor can be controlled, and the transport and release of cargo can also be achieved by adjusting the electric field.

Ultrasonic Drive

When an ultrasonic is applied to the liquid, there are different pressure nodes in the liquid, forming a pressure difference, which will drive the micro-nano motor to move towards the pressure node. The asymmetric structure will increase the pressure difference experienced by the micro-nano motor, correspondingly obtaining a faster movement speed. Wang and others used MHz frequency ultrasound to drive rod-shaped metal robots. The ultrasound emitted from the top interacts and superimposes with the bottom ultrasound, forming a standing wave nodal plane in the liquid. The micro-nano motor can move within the standing wave nodal plane, and the speed of the micro-nano motor can be adjusted by different ultrasound frequencies. Similarly, by applying different ultrasound waves, a pressure gradient is formed in the liquid, forcing the micro-nano motor to gather in the low-pressure area. When the ultrasound is turned off, the pressure gradient disappears, and the micro-nano motor quickly disperses due to Brownian motion, thereby achieving the aggregation and dispersion of the robot. In addition, ultrasound can be used to trigger volatile substances to form bubbles, indirectly driving the movement of the micro-nano motor, and controlling the movement of the micro-nano motor by turning the ultrasound on and off.

Light Drive

Light is ubiquitous in nature, so light-driven micro-nano motors have a wide range of applications. The driving mechanisms of light-driven micro-nano motors are mainly divided into three categories: photocatalytic self-diffusiophoresis, photocatalytic self-electrophoresis, and photothermal effect self-thermophoresis. (1) Self-diffusiophoresis: When light shines on the surface of the micro-nano motor, the surface material will undergo photocatalytic reactions or photodecomposition reactions to release product ions. At the same time, due to the different diffusion speeds of ions, a local concentration gradient field will form around the micro-nano motor, thereby forming osmotic flow and driving the micro-nano motor to move. (2) Self-electrophoresis: When light is irradiated on a micro-nano motor with photocatalytic capabilities (for example: TiO_2-Au), hole-electron pairs will be generated, and the holes are distributed on one side of the semiconductor (TiO_2) to undergo oxidation reactions with water, while the electrons will migrate to the metal side (Au), reducing H^+ to H_2. However, because the reduction rate is less than the oxidation rate, some H^+ will be surplus, eventually leading to the formation of a local electric field around the robot. The charged micro-nano motor will move under the local electric field. (3) Self-thermophoresis: When the surface of the micro-nano motor is coated with photothermal materials (for example: Au), after being illuminated, due to the photothermal effect, a local temperature gradient field will form around the micro-nano motor, thereby forming local convection and driving the micro-nano motor to move in the opposite direction.

9.1.2.3 Biomass Drive

Naturally existing micro-nano structures with self-driving capabilities, such as sperm cells, bacteria, and some microorganisms, can be combined with natural biological bodies with motion functions to prepare biomass-driven micro-nano motors. Because naturally existing biological structures have good biocompatibility, they have good application prospects in the fields of drug delivery and medical detection. However, cells and other entities with natural motion capabilities have poor motion control and cannot effectively complete predetermined targets such as targeted transport. Therefore, it is necessary to combine with external auxiliary structures to improve their control performance and better complete the predetermined tasks. Researchers have combined sperm cells with magnetic nano caps to form biomass hybrid micro-nano motors, controlling their movement direction through an external magnetic field, and using the movement of sperm flagella to provide power for the overall motor. It is also possible to prepare magnetic micron cones by self-curling, and sperm cells drill into the tube to form a biomass hybrid micro-nano motor. It is also possible to prepare magnetic flower-like structures by photolithography, and because the head of the sperm cell has a large specific surface area, it can achieve drug loading, and then reach the target location for controlled drug release.

9.1.3 Applications of Micro-Nano Motors

9.1.3.1 Drug Delivery

Drug therapy is one of the important treatment methods in clinical medicine. There are two traditional ways of administering drugs: one is oral intestinal administration, and the other is injection into the blood. However, both methods belong to passive drug transport in the body, which may lead to excessive drug dosage and off-target administration, producing side effects on the human body. In the diagnosis and treatment of tumor diseases, targeted drug delivery treatment is considered to have fewer side effects than traditional chemotherapy and radiotherapy. Because the disease microenvironment is different from normal tissues, such as higher temperature in inflamed organs, lower pH around cancer tissues, etc., intelligent controllable drug delivery systems can respond to external environmental stimuli such as pH, temperature, light, oxidants, etc., and selectively release drugs to the target disease site, thereby improving the efficiency of drug administration. Due to the biocompatibility and controllable mobility of micro-nano motors, they bring the possibility of transformative technological innovation for biomedical applications. From targeted drug delivery to photodynamic therapy and other applications, it has been proven that micro-nano motor-mediated targeted drug delivery can effectively find specific lesion sites and effectively deliver therapeutic drugs while avoiding adverse side effects. On the other hand, the drug delivery system needs to overcome various physiological and pathological obstacles, and the active and controllable movement of micro-nano motors to break through these biological barriers can further optimize the effect of drug delivery (Fig. 9.2a).

9.1.3.2 Minimally Invasive Surgery

Accurate positioning of the incision is key to the success of minimally invasive surgery, however, performing minimally invasive surgery with traditional surgical knives and other medical instruments is very challenging. Micro-nano motors, with their autonomous movement capabilities and miniaturized structures, are considered to have broad application prospects in various biomedical applications, especially in minimally invasive surgery. The natural advantage of external magnetic drive mode can penetrate biological tissues to provide forward propulsion or rotational torque to micro-nano motors. In addition, some microstructures may help perform precise navigation or cell transport and other fine operations in complex

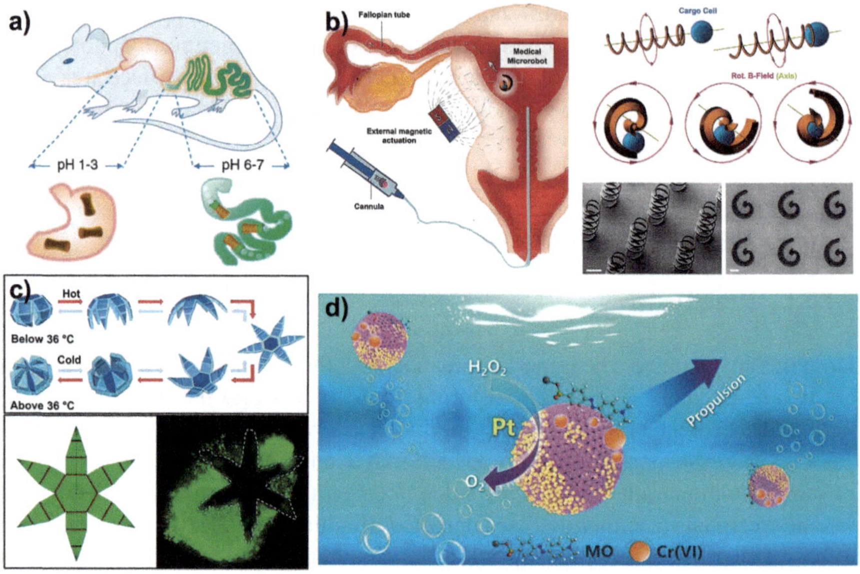

Fig. 9.2 Applications of micro-nano motors in biomedicine. (**a**) Schematic diagram of micro-nano motors moving in the digestive tract and drug delivery. (**b**) Rotating helical micro-nano motors for minimally invasive surgery in vivo. (**c**) Temperature-sensitive micro-nano motors for live tissue sampling. (**d**) Schematic diagram of the movement and pollutant removal mechanism of self-assembled organic framework micro-nano motors

restricted internal environments. By changing the thickness and stress of the material, different diameter micro-nano instruments can be easily manufactured to meet different ultra-micro surgical requirements. With the help of this controllable new type of micro-tool, the research of minimally invasive surgery will achieve new progress and breakthroughs (Fig. 9.2b).

9.1.3.3 In Vivo Sampling

Surgical procedures are used to cut, clamp, or puncture live tissue samples or cell samples from the patient's body for pathological examination by pathologists. Biopsies and pathological tests are usually conducted to gain a deeper understanding of whether a lesion/tumor is cancerous, to determine the type of cancer, and the nature of the cancer, etc., thereby guiding the patient's subsequent overall treatment plan, including surgery, radiation, and drug treatment. Conventional biopsies are invasive and destructive, having a significant impact on the patient. Micro-nano motors, with their small size and task execution capabilities, have a wide range of potential applications in the field of live tissue inspection. Micro-nano motors can capture individual cells or collect extremely small amounts of tissue samples from healthy or diseased organs (including breasts, lungs, liver, skin, prostate, etc.) for further disease diagnosis, with high specificity and selectivity. The application of micro-nano motors in cell manipulation refers to the precise mechanical operations such as cell separation, capture, transmission, and imaging without changing the normal physiological characteristics of the cells. Magnetic micro-robots with these functions, also known as micro-grippers, can perform functions similar to human hands under small size conditions, such as picking up and putting down, for example, thermally responsive flexible magnetic micro-robots (Fig. 9.2c).

9.1.3.4 Environmental Governance

Self-propelled micro-nano motors have a wide range of application prospects in the field of environmental governance, such as catalytic degradation of organic pollutants, adsorption of heavy metal ions, killing of pathogens, oil-water separation, etc. Micro-nano motors can provide a solution for dealing with these toxic and persistent pollutants. The polluted wastewater usually contains cations and anions, organic matter, and oils, which are toxic to the ecosystem. Adsorption is a widely used, practical, low-cost, and easy-to-operate method for purifying water pollutants. Traditional adsorption purification relies on the passive diffusion process of the adsorbed material and the physicochemical properties of the adsorbent, which is usually a time-consuming process. With the development of nanoscience and nanotechnology, researchers have designed moving micro-nano motor cleaners to accelerate their adsorption and removal of pollutants in wastewater. The newly developed motion purification technology uses the self-propulsion characteristics of the adsorbent to break through the limitations of the natural diffusion process on activated carbon adsorption purification. Accelerating the mixing of the solution is crucial for improving the efficiency and speed of many physicochemical processes, including adsorption. Therefore, the efficiency of active micro-nano motors in removing pollutants is higher, and the treatment time is shorter. In addition, bubble-propelled micro-nano motors can provide additional stirrers, and the generated bubbles will float to the surface, enhancing fluid convection. Furthermore, by decomposing H_2O_2 fuel to produce highly reactive free radicals, the degradation of organic pollutants can also be achieved. The Fenton reaction uses ferrous ions as catalysts and hydrogen peroxide as oxidants to produce oxidizing free radicals. Due to the advantages of the above-mentioned micro-nano motors in water purification, different types of pollutants, such as heavy metals, nerve toxins, biochemical toxins, and dyes, can achieve efficient and rapid decontamination or degradation treatment through the good adsorption performance and active movement characteristics of micro-nano motors. However, the high cost of manufacturing micro-nano motors and the small volume of wastewater sample treatment have imposed many restrictions on their practical applications (Fig. 9.2d).

9.2 Principles and Applications of Micro-Nano Motors

9.2.1 Intelligent Sensing Based on Micro-Nano Motor Motion Behavior Analysis

Self-propelled motion is one of the most important characteristics of micro-nanomotors. Since the invention of micro-nano motors, researchers have conducted a large amount of research on their motion behavior and mechanisms. During this period, researchers found that different substances have inhibitory effects on the movement of certain micro-nano motors, and have developed sensing methods based on changes in micro-nano motor motion behavior. In this section, we will introduce the sensing applications of micro-nano motors in metal ion, organic or polymer, and microbial detection according to different response mechanisms.

9.2.1.1 Ion Response

As shown in Fig. 9.3, taking the bimetallic gold-platinum nanomotor as an example, due to the different catalytic reactions of platinum and gold to hydrogen peroxide oxidation, the motor will move axially. Researchers found that in the presence of silver ions, the gold-platinum nanomotor will show a significant acceleration phenomenon, while in the presence of other ions such as lead, manganese, nickel, copper, etc., the speed of the nanomotor will decrease. This is because silver ions will undergo underpotential deposition on the surface of the nanomotor under the action of hydrogen peroxide, thereby enhancing the catalytic activity of the motor. On the contrary, for some heavy metal ions mentioned above, their presence will inhibit the

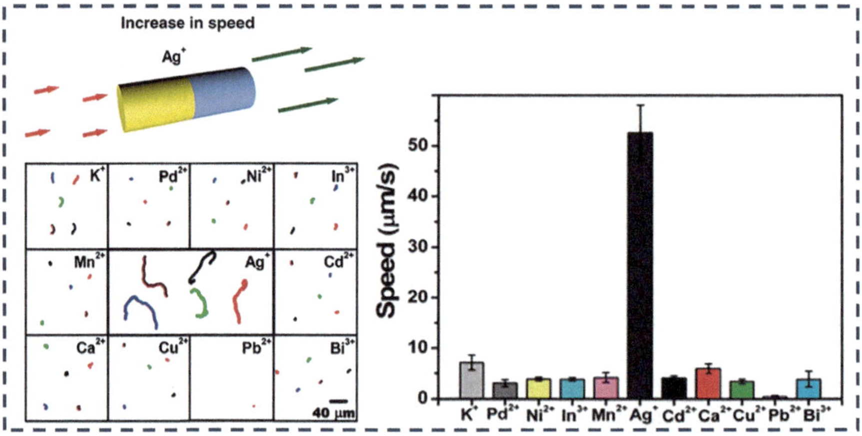

Fig. 9.3 Metal ion response of a bimetallic gold-platinum nanomotor based on motion changes

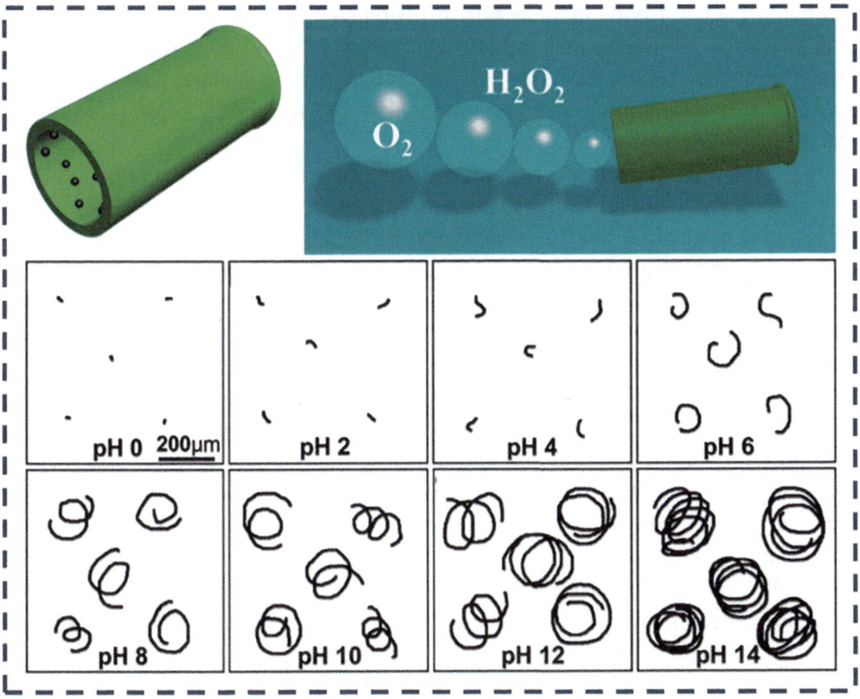

Fig. 9.4 pH-responsive bullet-shaped micro-nano motor based on motion changes

catalytic activity of the metal atoms on the motor surface. According to the acceleration or deceleration response of micro-nano motors to different ions, corresponding ion sensors can be prepared. By analyzing the motion trajectory images of micro-nanomotors, the increase or decrease in the speed of nanomotor motion can complete the quantitative detection of these ions to be tested.

In addition to metal ions, changes in the motion behavior of micro-nano motors can also be applied to hydrogen ion detection, i.e., pH sensing, as shown in Fig. 9.4. The bullet-shaped micro-nano motor, whose shell is composed of gelatin and loaded with platinum nanoparticles inside, can undergo self-propelled motion in a hydrogen peroxide solution based on the bubble propulsion mechanism. At the same time, research has found that the motion of this type of micro-nano motor will be affected by pH. Within the range of 0–14, as the pH continues to increase, the speed of this type of micro-nano motor will also continue to increase. This is due to the different structures of gelatin at different pH and the different catalytic effects of platinum nanoparticles. As the pH continues to increase, the opening of the gelatin shell will continue to increase, the frequency of bubble production will continue to accelerate, and the propulsion effect will continue to strengthen. In addition, platinum nanoparticles will catalyze hydrogen peroxide molecules directly into two hydroxyl radicals at low pH, and form a concentration difference at both ends of the bullet shell motor, at which time the driving mode is self-diffusion swimming; while at high pH, platinum nanoparticles will catalyze hydrogen peroxide to produce oxygen, at which time it is a better bubble recoil drive. Therefore, under the dual pH response of gelatin structure and platinum nanoparticle catalytic performance, micro-nano motors can achieve pH sensing detection.

In recent years, enzyme-driven micro-nano motors have gradually become one of the important directions in the research of micro-nano motors due to their excellent driving ability and biocompatibility. General micro-nano motors can use hydrogen peroxide enzyme and urease, respectively using hydrogen peroxide and urea as fuel, to achieve self-propelled motion. Common heavy metal ion pollutants (such as mercury ions, copper ions, etc.) often affect the activity of the enzyme. Taking hydrogen peroxide enzyme as an example, the thiol and amino groups of the enzyme can bind with mercury ions and copper ions respectively, forming complexes, changing the structure of the bioenzyme, leading to a significant decrease in enzyme activity, and thus causing a decrease in the speed of the micro-nano motor. Therefore, the amount of heavy metal ions in the solution can be quantitatively detected based on the change in the speed of the enzyme-driven micro-nano motor.

In addition to the above-mentioned sensing methods that change the motion behavior of micro-nano motors by affecting their chemical reaction activity, some micro-nano motors with novel structures and material designs can also be used for sensing detection. As shown in Fig. 9.5, researchers have used electrospinning to create a spindle-shaped nanomotor composed of polycaprolactone and sodium dodecyl sulfonate (an anionic surfactant). This motor will slowly release the surfactant in the solution, causing a change in the local solution surface tension. Due to the presence of the Marangoni effect, the micro-nano motor moves towards the side with higher surface tension. This movement highly depends on the surface tension of the liquid, so the liquid environment will have

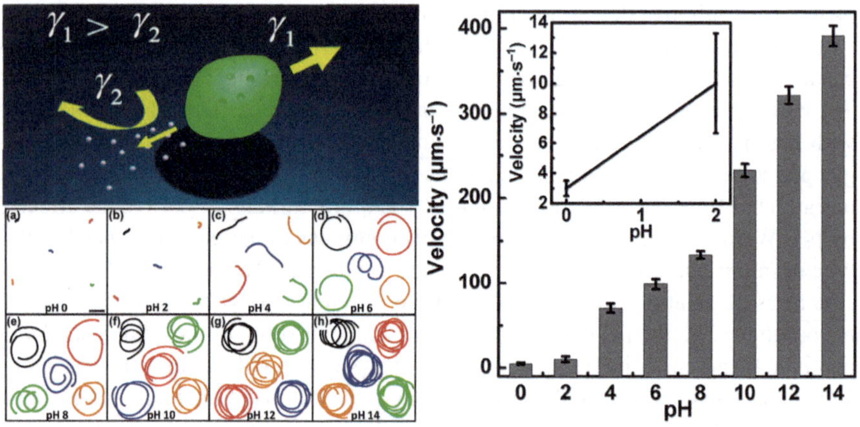

Fig. 9.5 pH-responsive spindle-shaped micro-nano motor based on motion changes

a huge impact on it. In this study, the surface tension of water is independent of pH, but the surface tension of the surfactant solution is affected by pH. As the pH increases, the surface tension between the surfactant solution and the micro-nano motor will decrease. In other words, as the pH increases, the difference in surface tension between the micro-nano motor and the liquid will be greater, the driving force of the motor will be greater, and ultimately the speed of the motor will increase, thereby completing the pH sensing detection based on motion speed analysis.

9.2.1.2 Organic/Macromolecule Response

Micro-nano motors, due to their small size and self-propelled motion characteristics, are considered to have important application prospects in the fields of biomedical detection and targeted treatment, which has aroused the research interest of many scholars. During this period, researchers have developed sensing detection methods and devices for certain organic or high molecular weight test substances based on the motion behavior characteristics of micro-nano motors.

Uric acid is a major metabolic product in the blood and is also one of the important markers in biomedical testing. Researchers have found that the presence of uric acid inhibits the motility of micro-nano motors driven by hydrogen peroxide as fuel. This is because during the process of catalyzing the decomposition of hydrogen peroxide, the hydroxyl free radical, an important intermediate product, is cleared by the reaction with uric acid in both the bubble-driven mechanism and the self-diffusion swimming-driven mechanism, making the micro-nano motor unable to generate sufficient driving power. Therefore, as the concentration of uric acid increases, the speed of the micro-nanomotor gradually decreases, and above a certain concentration, it loses its motility. In addition, there are a large number of blood proteins in the blood, which also have a significant impact on the movement of micro-nano motors. This phenomenon is particularly evident in micro-nano motors with platinum as a catalyst. Blood proteins bind to platinum nanoparticles through their contained thiol groups, directly adsorbing on the surface of the micro-nanomotor, thereby hindering the contact between the surface catalyst and the driving fuel, reducing the active sites of the micro-nanomotor, and hindering the self-driving movement of the micro-nanomotor. Based on the movement response characteristics of micro-nanomotors to these organic substances or large biological molecules, corresponding intelligent sensor devices can be developed. Of course, the mutual influence between different molecular tests needs to be considered, and the specificity and accuracy of detection can be improved through specific pretreatment or the design of micro-nanomotor structure and materials. The catalytic activity of biological enzymes strongly depends on their special protein conformation structure, so the activity of enzymes will change with the change of their conformation structure. At the same time, factors such as temperature, pH, inorganic molecules, and organic molecules will cause changes in the spatial conformation structure of biological enzymes. Therefore, micro-nanomotors driven by enzymes can

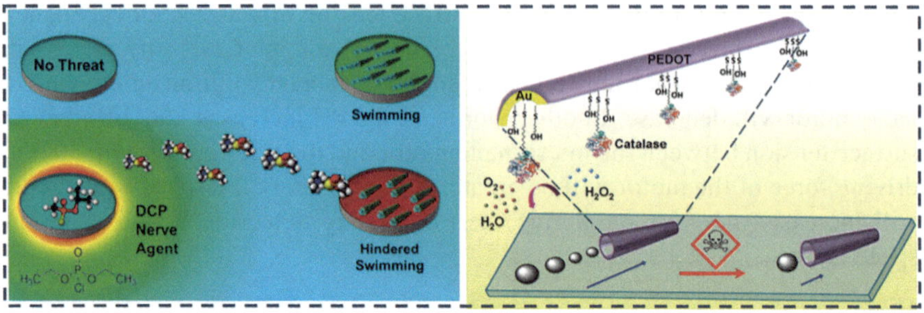

Fig. 9.6 Nanomotor for large molecule detection based on motion changes. (**a**) Detection of diethyl chlorophosphate by a polyethylene dioxythiophene/gold nanoparticle composite tubular micro-nanomotor driven by hydrogen peroxide enzyme; (**b**) Detection of pesticide residues in vegetables by a hydrogen peroxide-driven motor

respond to environmental changes and specific substances, and ultimately reflect through changes in the behavior of micro-nanomotor movement. As shown in Fig. 9.6a, researchers have prepared a polyethylene dioxythiophene/gold nanoparticle composite tubular micro-nanomotor driven by hydrogen peroxide enzyme. This micro-nanomotor can detect diethyl chlorophosphate (DCP, a nerve toxin) in the gas environment. This gaseous substance can dissolve in water and cause damage to the hydrogen peroxide enzyme, thereby reducing the speed of the micro-nanomotor. The nerve toxin detection device based on this micro-nanomotor has a lower detection limit, and its selectivity and sensitivity are also high, and it has good application potential in both military and civilian fields. As shown in Fig. 9.6b, the same hydrogen peroxide enzyme-driven nanomotor can also be used to detect pesticide residues in vegetables, which has application prospects in daily food safety monitoring.

Furthermore, changes in the motion behavior of micro-nanomotors can also be used for DNA detection. As shown in Fig. 9.7, similar to the micro-nanomotors in Fig. 9.4, researchers have developed a tubular micro-nanomotor composed of polyethylene dioxythiophene/gold particles driven by hydrogen peroxide enzyme. However, in this micro-nanomotor, researchers introduced a designed nucleic acid chain group. Due to the competition between specific nucleic acid chains, the target nucleic acid chain will cause the nucleic acid chain modified with hydrogen peroxide enzyme to fall off from the inside of the micro-nanomotor, thereby reducing the driving force of the micro-nanomotor. By testing the degree of reduction in the speed of the micro-nanomotor, the content of the target nucleic acid chain can be analyzed.

In summary, changes in the motion rate of micro-nanomotors can be used in the sensing applications of certain specific ions, molecules, and even biomolecules. The reasons for the regulation of the motion rate of micro-nanomotors can generally be divided into changes in catalytic activity (including enhancement or weakening of catalytic ability), changes in the number of catalytic sites (mainly the target will cause a reduction in catalytic sites), and rate changes caused by specific structural

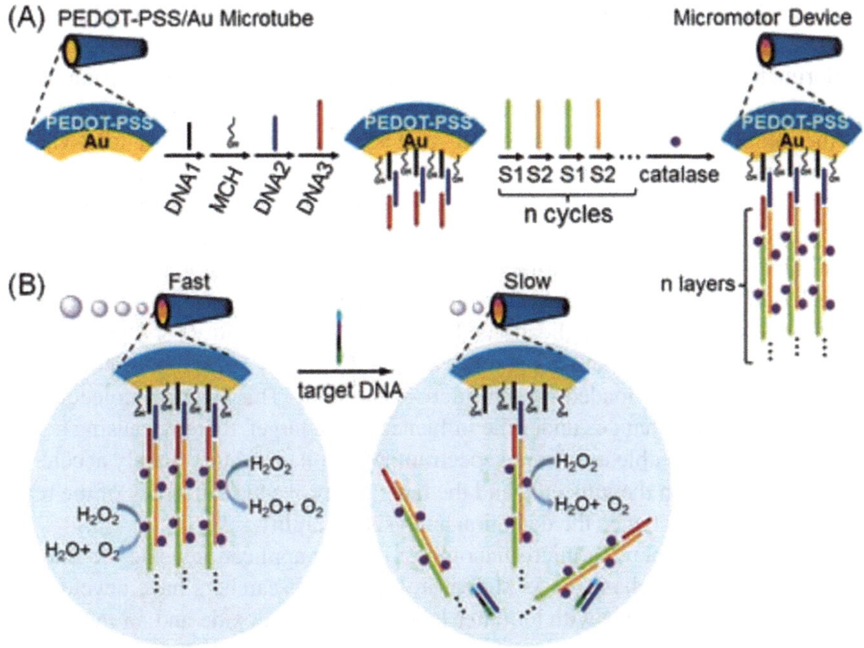

Fig. 9.7 DNA detection nanomotor based on motion changes

design. Different micro-nanomotor designs can be made for different target detection objects. Sensing technology based on changes in the motion behavior of micro-nanomotors has application value in environmental detection, biomedical sensing, and food hygiene safety, but the accuracy and signal reproducibility of its detection are challenges for further application in the future.

9.2.2 Intelligent Sensing of Micro-Nano Motors Based on Optical Detection Methods

Optical detection methods are a type of sensing method that relies on spectroscopic information such as spectral bands, light intensity, or lifetime to qualitatively or quantitatively detect the target. With the continuous development of optical instruments in recent years, the specificity, accuracy, and sensitivity of optical detection methods have greatly improved, playing an important role in the detection of inorganic small molecules, biological macromolecules, and microorganisms. Next, starting from the specific principles of optical detection methods, we will introduce the specific applications of micro-nanomotors in colorimetry and fluorescence quenching methods, and introduce the sensing applications of micro-nanomotors combined with electrochemiluminescence detection and Raman enhanced spectroscopy detection.

9.2.2.1 Micro-Nano Motor Sensing Applications Based on Colorimetry

Colorimetry is a detection method that determines the type and concentration of the test substance by comparing the color of the solution or measuring the absorbance of the solution. Colorimetry can generally be achieved by testing ultraviolet-visible absorption spectra, fluorescence spectra, and infrared spectra. Since micro-nanomotors can frequently actively contact the target to be tested, the colorimetric detection technology based on micro-nanomotors has higher sensitivity and shorter detection time than traditional detection methods. In the micro-nanomotor sensing applications based on colorimetry, the principle used is mostly fluorescence quenching. In addition, there are some applications that use the shift (quantity) or change of the spectrum on the micro-nanomotor for detection. Generally speaking, a special type of probe molecule is loaded on the micro-nanomotor. This type of molecule can undergo structural changes under the influence of the target, thereby causing a shift in the ultraviolet-visible or infrared spectrum. Micro-nanomotors mainly accelerate the contact between the molecule and the target, improve the efficiency of the reaction, and thereby enhance the detection analysis capability.

Here is an example of a micro-nanomotor sensing application using the fluorescence quenching mechanism. As shown in Fig. 9.8, researchers have developed a tubular micro-nanomotor with an outer layer of graphene oxide and an inner layer

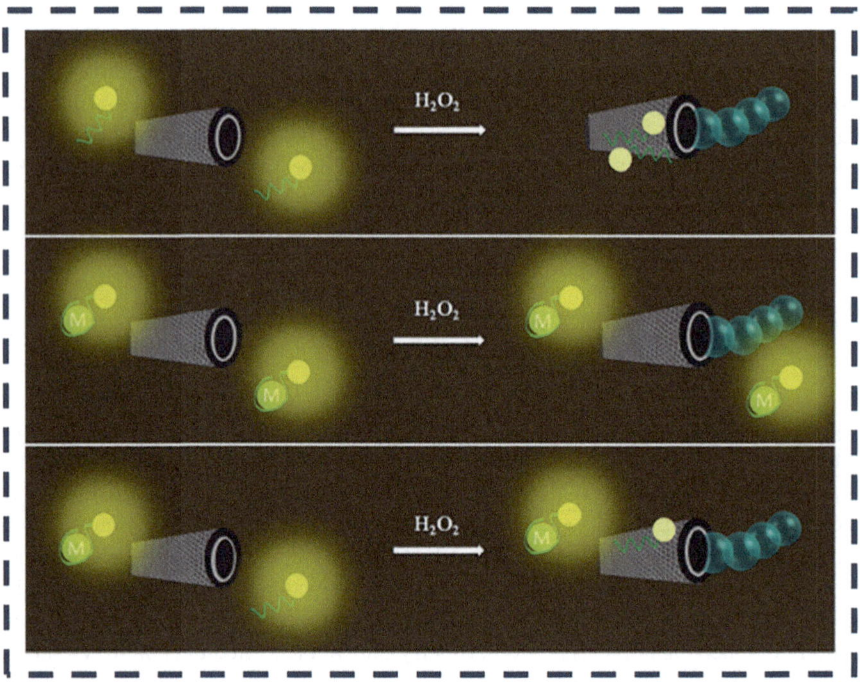

Fig. 9.8 Fluorescent sensing nanomotor based on colorimetry

of platinum nanoparticles. The outer layer of graphene oxide first adsorbs an aptamer with a fluorescent group. Due to the quenching effect of graphene oxide on the fluorescent group, the fluorescent group does not fluoresce at this time. However, the aptamer connected to the fluorescent group can bind with high specificity to mycotoxins (target molecules), thereby "dragging" the fluorescent group away from the surface of the graphene oxide, causing the fluorescent group to recover fluorescence. By testing the intensity of its fluorescence, the content of mycotoxins can be quantitatively detected.

In addition to generating spectral changes through changes in chemical molecular structure, some structural color materials also have color distinguishability, so they can also be combined with micro-nanomotors and used in colorimetric sensing. Structural color is formed due to the periodic arrangement of the microscopic structure of the material, which can not only be directly observed through an optical microscope, but also be accurately quantitatively analyzed through a spectrometer. As shown in Fig. 9.9, researchers have prepared Janus structural color microspheres with graphene oxide and silica nanoparticles on either side through a microfluidic device, using the phase separation of a mixed solution of graphene oxide and silica nanoparticles.

Because graphene oxide is pre-mixed with platinum nanoparticles and ferrite particles, this structural color micro-nanomotor can achieve two modes of motion in a hydrogen peroxide solution: bubble drive and magnetic field drive. In addition, the surface of the structural color micro-nanomotor is also coated with a hydrogel with an aptamer, which can bind with the target ion, causing a change in structural color. This detection method has the advantage of being label-free. In this detection method, the spectral change of the structural color micro-nanomotor can reflect the presence or absence of ions, and the amount of ions can be quantitatively described

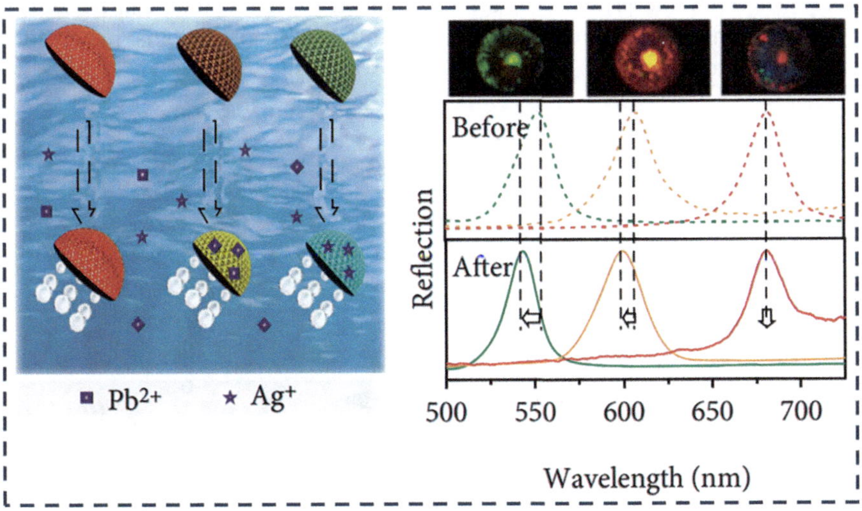

Fig. 9.9 Structural color sensing nanomotor based on colorimetry

by the degree of spectral shift. On this basis, by modifying the type of aptamer, it is possible to respond to different ions. Furthermore, if aptamers for different ions are modified on structural color micro-nanomotors with different colors, parallel label-free detection of multiple ions can be achieved. In summary, using structural color micro-nanomotors, simple and convenient multi-element detection with high sensitivity can be achieved through simple spectral testing, which has certain application prospects in high-throughput testing.

9.2.2.2 Micro-Nano Motor Sensing Application Based on Fluorescence Quenching Effect

Fluorescence quenching is actually still within the scope of colorimetry, and in recent years it has gradually become a new trend in the research of micro-nanomotor sensing applications, so it is listed as a separate section for introduction. Fluorescence quenching refers to the phenomenon where the fluorescence intensity or fluorescence lifetime of fluorescent molecules or quantum dots and other luminescent materials decreases under specific conditions. This phenomenon has been applied in sensing detection. For example, mercury ions can cause carbon quantum dots to undergo fluorescence quenching, so some researchers use carbon quantum dots as probes to detect the content of mercury ions in the water environment. Similarly, by modifying specific fluorescent molecules or quantum dots on micro-nanomotors, the micro-nanomotors can have the function of detecting certain ions or molecules.

As shown in Fig. 9.10, researchers modified one side of the silicon dioxide/platinum nanomotor with fluorescein (an organic molecule with green fluorescence), which responds to nerve toxins such as sarin and soman, losing its fluorescence characteristics within ten seconds. This “on-off” fluorescence design can qualitatively determine the presence of the test molecule. This method can be used to detect substances such as pesticides, plant toxins (proteins), microbial metabolites, and even bacteria.

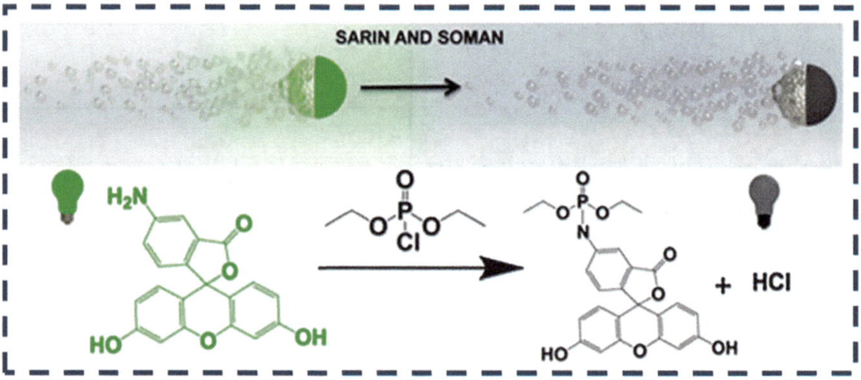

Fig. 9.10 Fluorescence quenching sensor nanomotor based on colorimetry

In addition, for micronanomotors using quantum dots as fluorescent materials, the degree of fluorescence quenching of the quantum dots is directly proportional to the content of the corresponding quenching substance. Therefore, this type of micronanomotor can not only qualitatively detect the presence of the target substance, but also quantitatively analyze the target substance within a certain range. Micronanomotors based on fluorescence quenching have become a new direction in the field of sensor detection. The motion characteristics of micronanomotors can enable rapid and effective reactions between fluorescent substances and target substances, greatly reducing the detection limit of this sensing method, improving detection sensitivity, and having practical application value in food safety and environmental monitoring.

9.2.2.3 Micro-Nano Motor Sensing Applications Based on Electrochemiluminescence

Electrochemiluminescence refers to the generation of new substances in an electrolyte solution through electrochemical methods, and these new substances can further react with each other or with existing substances to cause luminescence. As a detection analysis technique that combines electrochemical methods and chemical methods, electrochemiluminescence has the advantages of strong selectivity and high sensitivity, and has developed rapidly in fields such as immunoanalysis and nucleic acid detection in recent years. With the deepening of researchers' functionalization research on micronanomotors, electrochemiluminescence has also been integrated into the sensing applications of micronanomotors.

As shown in Fig. 9.11, researchers designed a glucose detection device composed of a glass carbon microbead and a vertical capillary with electrodes at both ends. The capillary contains an inorganic complex containing ruthenium for electrochemiluminescence and an auxiliary reactant tripropylamine. During the detection process, the glass carbon microbeads will undergo asymmetric chemical reactions under the action of an electric field. Hydrogen ions are reduced to produce hydrogen gas at one end of the microbead cathode, thereby generating driving force. At one end of the microbead anode, glucose oxidation and coenzyme NAD^+ is converted to reduced coenzyme I (NADH), and NADH participates in the electrochemiluminescence process of the ruthenium-containing complex, thereby completing the luminescence and coenzyme cycle. In electrochemiluminescence testing, the luminescence intensity is highly related to the initial concentration of glucose, so glucose content monitoring can be completed.

The sensing method based on electrochemiluminescence not only has the high sensitivity, wide linear range, and simple equipment advantages of traditional colorimetric sensing, but also has the advantages of good repeatability and easy control of electrochemical methods. It has great application potential in the next generation of portable biosensing devices. However, due to the complex principle of electrochemiluminescence testing, further research and development are needed for its application in the field of intelligent sensing of micronanomotors.

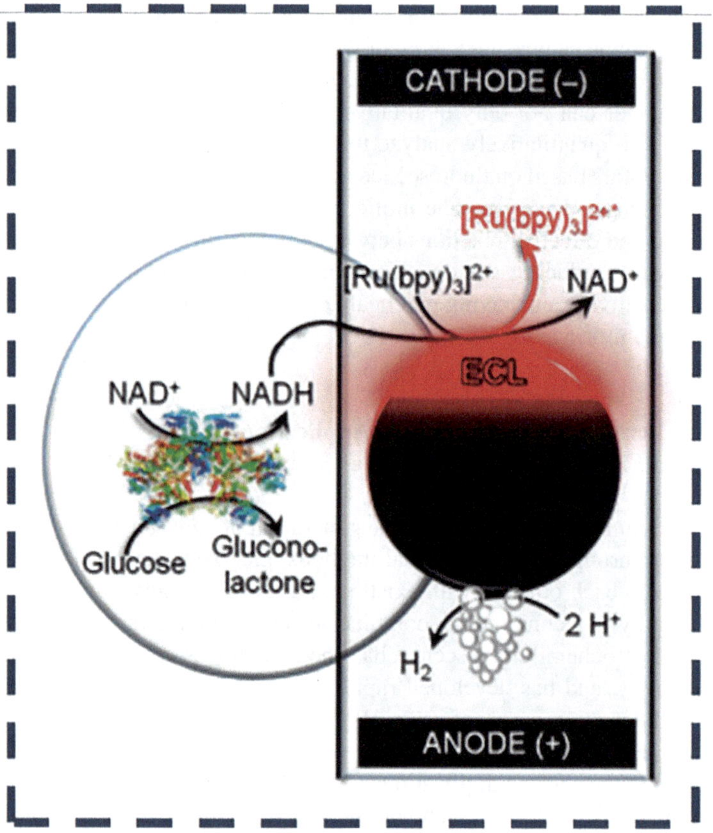

Fig. 9.11 Glucose sensing micronanomotor based on electrochemiluminescence

9.2.2.4 Micro-Nano Motor Sensing Applications Based on Surface-Enhanced Raman Spectroscopy

The surface Raman enhancement effect refers to the ability to enhance Raman scattering signals in the gaps between certain special metals (such as gold, silver, copper) or rough surfaces or metal nanoparticles. Surface-enhanced Raman spectroscopy inherits the advantages of Raman spectroscopy, which is simple, fast, non-destructive, and can reflect the specific structural features of molecules, thus enabling ultra-sensitive sensing of low-concentration target molecules. The emergence of surface-enhanced Raman spectroscopy has greatly expanded the application of Raman spectroscopy in medical detection, environmental monitoring, and material analysis. Micro-nano motors, which can move under external fields, can be modified with Raman enhancement "hot spot" probes on their surfaces, enabling dynamic enrichment of Raman signals and targeted micro-area detection.

Researchers have constructed an active surface-enhanced Raman scattering (SERS) probe based on micro-nano robots with autonomous movement and precise control. Firstly, using nanosilver wires as the main structure, a silica shell is used to coat the nanosilver wires and form a silver chloride power drive part, constructing the SERS enhancement probe. The ingenious integration of micro-nano robot drive and detection is realized, and the phototactic characteristics of the micro-nano robot are used to achieve controllable aggregation at the target location, and the feasibility of early tumor detection with this probe is systematically verified for the first time (Fig. 9.12a). Furthermore, to optimize the control precision of the micro-nano robot, researchers designed and developed a magnetic rod-shaped micro-nano robot, which grows silver nanoparticles in situ on the rod-shaped structure surface through a silver mirror reaction, serving as an active intelligent SERS detection probe. At the micro-nano scale, it has the ability to actively and precisely target, achieving high sensitivity detection, achieving self-cleaning and targeted intracellular signal detection (Fig. 9.12b). By in situ rapid growth of Au nano-spikes, a surface-enhanced Raman spectroscopy substrate is constructed, achieving rapid detection of Rhodamine 6G and crystal violet (Fig. 9.12c).

This section mainly introduces some sensing applications of micro-nano motors based on optical detection methods. Optical detection methods, due to their simplicity and accuracy, have been widely used in portable detection equipment. Combining

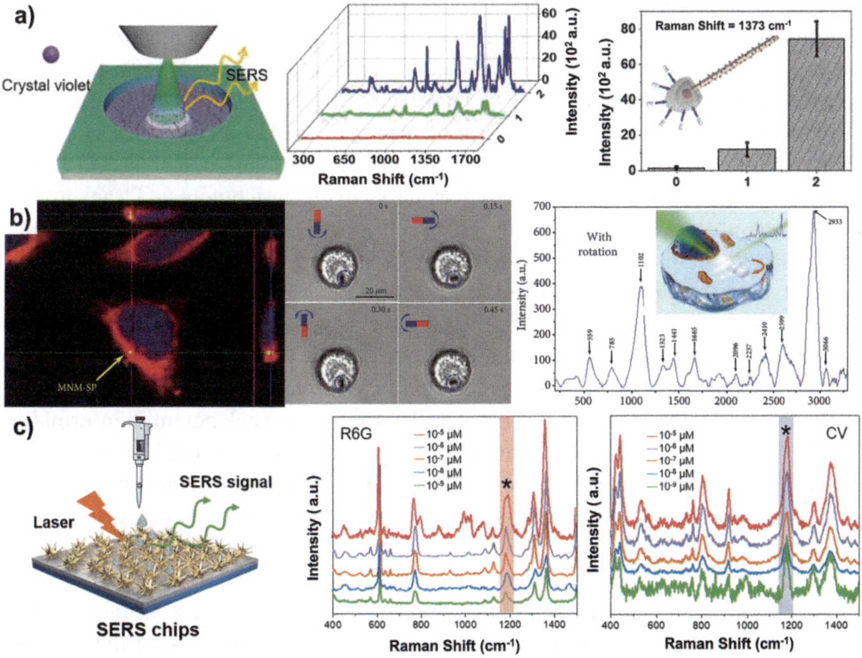

Fig. 9.12 Small molecule sensing micro-nano motor based on surface-enhanced Raman. (**a**) One-dimensional photocatalytic micro-nano robot for ex vivo Raman sensing detection. (**b**) Magnetic rod-shaped micro-nano robot used for intracellular signal detection. (**c**) Schematic diagram of Au nano-spike detection and Raman detection spectra of Rhodamine 6G and crystal violet

the motion characteristics of micro-nano motors with optical sensing methods greatly enhances the detection capability of optical detection methods. In addition, some novel micro-nano motor structure designs have made the application range of optical detection methods even wider.

9.2.3 Intelligent Sensing of Micro-Nano Motors Based on Electrochemical Analysis Methods

Electrochemistry is a sensing method that relies on measuring the transfer of electron charge during the oxidation-reduction reaction process on the electrode surface. In an electrolyte solution environment, changes on the electrode surface can be easily obtained by measuring microcurrents, impedance, and photoelectric conversion signals on the electrode surface. In addition, because electrochemical testing often does not require complex testing equipment and has high adaptability to the environment, sensing methods based on electrochemical principles have great potential in the development of miniaturized portable sensing devices. With the continuous development of micro-nano motors in recent years, sensing methods that use the motion behavior of micro-nano motors to enhance electrochemical detection performance have gradually appeared in the field of electrochemical sensing.

Micro-nano motor sensing based on electrochemical analysis methods is a new testing method that has emerged in recent years. The principles are similar, generally through Cyclic Voltammetry (CV) or Electrochemical Impedance Spectroscopy (EIS) to carry out qualitative analysis of electrode surface reactions. Differential Pulse Voltammetry (DPV) can be used for precise quantitative analysis of the test substance. Micro-nanomotors mainly play a role in the enrichment and transmission of the test substance during testing. This section will mainly introduce it based on the development history of micro-nanomotors in electrochemical testing.

First, a simple introduction to the three common testing principles in electrochemical testing. As shown in Fig. 9.13a, cyclic voltammetry applies a constant scanning speed voltage signal to the electrode within a certain range, while testing the relationship between the surface current of the electrode and the applied voltage. In the CV curve, the shape of the curve, peak current, and peak potential information can be used to judge the mechanism of the redox reaction occurring on the electrode surface. In sensor testing, the CV curve is generally used to obtain the oxidation peak potential interval, paving the way for the subsequent differential pulse voltammetry curve. As shown in Fig. 9.13b, differential pulse voltammetry applies a gradually rising pulse voltage to the electrode, and then tests the relationship between the electrode current and the pulse voltage. In the DPV curve, the size of the oxidation peak current will show a linear relationship with the concentration of the reactant, so that the content of the substance can be calculated. Due to the sensitivity of electrochemical sensing tests, the combined testing method of CV curve and DPV curve can often be applied to the testing of trace substances. As shown in Fig. 9.13c, electrochemical impedance spectroscopy can generally be used to characterize the

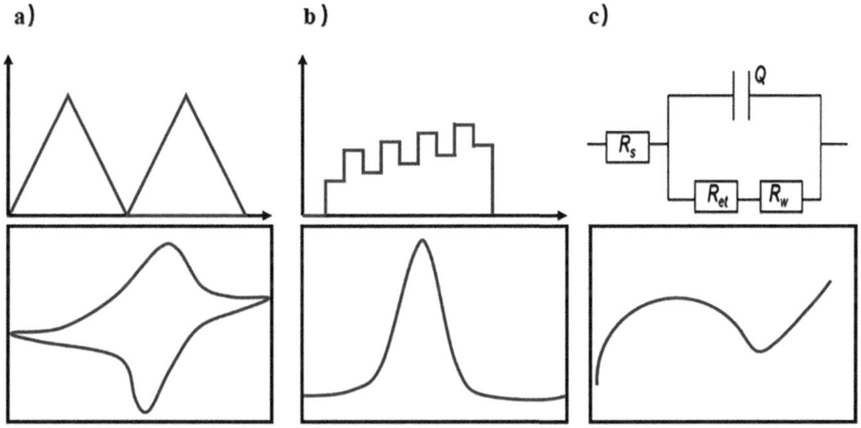

Fig. 9.13 Micro-nano motors combined with different electrochemical detection methods to achieve biochemical sensing. (**a**) Cyclic voltammetry; (**b**) differential pulse voltammetry; (**c**) electrochemical impedance

material properties of the electrode surface. In the intelligent sensing application of micro-nanomotors, micro-nanomotors often need to undergo multiple steps of modification to achieve their sensing characteristics. Electrochemical impedance spectroscopy can be used to judge whether each step of the reaction is successful, and it is often used in the characterization of micro-nanomotors. In addition, electrochemical impedance spectroscopy also has the potential for quantitative testing. For the same molecule, the higher the content, the greater the measured electrochemical impedance, which also makes electrochemical impedance spectroscopy have good application potential in micro-nanomotor intelligent sensing.

Since the invention of micro-nanomotors, researchers have conducted extensive and in-depth research on their motion behavior. Among them, bubble-propelled self-driven micro-nanomotors have been proven to improve the mixing of fluids and the transport of substances in fluids, thereby improving the efficiency of conventional chemical reactions, which also has a good performance promotion effect on electrochemical sensing. Taking magnesium-based Janus micro-nanomotors as an example, micro-nanomotors will produce obvious microbubbles in water. The existence of microbubbles will greatly promote mass transfer and local convection in the solution. At this time, substrates that can interact with the test substance are modified on the working electrode (for example, glucose oxidase is modified on the glass carbon electrode, which can be used to detect glucose in the solution). The position of the substrate oxidation peak can be judged by the cyclic voltammetry curve, and then the peak value of the differential pulse voltammetry curve can be used to quantitatively analyze the concentration of the test substance. Because the micro-nanomotor produces the enhancement effect of convection and mass transfer, more glucose molecules will contact the glucose oxidase on the electrode, so as to detect as much glucose as possible. In this process, the participation of micro-nanomotors can improve the sensitivity of electrochemical detection.

In recent years, researchers' studies on micro-nano motors have not only focused on improving their mobility, but also on functionalizing these motors, which is one of their important development directions. By modifying the surface of the micro-nano motor, it can accomplish more complex tasks. As shown in Fig. 9.14a, by modifying

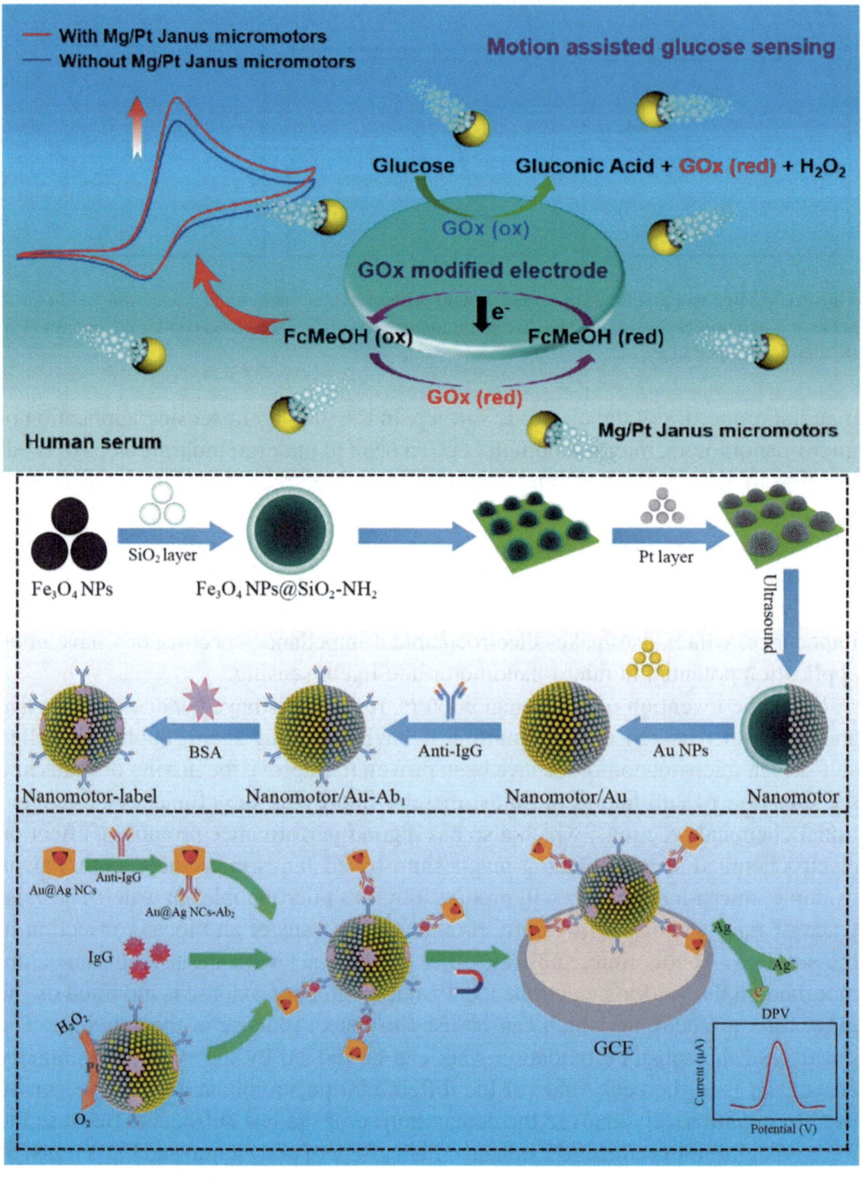

Fig. 9.14 Signal amplification in the process of electrochemical sensing by micro-nano motors. (**a**) Enhancing the electrochemical signal by constructing a binding-lysis cycle based on magnetic motors; (**b**) micro-nano motors catalyze the substrate reaction, improve electrochemical activity, and enhance the electrochemical signal

the surface of the micro-nano motor with an antibody of a certain antigen substrate, when this functionalized micro-nano motor is added to the substrate solution, it can actively capture the antigen through its own motion behavior. This active capture efficiency is much higher than the passive capture relying on molecular diffusion. Against this background, a large number of micro-nano motors with magnetic nanoparticles as the core have been applied in the field of electrochemical sensing. Researchers have prepared silica-coated magnetite nano motors driven by a magnetic field. By modifying the surface of silica with IgG antibodies, under the drive of a magnetic field, the micro-nano motor can capture free antigens in the solution, and then the micro-nano motor can be collected on the surface of the working electrode under the action of the magnetic field, and further complete the test. Through the two-step test of cyclic voltammetry and differential pulse voltammetry, the content information of the antigen in the solution can be obtained. This kind of micro-nano motor that actively captures and transports to the electrode surface will have higher sensitivity and detection accuracy in electrochemical tests compared to the previous micro-nano motors that only improve convection and mass transfer.

Although the active capture micro-nano motors have already achieved good test results in electrochemical sensing, researchers are still committed to exploring other characteristics of micro-nano motors in the hope of further improving the sensing detection capability. As shown in Fig. 9.14b as shown, researchers first used polydopamine to coat magnetite nanoparticles and modified their surface with heparin (i.e., introduced amino groups on their surface). Due to the porosity of polydopamine, it can adsorb a large number of biomolecules. Researchers loaded a magnesium ion-responsive DNA hydrolase into it, which can break at specific nucleic acid sites under the action of magnesium ions. Subsequently, researchers modified one side of the micron-sized magnesium particles with amino groups. With the participation of glutaraldehyde, the magnesium particles were connected to the aforementioned magnetite particles. Since only one side of the magnesium particle surface is covered with magnetite particles (i.e., a Janus structure is formed), the exposed magnesium on the other side will react with water to generate bubbles and cause the micro-nano motor to move. When using such micro-nano motors for specific nucleic acid chain detection, the target nucleic acid chain will specifically bind with the DNA hydrolase, and then split into a marker chain that can be used for testing, a useless free chain, and an original nucleic acid chain (target nucleic acid chain) under the action of magnesium ions. The original nucleic acid chain that has been split out will further bind with the next DNA hydrolase and split again. Through multiple binding-splitting cycles, multiple marker chains can be produced from a target nucleic acid chain to be tested, and then the marker chain is combined with a complementary chain with electrochemical active substances to enhance the signal strength. All the above nucleic acid chain reactions occur within the mesopores of polydopamine on the surface of magnetite. As the magnesium particles are continuously consumed by the reaction, the magnetite particles are released, and then the magnetite particles can be attracted to the electrode surface by a magnetic field, and the detection is finally completed. In this work, the micro-nano motor significantly reduces the detection limit of specific nucleic acid chains through the amplification effect and has high sensitivity. Similarly, some micro-nano motors can also improve the sensitivity of sensing by catalyzing the substrate to be tested from low electrochemical active substances to high electrochemical active substances. These all reflect the important role and function of micro-nano motors in electrochemical sensing.

In addition to the sensing applications of the above micro-nano motors, electrochemical sensing methods also have important applications in the study of the driving mechanism of micro-nano motors themselves. For example, the movement of micro-nano motors will cause convection in the solution, which will disturb the diffusion layer on the electrode surface. Using electrochemical testing methods can obtain disturbance signals, and then the number and speed of micro-nano motors in the solution can be judged by analyzing the disturbance signals. Inspired by this, researchers have also developed sensing systems that can be used for microbial detection. As shown in Fig. 9.15, researchers have designed a detection device for E. coli, where the presence of E. coli will affect the diffusion of hydroxyl to the electrode surface, thereby affecting the strength of the electrode detection signal. Based on this indicator, the concentration of E. coli can be judged.

Compared with other detection methods, electrochemical sensing methods have higher sensitivity and reliability, and the testing equipment is simple and the testing cost is relatively low. In recent years, it has become one of the important research directions for the application of micro-nano motors in the field of sensing. However, electrochemical testing methods generally have poor selectivity in

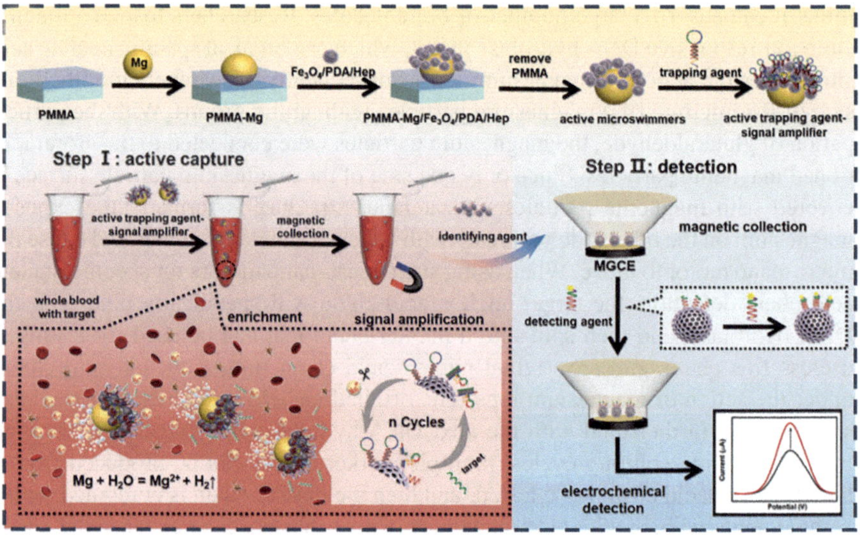

Fig. 9.15 Micro-nano motors with signal amplification function based on electrochemical detection methods

testing, which requires micro-nano motors using electrochemical sensing detection to have special structural design. This increases the difficulty of its subsequent development and application, and also makes it difficult to apply in complex liquid environments.

9.2.4 Other Micro-Nano Motor Intelligent Sensing

In addition to the above-mentioned applications of micro-nanomotor sensing based on the motion behavior of micro-nanomotors, combined with optical detection methods and electrochemical detection methods, there are also some other detection methods that are also applied to intelligent sensing of micro-nanomotors.

As shown in Fig. 9.16, researchers have prepared a magnetically driven nanorod, which can achieve rotational motion under a rotating magnetic field. Since the nanorod has anisotropy, collective plasmon oscillations will occur on the surface during the rotation, thereby generating periodic extinction signals. This signal can be output as a frequency signal through Fourier transform. After connecting the antibody used to capture the antigen and finally capturing the antigen, due to the change in surface structure, the measured frequency signal will decrease. Based on this phenomenon, a relationship can be established between frequency changes and antigen concentration, thereby creating intelligent sensing devices.

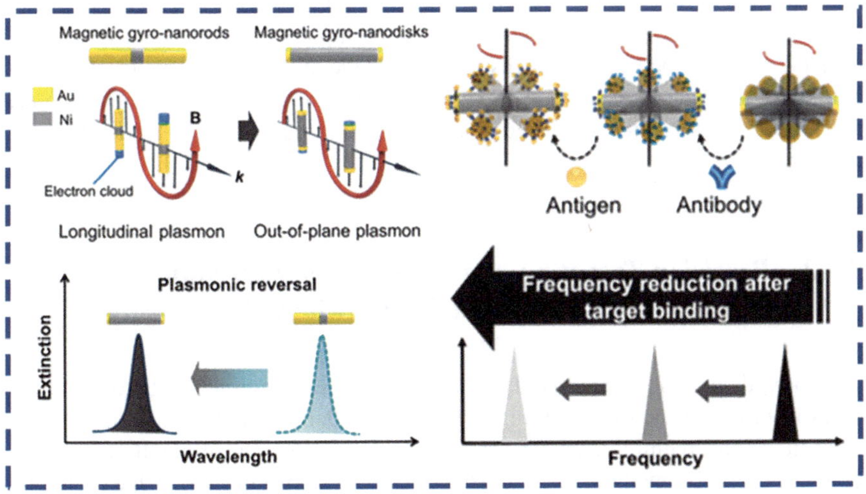

Fig. 9.16 Periodic changes in extinction signals after antigen-antibody binding

9.3 Advantages of Micro-Nano Motor Intelligent Sensing

9.3.1 Enhanced Material Exchange in Low Reynolds Number Environment by Micro-Nano Motor Motion

Efficient and uniform fluid mixing is crucial for complex chemical reactions, small molecules, and various applications involving microfluidic interactions, such as biochemical detection. However, fluids at the micro and nano scales are usually in a laminar state, characterized by a low Reynolds number $Re = \frac{\rho UL}{\mu} << 1$, where ρ is the fluid density, U is the flow velocity, L is the characteristic length, μ is the dynamic viscosity of the fluid. Therefore, without external influence, the mixing of solutions at the micro-nano scale mainly relies on passive diffusion, which progresses slowly, leading to low detection efficiency and inability to achieve rapid real-time detection.

In order to reduce the solution mixing time and improve the mixing efficiency, many strategies to enhance material mixing have been studied. They can generally be divided into two categories: passive methods and active methods. Each method has its own advantages and limitations in terms of effect, mixing time, control, and ease of operation. Since passive mixers do not require any external equipment or power source, they are very simple to operate and can be integrated with different detection applications. However, the mixing efficiency is low and the time used for mixing is long. The other category is active mixing, which uses external ultrasound, magnetic fields, electric fields, etc., to drive particles, making the mixing more efficient and uniform.

Micronano motors have the ability to move under low Reynolds numbers. Intelligent detection probes based on micronano motors can achieve different controllable motion modes, generate disturbances at the micronano scale, accelerate the binding of the molecules to be tested with the probe detection target, quickly reach the kinetic equilibrium state, and complete processes such as specific immune binding.

9.3.2 Precision Targeting and Controllable Detection of Substances to Be Tested at the Micro-Nano Scale

Currently, detection probes mainly rely on passive diffusion to the detection location for sensing detection, resulting in poor controllability and precision, unable to meet the future requirements for precise targeting detection of small lesion locations. Microprobe methods can achieve detection at specific target locations, but require complex control platforms and have tissue destructiveness. Intelligent detection probes based on micronano motors can achieve controllable targeting of the parts to be tested, and have the advantages of being able to achieve dynamic detection, small destructiveness to the tissues to be tested, etc., and have important potential in the future development of intelligent detection probes.

9.3.3 Achieving Automated Detection by Controlling Detection Probes

Currently, the market has seen the use of automated equipment to replace manual operations, and the probes used in this solution are mostly passive probes. Passive probe immunological detection requires complex mechanical control equipment and systems to achieve automatic cleaning processes and solution transfers, which increases the cost of the instrument, which is not conducive to further development of point-of-care testing (POCT) miniaturized devices. Another strategy is to directly control the probe material, simplify external operation complex equipment, and reduce costs. Intelligent detection probes based on micronano motors can be directly operated by external magnetic fields to complete different detection process tasks. Therefore, based on the research foundation of controllable motion of micronano motors, it is possible to directly operate detection probes at the micronano scale and achieve automated detection.

9.3.4 Combination of Micro-Nano Motors and Smartphones

Using micronano motors as a micro-nano operation platform and combining with smartphones to form a digital intelligent diagnosis and treatment system. Using surface-modified antibody micronano motors and the immune response to the Zika virus, a simple and rapid detection of the Zika virus can be achieved. Kristin Weidemaier, based on magnetic nanorobots, used surface-enhanced Raman detection probes to detect antigens on the surface of the Ebola virus, achieving rapid diagnosis of patients with fever caused by the Ebola virus. Harvard Medical School uses bubble-driven micronano motors combined with mobile phones to detect the substances to be tested by analyzing the changes in the bubbles produced by the micronano motors. In the future, the microscopic photography optical system of smartphones can be further utilized, and matching detection software can be developed to achieve instant rapid detection of various substances to be tested.

9.4 Chapter Summary

Micro-nano motors have the characteristics of micro/nano scale size and high specific surface area, which can achieve efficient adsorption and specific binding, making them ideal carriers for detection probes. Through surface material and bio-functionalization, micro/nanomotors can be prepared into dynamic microbiological sensors, achieving real-time, rapid, and highly sensitive detection of biomolecules (such as nucleic acids and proteins). At the same time, micro-nano motors have the advantages of autonomous movement and precise control, which can achieve efficient fluid mixing and accelerate the binding efficiency of the test substance and the detection probe. In the future, micro-nano motors can serve as a manipulable universal micro-nano platform, which can be extended to other in vitro diagnostic (IVD) applications, or combined with electrochemical sensing technology to explore a wider range of application scenarios, providing an alternative strategy for innovating traditional biochemical sensing methods.

Bibliography

1. Zhou C, Gao C, Lin Z, Wang D, Li Y, Yuan Y, Zhu B, He Q, Autonomous motion of bubble-powered carbonaceous nanoflask motors [J]. Langmuir, 2020, 36: 7039-7045.
2. Wang W, Castro L A, Hoyos M, Mallouk T E, Autonomous motion of metallic microrods propelled by ultrasound [J]. ACS Nano, 2012; 6: 6122-6132.
3. Guo J, Gallegos J J, Tom A R, Fan D, Electric-field-guided precision manipulation of catalytic nanomotors for cargo delivery and powering nanoelectromechanical devices [J]. ACS Nano, 2018, 12: 1179-1187.
4. Tang S-Y, Sivan V, Khoshmanesh K, O'Mullane A P, Tang X, Gol B, Eshtiaghi N, Lieder F, Petersen P, Mitchell A, Kalantar-zadeh K, Electrochemically induced actuation of liquid metal marbles [J]. Nanoscale, 2013, 5: 5949-5957.
5. Villa K, Pumera M, Fuel-free light-driven micro/nanomachines: Artificial active matter mimicking nature [J]. Chemical Society Reviews, 2019, 48: 4966-4978.
6. Hu W, Lum G Z, Mastrangeli M, Sitti M, Small-scale soft-bodied robot with multimodal locomotion [J]. Nature, 2018, 554: 81-85.
7. Yan M, Liu T, Li X, Zhou S, Zeng H, Liang Q, Liang K, Wei X, Wang J, Gu Z, Jiang L, Zhao D, Kong B, Soft patch interface-oriented superassembly of complex hollow nanoarchitectures for smart dual-responsive nanospacecrafts [J]. Journal of the American Chemical Society, 2022, 144: 7778-7789.
8. Xu H, Medina-Sánchez M, Magdanz V, Schwarz L, Hebenstreit F, Schmidt O G, Sperm-hybrid micromotor for targeted drug delivery [J]. ACS Nano, 2018, 12: 327-337.
9. Li J, Thamphiwatana S, Liu W, Esteban-Fernández de Ávila B, Angsantikul P, Sandraz E, Wang J, Xu T, Soto F, Ramez V, Wang X, Gao W, Zhang L, Wang J, Enteric micromotor can selectively position and spontaneously propel in the gastrointestinal tract [J]. ACS Nano, 2016, 10: 9536-9542.
10. Huang H, Li J, Yuan M, Yang H, Zhao Y, Ying Y, Wang S, Large-scale self-assembly of mofs colloidosomes for bubble-propelled micromotors and stirring-free environmental remediation [J]. Angewandte Chemie International Edition, 2022, 61: e202211163.
11. Schwarz L, Karnaushenko D D, Hebenstreit F, Naumann R, Schmidt O G, Medina-Sánchez M, A rotating spiral micromotor for noninvasive zygote transfer [J]. Advanced Science, 2020, 7: 2000843.
12. Breger J C, Yoon C, Xiao R, Kwag H R, Wang M O, Fisher J P, Nguyen T D, Gracias D H, Self-folding thermo-magnetically responsive soft microgrippers [J]. ACS Applied Materials & Interfaces, 2015, 7: 3398-3405.
13. Reisoval, Hermanova S, Pumera M. Micro/Nanomachines: What is needed for them to become a real force in cancer therapy [J]. Nanoscale, 2019, 11(14): 6519-32.
14. Xing Y, Du X, Xu T, et al. Janus dendritic silica/carbon@Pt Nanomotors with multiengines for H_2O_2, near-infrared light and lipase powered Propulsion [J]. Soft Matter, 2020, 16(41): 9553-58.

15. Gao W, Dong R, Thamphiwatana S, et al. Artificial micromotors in the mouse's stomach: a step toward in *vivo* use of synthetic motors [J]. ACS Nano, 2015, 9(1): 117-23.
16. Chatzipirpirdis G, De Marco C, Pellicer E, et al. Template-assisted electroforming of fully semi-hard-magnetic helical microactuators [J]. Advanced Engineering Materials, 2018, 20(9): 1800179.
17. Pal M, Somalwar N, Singh A, et al. Maneuverability of magnetic nanomotors inside living cells [J]. Advanced Materials, 2018, 30(22): 1800429.
18. Yu J, Wang B, Du X, et al. Ultra-extensible ribbon-like magnetic microswarm [J]. Nature Communications, 2018, 9(1): 3260.
19. Wang B, Chan Kai F, Yuan K, et al. Endoscopy-assisted magnetic navigation of biohybrid soft microrobots with rapid endoluminal delivery and imaging [J]. Science Robotics, 2021, 6(52): eabd2813.
20. Kagan D, Calvo-Marzal P, Balasubramanian S, et al. Chemical sensing based on catalytic nanomotors: motion-based detection of trace silver [J]. Journal of the American Chemical Society, 2009, 131(34), 12082–083.
21. Su Y, Ge Y, Liu L, et al. Motion-based pH sensing based on the cartridge-case-like micromotor [J]. ACS Applied Materials & Interfaces, 2016, 8(6), 4250–4257.
22. Liu L, Dong Y, Sun Y, et al. Motion-based pH sensing using spindle-like micromotors [J]. Nano Research, 2016, 9(5), 1310–1318.
23. Singh V. V, Kaufmann K, Esteban-Fernandez De Ávila B, et al. Nanomotors responsive to nerve-agent vapor plumes [J]. Chemical Communications, 2016, 52, 3360–63.
24. Orozco J, Garcia-Gradilla V, D Agostino M, et al. Artificial enzyme-powered micro fish for water-quality testing [J]. ACS Nano, 2013, 7(1), 818–24.
25. Fu S, Zhang X, Xie Y, et al. An efficient enzyme-powered micromotor device fabricated by cyclic alternate hybridization assembly for DNA detection [J]. Nanoscale, 2017, 9, 9026–33.
26. Molinero-Fernandez Á, Moreno-Guzman M, Lopez M, et al. Biosensing strategy for simultaneous and accurate quantitative analysis of mycotoxins in food samples using unmodified graphene micromotors [J]. Analytical Chemistry, 2017, 89(20), 10850–57.
27. Wang H, Cai L, Zhang D, et al. Responsive Janus structural color hydrogel micromotors for label-free multiplex assays [J]. Research, 2021, 1–9.
28. Singh V, Kaufmann K, Orozco J, et al. Micromotor-based on-off fluorescence detection of sarin and soman simulants [J]. Chemical Communications, 2015, 51, 11190–93.
29. Sentic M, Arbault S, Goudeau B, et al. Electrochemiluminescent swimmers for dynamic enzymatic sensing [J]. Chemical Communications, 2014, 50, 10202–05.
30. Fan X, Hao Q, Li M, et al. Hotspots on the move: active molecular enrichment by hierarchically structured micromotors for ultrasensitive SERS sensing [J]. ACS Applied Materials & Interfaces, 2020, 12(25), 28783–91.
31. Kong L, Rohaizad N, Nasir M, et al. Micromotor-assisted human serum glucose biosensing [J]. Analytical Chemistry, 91(9), 5660–66.
32. Ma E, Wang K, Wang H, et al. An immunoassay based on nanomotor-assisted electrochemical response for the detection of immunoglobulin [J]. Microchimica Acta, 2022, 189(47), 1–9.
33. Wang Q, Tang X, LV R, et al. Microswimmer-based electrochemical platform with active capturer/signal amplifier/funnel-type device for whole blood detection [J]. Chemical Engineering Journal, 2022, 430(4), 132665.
34. Ramanujam A, Neyhouse B, Keogh RA, et al. Rapid electrochemical detection of Escherichia coli using nickel oxidation reaction on a rotating disk electrode [J]. Chemical Engineering Journal, 2021, 411(1), 128453.
35. Jung I, Ih S, Yoo H, et al. Fourier transform surface plasmon resonance of nanodisks embedded in magnetic nanorods [J]. Nano Letters, 2018, 18(3), 1984–92.
36. Liu X, Yang S, Lyu X, Liu S, Wang Y, Li Y, Wang B, Chen W, Wang W, Guo J, Ma X, Instant preparation of ultraclean gold nanothorns under ambient conditions for SERS kit-enabled mobile diagnosis [J]. Analytical Chemistry 2021, 93: 16628-16637.

37. Wang Y, Liu Y, Li Y, Xu D, Pan X, Chen Y, Zhou D, Wang B, Feng H, Ma X, Magnetic nanomotor-based maneuverable SERS probe [J]. Research, 2020, 7962024.
38. Wang Y, Zhou C, Wang W, Xu D, Zeng F, Zhan C, Gu J, Li M, Zhao W, Zhang J, Guo J, Feng H, Ma X, Photocatalytically powered matchlike nanomotor for light-guided active SERS sensing [J]. Angewandte Chemie International Edition, 2018, 57: 13110-13113.

GPSR Compliance

The European Union's (EU) General Product Safety Regulation (GPSR) is a set of rules that requires consumer products to be safe and our obligations to ensure this.

If you have any concerns about our products, you can contact us on ProductSafety@springernature.com

In case Publisher is established outside the EU, the EU authorized representative is:

Springer Nature Customer Service Center GmbH
Europaplatz 3
69115 Heidelberg, Germany

Batch number: 10725825

Printed by Printforce, the Netherlands